Law Librarianship in the Digital Age

Edited by Ellyssa Kroski

THE SCARECROW PRESS, INC.
Lanham • Toronto • Plymouth, UK
2014

Published by Scarecrow Press, Inc.
A wholly owned subsidiary of The Rowman & Littlefield Publishing Group, Inc.
4501 Forbes Boulevard, Suite 200, Lanham, Maryland 20706
www.rowman.com

10 Thornbury Road, Plymouth PL6 7PP, United Kingdom

British Library Cataloguing in Publication Information Available

Library of Congress Cataloging-in-Publication Data

Law librarianship in the digital age / Edited by Ellyssa Kroski.
pages cm
Includes index.
ISBN 978-0-8108-8805-0 (cloth : alk. paper) — ISBN 978-0-8108-8806-7 (pbk. : alk. paper) — ISBN 978-0-8108-8807-4 (ebook)
1. Law libraries—Handbooks, manuals, etc. 2. Libraries and the Internet. 3. Digital communications—United States. I. Kroski, Ellyssa, editor of compilation.
Z675.L2L3839 2013
026.34—dc23

2013021494

∞™ The paper used in this publication meets the minimum requirements of American National Standard for Information Sciences—Permanence of Paper for Printed Library Materials, ANSI/NISO Z39.48-1992.

Printed in the United States of America.

Contents

Part II. Technologies

Part III. Reference Services

Part IV. Instruction

Part V. Technical Services

Figures

Tables

Foreword

My first reaction to *Law Librarianship in the Digital Age* was, "I wish there had been a book like this when I was in graduate school." But e-books, iPads, virtual reference, webinars, cloud computing, Web scale discovery, apps, and avatars were the stuff of fantasy. If there had been such a book, it would have been classified as "science fiction." In the thirty years since I entered the profession, the externals of librarianship have been wildly transformed. But the core mission of the profession—matching people to knowledge—remains intact and drives a vision of the future that distinguishes our profession from all others. We are charged with preserving and optimizing access to knowledge. We wear all the hats: grinding, finding, minding, and connecting. For those who still think of the "bun-headed" cartoon librarian stereotype, this book will be a rude awakening. The editor has assembled an impressive lineup of thought leaders to provide strategic insights into the various facets of digital information preservation, presentation, access, and management in a variety of contexts.

This is the proverbial best of times and worst of times. Since the collapse of Lehman Brothers in 2008, the law firm market has been thrown forward into an altered world where technology, client pricing pressures, and globalization have generated a truly "disrupted market." New kinds of legal practices are emerging, such as virtual law firms, coupled with the emergence of offshored legal drafting and e-discovery centers. As law firms have retrenched, their hiring of new associates has reduced. Law schools are faced with falling enrollments due both to the high cost of legal education and the uncertain job prospects. We will all adapt what we do to address the changed environments and reinvent ourselves and our services, as we have done before.

Graduate curriculum for legal information professionals needs to focus on cultivating the wide range of professional competencies outlined in the book. The word "librarian" hardly covers the breadth of our universe. We are strategic leaders, research analysts, taxonomists, teachers, digital pioneers, app developers, knowledge managers, information literacy evangelists, and competitive intelligence gurus. In short, we are both educators and digital cartographers who build the bridges and help researchers chart the course between knowledge from the past and data that will become knowledge of the future.

But this book isn't just a practical handbook for students; it also contains a wealth of state-of-the-art information for practitioners and those thinking of a career change into law librarianship.

The editor has selected a range of topics that offer an exploration of both the core practice issues and the transformational initiatives in law school, government, and law firm library environments. Each kind of library may adopt new practices ahead of each other, and then inspire and cross-fertilize new initiatives in another environment. Seeing how an academic library promotes distance learning provides inspiration for law firms that are increasingly globalized. The competitive intelligence initiatives in law firms may be adapted to customized faculty research needs.

Steve Jobs attributed the success of Apple to the fact that it existed at the intersection of technology and humanities. Jobs was referring to the kind of multidisciplinary thinking that our professional excels at. We connect technology, law, and the multitude of social, literary, technical, and scientific issues that stream through legislation, case law, and commercial activities.

I have to admit I was someone who entered the profession because I liked books. I especially loved breathing the air in a cloistered alcove of a research library surrounded by aging volumes. And yet I entered the profession as it was about to tumble into decades of technological change and professional uncertainty. Libraries as places are shrinking. As we all know, change leads to opportunity, and we face the opportunity to radically transform both libraries and our profession. *Law Librarianship in the Digital Age* provides a panorama of how traditional functions such as research, collection development, technical services, and administration have been and will continue to be transformed by the innovative professionals who contributed to this book.

Jean O'Grady, JD, MLS
Washington, D.C., June 2013

Preface

Although not all LIS professionals need to be digital natives, it is absolutely essential that today's law librarians are digitally literate as well as possess an understanding and awareness of recent advancements and trends in information technology as they pertain to the library field. *Law Librarianship in the Digital Age* offers a comprehensive guidebook to achieving both of those goals.

This resource covers the most cutting-edge developments that face today's modern law libraries, including e-books, mobile device management, Web scale discovery, cloud computing, social software, and much more. These critical issues and concepts are approached from the perspective of tech-savvy library leaders who discuss how forward-thinking libraries are tackling traditional library practices such as reference, collection development, technical services, and administration in this new "digital age."

Each chapter explores the key concepts and issues that are that are currently being discussed at major law library conferences and events today, and looks ahead to what's on the horizon for law libraries in the future. Chapters have been written by the field's top innovators from all areas of law librarianship, including academic, government, and private law firm libraries who have strived to provide inspiration and guidance to tomorrow's law library leaders.

Ellyssa Kroski
The New York Law Institute

Acknowledgments

I would like to acknowledge the hard work and exemplary insight contributed by the stellar team of authors and editorial board members who collaborated to create this excellent resource. Thank you, everyone!

—Ellyssa Kroski

Part I

MAJOR INTRODUCTORY CONCEPTS

1

Law Librarianship 2.0

Jennifer Wertkin
THE ARTHUR W. DIAMOND LAW LIBRARY, COLUMBIA UNIVERSITY

HISTORY

Law libraries have a rich tradition in the United States. The relationship between law libraries and legal information has its roots in our common law legal system. Strictly speaking, the first law libraries, dating back as far as the American republic itself, consisted of "reporters" that contained judicial opinions. The principle of *stare decisis*[1] guided the way legal research was conducted and, as such, the only necessary resources were judicial opinions organized into sets of volumes called reporters. These collections of reporters were all indexed separately and published by various private publishers. The first libraries that collected this material were a select number of law schools and a few specialized libraries in Boston, New York, and Philadelphia.

By the beginning of the twentieth century, the nature of legal practice had changed. Attorneys began looking beyond judicial opinions and started bringing other materials, outside of case law, into their court documents. These materials included scholarship from many fields and government documents. It was at this point that law libraries began to flourish. As new materials were added to collections, it became clear that this influx required curators to organize this information, index it, and assist patrons in using it. The professional field of law librarianship was born.[2]

The first prestigious law libraries were in academic settings and in specialized subscription libraries in large cities. Before the advent of electronic resources, it is easy to envision the law library as the heart of the law school. All research had to physically take place in the library. Many roles for librarians emerged as students and faculty gravitated to the library.

First, the information needed to be organized and cataloged. This created specialized roles for those traditionally charged with these tasks in other libraries. Maintaining serials was challenging, as the law changed quickly.

Second, the patron needed actual assistance finding information. When a library consisted only of case reporters, it was easy for the researcher to find the necessary materials. Now, with the proliferation of a large variety of material, librarians became experts in helping patrons locate these specialized resources.

Finally, the library needed a strong director and a team of administrators to form a collection development plan, to continue to work with the parent institution in creating a solid collection, and to create legal bibliographies that were useful to the growing number of legal researchers.

In private law firms, libraries also held a prestigious allure. As law firms grew in the middle of the twentieth century, large, impressive law libraries became the centerpiece of the firm. At first, law firms only collected reporters and other materials related to case law research and legislation.[3] They left the broader spectrum of information to academic institutions. However, firms did start acquiring specific practice-related secondary sources as these materials developed.

The need for specialized law librarians grew alongside the field. The American Association of Law Libraries (AALL) was founded in 1906. Its membership grew as the profession blossomed. AALL allowed librarians across the spectrum to participate in meaningful discussions of the issues that were specific to working with legal information. AALL still thrives today as more members enter the ever-changing field of law librarianship.

TYPES OF LAW LIBRARIES

Law librarians have been inextricably linked to both the study and the practice of law. Where law libraries used to be reserved for the elite, law libraries currently flourish in many different settings, serving a variety of patrons. Students, scholars, attorneys, *pro se* patrons,[4] government employees, and prisoners alike all have access to legal information. Libraries vary in type, but, in general, all libraries are charged with helping patrons find legal information efficiently.

Academic Law Libraries

Academic law libraries require a lot of personnel. Not only do they provide numerous types of public service assistance to the law school community, but librarians also work behind the scenes in maintaining the integrated library system (ILS), cataloging both the monographs and serials, working with and troubleshooting electronic resources, and doing other, specialized work.

Academic law librarians in public services usually possess both a JD and an MLS.[5] This lends assurances to the student researcher that the librarian has gone through the same process. Law students are introduced to the concept of legal research during their first year of law school. Ideally, law librarians are charged with teaching basic legal research, as they are the experts in the field. In most schools, librarians teach advanced classes in legal research.

Law libraries typically have a busy reference desk. Although there is more legal information available now than ever before, it can still be challenging for the student to conduct research in a sophisticated manner. As is the case in any library, legal reference librarians conduct sophisticated reference interviews to gain insight into the resources that might be helpful to the student who is writing a paper, working in a law school clinic, or conducting research for a faculty member. Unraveling some of the mysteries of the research process provides many teachable moments for reference librarians. The more students come to the reference office, the better equipped they will be for their post–law school career.

However, law librarians differ from their other academic counterparts in one important way: In law schools, most students are learning research in order to apply it in a professional context. Therefore, the skills the librarian teaches are necessary to make the student ready to practice. Although all librarians work to make students more efficient researchers, law students must be efficient in order to perform cost-effective research once they are in practice. Most importantly, students must learn how to make sure their research is up to date and that the sources upon which they are relying are still "good law." In addition to providing reference services, academic librarians work directly with faculty in providing research assistance, creating bibliographies, collecting library materials related to the faculty member's area of scholarship, and assisting the faculty in the digitization and preservation of their work.[6]

Traditionally, for law schools within a larger university setting, law libraries maintained their own catalog. As the online public access catalog (OPAC) has emerged in the past two decades, many law libraries have continued this tradition.[7] The justification for maintaining a separate catalog and technical services team has been that that legal information, particularly serials and foreign materials, is different or specific enough to warrant specialists.

Law Firm Libraries

The library has traditionally been a central part of the law firm environment, particularly in large firms. With reporters lining the walls, and large tables set up for research, lawyers and their clerks, assisted by librarians, would come to the library for their research tasks. Libraries often focused on beauty as well as functionality.[8]

The library's space and the needs of the attorneys dictate the size of the collection. Many large firms have a main library and satellite libraries in other branches of the firm worldwide. Different offices may keep separate parts of the collection and material is shared throughout. The roles of the librarians were to select and maintain the materials, answer queries, assist with research, and provide cataloging functions.

However, this all changed when the economic downturn began in 2007. It hit law firms particularly hard, changing the face of their libraries forever. The libraries that remained were severely downsized. Some branch offices were eliminated, and almost all of the decorative library space was removed. Libraries had become more than "library as place" as the concentration focused solely on function.

Currently, many print resources have been canceled except for those that are absolutely necessary. Most firms operate using mainly electronic databases for research. This is a big change, particularly for those not familiar with the various electronic resources. Librarians are often the members of the firm that first receive the bills from the electronic services (such as Westlaw and Lexis). When an associate runs up a large bill, it is often the librarian's job to speak with the associate and insist that the associate receive extra training.

Librarians in law firms have many of the same tasks as academic librarians, especially in terms of assisting patrons in retrieving information. But firm libraries move at a different speed. The activity in the law firm drives the workflow of the library. If an attorney needs a document, the librarian retrieves it as quickly as possible (there are few "teachable moments" in a firm setting). Librarians are usually part of the billable hours scheme at firms, so finishing tasks quickly is of utmost importance.

Few attorneys visit the library in person. Rather, they use e-mail as their main source of interaction. The vast majority of reference interviews are therefore conducted virtually, rather than in person.

Law firm librarians also handle specialized tasks for their patrons. For example, librarians work with attorneys to set up "current awareness" tools that keep attorneys abreast of the issues in their particular fields. In large firms, librarians may assist with large "Knowledge Management" (KM) databases.[9] These databases contain all of the documents produced by attorneys at the firm, such as pleadings, interrogatories, and briefs, so other attorneys can find them by keyword searching.

Since the economic crisis, many law firm librarians had to become experts in researching competitive corporate intelligence information in order to assist attorneys in generating new business for the firm.[10] Some firms have even outsourced their library work or have only one professional librarian on staff.

Government Information and Libraries

Government libraries are part of librarianship's rich history in the United States. The Library of Congress is a prime example. Libraries within the governmental bodies are an obvious fit because the of the government documents that come from the Government Printing Office. Much of this information is becoming more freely available online, so the law librarian's job (in any type of library) is to assist patrons in finding what they need from the enormous amount of government information that comes in every day.[11]

The U.S. government has a printing office that publishes our primary materials. The proliferation of government documents throughout the years has required that librarians become specialists in how to find and use this information. Primary government documents flow from the three branches of government—Congress (legislation), the administration (regulations), and the judiciary (case law). This is the case with the federal government, but the states and U.S. territories also publish much of their primary materials in various formats.

The proliferation of government materials makes libraries associated with the government a natural fit. One can find law libraries in many governmental bodies, including both federal and state courts, county and state buildings, offices of both the district attorney and public defender, and offices within Congress and the various federal agencies.

Government libraries serve two different sets of patrons. First, they serve employees of the various government departments who work in the parent institution of the libraries. However, many county and state libraries act as libraries for attorneys and the public alike. This presents certain specific challenges for the librarian.

Assisting government employees is the same as assisting any patron who needs information. Like a law firm, the information is needed quickly as a judge may be making a decision or a member of Congress may be ready to vote on an issue. The librarian must help gather the necessary resources so the business of government can run smoothly.

But unlike other types of law libraries, many government libraries are open to the public. Serving the public presents certain opportunities and challenges for librarians who work in this field. Because most public patrons are not well-versed in legal research, the librarian has opportunities for "teachable moments," which is not usually a task outside of an academic setting. It is really satisfying to teach basic legal research to people representing themselves *pro se* or conducting legal research on behalf of another.

Assisting public patrons also presents problems that a librarian may not anticipate. Namely, public patrons may want "legal advice." Librarians may not give legal advice. They are only able to guide patrons to the resources they may need and assist them in using those resources.[12] Sometimes it is hard to find the line between giving advice and providing research help. Therefore, many law libraries open to the public have lists of free and low-cost legal services on hand. Public law librarians also publish pamphlets to assist patrons with simple research tasks.

Special Libraries

Law libraries are not only found in large institutional settings. There are other types of libraries as well. Stand-alone, subscription libraries were some of the first law libraries in this country. They still offer a variety of services (both walk-in and remote) to organizations of all sizes. Subscribers have access to databases as well as print materials through these libraries.

One can also find libraries in the nonprofit sector. It is not uncommon for advocacy groups, NGOs (nongovernmental organizations), and public interest organizations to have their own small law libraries focused on the specialized work they do. Most often, these libraries are staffed by a solo librarian who collects resources that are relevant to the organization.

CHALLENGES NOW AND INTO THE FUTURE

Shrinking Collections

Many factors are changing the practice of law librarianship across the profession. First, and most glaring, is the economy. As it was, libraries had begun weeding their collections as subscription prices increased. But beginning in 2007, when the economy hit law firms particularly hard, all law libraries had to make changes to their collections and personnel, and law librarians had to acquire new skill sets.[13]

The shrinking print collection is the most noticeable and obvious part of downsizing. The cost of renewing subscriptions and continuing serials has substantially increased, with some publishers asking for costs that are prohibitive, even to well-funded libraries. As a result, libraries are making different choices regarding their collections.[14]

One solution is for the library to involve itself in a resource-sharing consortium. In such consortiums, libraries can share the cost of electronic resources, thus giving each library a substantial discount. Libraries can also arrange for local or regional interlibrary loan programs that do not require long distance shipping charges. Finally, many libraries are opting out of print resources altogether, in favor of their electronic counterparts.

Due to the realities of the economy, new associates no longer have the luxury of learning research on the job. They need excellent legal research skills from day one. Although there has historically been tension between law firm librarians and academic librarians as to the readiness of new associates, everyone is now working harder to make new lawyers practice-ready. In law school settings, the trend is to offer more advanced legal research classes—even specialized ones—so new attorneys have the skills they need on day one.[15]

Changes in Technology

New technology has required law librarians to quickly adapt to the ever-expanding technological landscape. This can provide a challenge to librarians and patrons alike who have not been exposed to some of the newer platforms. Sometimes, there is a steep learning curve.

For the most part, however, technology has positively impacted the way we practice librarianship. Libraries are hiring new types of librarians such as those who are experts in emerging technologies and those who work with the integrated library systems (ILS). Workflow is changing and adapting to the needs of patrons. An example is "chat" reference. This is a new way to meet patrons in a medium in which they are comfortable.[16] Although much legal research no longer requires a trip to the library, developments such as chat keep reference traffic statistics up. The reference transaction is just as informative; it is only packaged in a different format.

By necessity, law libraries have adopted marketing techniques that attract younger digital natives. Some libraries communicate with their patrons through social media platforms. Many law libraries have successful blogs that keep patrons abreast of everything from the library's collection to changes in laws that affect the patron base. Other libraries have Twitter feeds that announce new faculty publications or research events in the library. Other libraries successfully use Facebook to announce changes in the library's schedule or encourage attendance at library-related activities.[17]

International and Interdisciplinary Research

Legal issues are becoming more global. Lawyers practice across countries and intergovernmental organizations like the European Union and the United Nations. Treaties and other instruments govern the laws of jurisdiction as well as subject matter. Librarians must be trained in how to locate the governing laws and this can be quite challenging. Some materials are available only in print, and others are contained in websites in other languages.[18] The learning curve may be steep, but law librarians need the skills necessary to convey this information.

Law librarians have had to quickly adapt to these changes. In some libraries, there is a foreign and international law specialist. In others, all librarians must be well versed in searching for complex and intertwining multinational laws.

In addition to foreign and international resources, more researchers are using statistics, as well as literature from the humanities, social sciences, and medicine.[19] Creating bibliographies or researching across disciplines is a necessary skill that may not have been as important in recent years. Now it is critical.

CONCLUSION

In spite of the proliferation of new technology and new tasks, the law library has remained unchanged in many ways. Law librarians are still charged with finding information even if the format has changed. Part of the job is creating bibliographies and finding aids, even if many of these are now online. Irrespective of the format, librarians who conduct collection development tasks still negotiate with publishers in a similar format. The medium has changed but not the process or the functions.

There is a misconception that because we live in a digital age, libraries and librarians are becoming obsolete. This could not be further from the truth. We are living in a time of information overload. But library professionals are able to weed through the vast amount of information and produce the correct, authoritative information for their patron base.[20] This is particularly true in the law library setting. Information that has not been vetted can cause someone to lose a case.

In order to remain relevant in the eyes of our parent institutions, law librarians need to develop skills to keep up with changing technology. Just as lawyers subscribe to current awareness tools, librarians too should take it upon themselves to read about the latest trends. Additionally, librarians should attend skill-building workshops and conferences. If librarians do not engage in professional development, the profession runs the risk of becoming irrelevant or obsolete. Law librarianship has a long and distinguished history. Today's librarians are more skilled than ever and can certainly usher the profession into the digital age.

FURTHER READING

Acosta, L. M., and A. M. Cherry. 2007. "Reference services in courts and governmental settings." *Legal Reference Services Quarterly* 26 (1–2): 113–134.

Balleste, Roy, Sonia Luna-Lamas, and Lisa Smith-Butler. 2007. *Law librarianship in the twenty-first century*. Lanham, Md.: Scarecrow Press.

The changing role of academic law librarianship: Leading librarians on teaching legal research skills, responding to emerging technologies, and adapting to changing trends. 2008. Boston, Mass.: Aspatore Books.

Danner, R. A., S. B. Kauffman, and J. G. Palfrey. 2009. "The twenty-first-century law library." *Law Library Journal* 101 (2): 143–156.

Estes, Mark. 2007. "Reference services in law firm environments." *Legal Reference Services Quarterly* 26 (1–2): 1–2.

Gasaway, Laura N., and Michael G. Chiorazzi. 1996. *Law librarianship: Historical perspectives.* Littleton, Colo.: F. B. Rothman.

Marke, Julius J., and Richard Sloane. 1982. *Legal research and law library management.* New York: Law Journal Seminars-Press.

Primary Research Group. 2002. *Law library benchmarks.* New York: Primary Research Group.

NOTES

1. The doctrine of precedent, under which a court must follow earlier judicial decisions when the same points arise again in litigation.

2. Robert C. Berring Jr., "A brief history of law librarianship," in *Law librarianship in the twenty-first century*, edited by Roy Balleste, Sonia Luna-Lamas, and Lisa Smith-Butler (Lanham, Md.: Scarecrow Press, 2007).

3. The West National Reporter System dominated the legal research field. A large team of legal experts, both lawyers and librarians, indexed all points of law into various key numbers. A sophisticated digest listed all cases by key number, so the researcher could find cases by subject within reporters. The West National Reporters are large sets of gold books, which helped make the law firm library so impressive.

4. *Pro se* patrons are individuals who are representing themselves in court.

5. There has historically been a debate as to whether or not librarians should have a dual-degree in order to be a public services librarian in an academic law library. There are arguments on both sides. See Karl T. Gruben "Working at the law library: A practical guide," in *Law librarianship in the twenty-first century*, edited by Roy Balleste, Sonia Luna-Lamas, and Lisa Smith-Butler (Lanham, Md.: Scarecrow Press, 2007).

6. Not all academic libraries perform the same tasks, but most do these basic tasks in varying degrees.

7. Some law schools, like Harvard, have integrated their catalog with that of the larger university. For most schools, however, this is not the case.

8. Law firms were proud of their libraries and showcased them. See Joan Batchen, "Life at a large U.S. law firm," *Legal Information Management* 6, 3 (2006): 190–193.

9. Elizabeth Nelson, "Knowledge management for libraries," *Library Administration & Management* 22, 3 (2008): 135.

10. Arthur Weiss and Ellen Naylor, "Part I: Competitive intelligence: How independent information professionals contribute to organizational success," *Bulletin of the American Society for Information Science and Technology* 37, 1 (2010): 30–34.

11. Some libraries act as federal depositories for government documents. This practice has become less prevalent as more official government documents become available online.

12. Amy Hale-Janeke and Sharon Blackburn, "Law librarians and the self-represented litigant," *Legal Reference Services Quarterly* 27, 1 (2008): 65–88.

13. T. Fitchett, P. Hazelton, J. Hambleton, A. Klinefelter, and J. Wright, "Law library budgets in hard times," *Law Library Journal* 103, 1 (2011): 91–111.

14. Alicia Brillon, "Collection analysis when the budget decreases," *Legal Reference Services Quarterly* 30, 4 (2011): 289–298.

15. See Patrick Meyer, "Law firm legal research requirements for new attorneys," *Law Library Journal* 101, 3 (2009): 297.

16. Yasmin Morais and Sara Sampson. "A content analysis of chat transcripts in the Georgetown Law Library," *Legal Reference Services Quarterly* 29, 3 (2010): 165–178.

17. Deborah Ginsberg, Meg Kribble, and Bonnie Shucha, "Inspiring innovation: Planning, implementing, and evaluating the Web 2.0 Challenge," *Law Library Journal* 101, 3 (2009): 355.

18. Mary Rumsey, "The skills they need: International and foreign legal research," *Legal Reference Services Quarterly* 28, 1–2 (2009): 133–149.

19. John Palfrey, "Cornerstones of law libraries for an era of digital-plus," *Law Library Journal* 102, 2 (2010): 171.

20. R. L. Sims and R. Munoz, "The long tail of legal information: Legal reference service in the age of the content farm," *Law Library Journal* 104, 3 (2012): 411–425.

2

Embedded Librarianship

Thomas J. Striepe
UNIVERSITY OF GEORGIA LAW LIBRARY

Mary Talley
TALLEY PARTNERS

> Across the globe, we see librarians leaving the library. Librarians are going to the conversation. Librarians are sitting in business meetings . . . are joining legal teams in court. . . . [Librarians are] showing up where they are needed and not waiting for someone to find them. They are proving their worth.
>
> —R. David Lankes[1]

The traditional model of library service is failing our organizations and us. It's an old model that depends on the initiative of information users to seek out a librarian and ask questions—if and when they decide a librarian can be of help. It presumes that information seekers know when to go through a librarian for their data. Couple this assumption with the physical nature of libraries and print resources, and librarians have had little incentive to seek out information users. That is, until information became unshackled from place and time, and unleashed information consumers from the library. With the rise of abundant, ubiquitous, fast information, former library users found they did not need librarians the same way they used to. For law librarians, the evidence of this is dramatic—from increasing predominance of self-serve digital resources, to vastly shrinking print collections and library space, to uncomfortable questions about relevance and value. Today, the problem is not only that they don't need us in the same ways, but that we continue to operate as if they do while information users still make the decision of when and whether to use us. And they often turn to Google instead of us.

For many law librarians, the question is no longer whether to replace the passive, transactional service model, but what to replace it with? And, of course, any new model must guarantee roles that are relevant, visible, and valuable to our organizations. A tall order—but one that is being filled by a radically new approach that is gaining traction in all types of libraries and frequently described as "embedded librarianship." What is it? How and why is it effective? What does it have to do with law libraries? How does one get started? In this chapter, we tackle these questions using firsthand research, professional literature, and real-life examples to explain the what, why, and how of embedded librarianship as it is occurring in private and academic law libraries.

UNDERSTANDING EMBEDDED LIBRARIANSHIP

The term "embedded librarian" (and "embedded librarianship") has been popularized in recent research studies,[2] professional literature,[3] and in several new books,[4] and used as a convenient framework for discussing a paradigm shift in the way librarians interact with information users and contribute to their organizations' goals and mission. At the same time, "embedded" is a somewhat misleading term. It is applied to a wide range of collaborative roles: from ad hoc collaboration with a committee, group, or other organizational community at one end, all the way to physically embedding with an organizational unit to provide exclusive, specialized work products—the relationship most commonly associated with it—at the other. Perhaps a clearer and more precise way to understand the concept is to think of it as "intentionally building strong, collaborative work relationships with communities[5] in our organizations." This approach can just as accurately be termed "collaborative" librarianship, and we use the terms "embedded" and "collaborative" interchangeably in this chapter.

Engaging in collaborative work with internal communities repositions librarians within their organization by changing the dynamic of "them and us" to "us" and, in the process, changes the perception of what librarians can and should do for their organizations—and of their value. When we work closer to an internal community's point of need, we have more influence over the decision to use our expertise and, by extension, a greater impact on our organizations' information-related challenges. To understand the importance of this shift in relationship, we have to first look hard at where traditional service places us within our organizations.

EMBEDDED VS. TRANSACTIONAL ROLES

Typically, our interactions with information consumers are transactional in nature: We provide a service or a good (information, training, resources) in response to an information consumer who determines a need and makes a request (asks a question, uses a resource, gets training). When we are asked for information or research, we are given only a specific piece of the information consumer's challenge, not the whole. Without full knowledge of the whole, our roles and our influence are always limited; it's almost impossible to understand the actual part our work product played in the outcome, to know what other information-related issues we might have solved but weren't asked because the information user was unaware of the full extent of our capabilities. Transactional relationships are superficial, preventing the sustained interactions necessary to develop deep knowledge of the information consumers' issues, goals, and objectives. Transactional roles, by definition, are adjunct roles: In them, we are always the supporting cast, never the main characters, and in the uncomfortable position of continually proving relevance and value.

Collaboration has the power to transform librarians' roles in their organizations, making it possible to actively participate in and contribute to the high-value work of their organizations. A look at the shared characteristics that set them apart from transactional librarians can be summed up in three words: relationship, results, and relevance.

- *Relationship:* The most successful embedded librarians—and those who enjoy the most job satisfaction—are the most adept at building strong relationships with organizational communities. They depend on these relationships for new and meaningful insight into their communities' work and information challenges. Strong relationships and insight open up opportunities to recognize information-related issues that might not otherwise be put to the librarian to solve or be recognized. New and unique opportunities provide a wider arena and demonstrate more capabilities than possible in a transactional relationship. Strong relationships, new insights, and a larger view of capabilities are the foundation for building knowledge, credibility, and trust: the ingredients for successful collaboration.
- *Results:* Work products tailored to community-specific problems are a hallmark of collaborative librarians. Taking on complex projects and customizing work products is not only time intensive but also requires the librarian to be thoroughly conversant with the issues. Those who are most successful at it have traded in lower-level, low-value tasks for complex work, such as competitive intelligence, market research, in-depth research and analysis, and creating technology-related solutions.
- *Relevance:* A librarian's work product may be exactly what was needed and invaluable to information users; but it—and the librarian—are only truly valued when the work product is recognized as a relevant contribution to

> the community's agenda and goals. Relevance, in this context, depends on the depth of the engagement with internal communities: Deeply engaged librarians are focused on the success of their teams. Consistently, collaborative librarians have replaced the concepts of service with that of collaboration, team member, and partnership.

How do these professionals build collaborative relationships? They approach collaboration as a sustained conversation in which information needs or problems might not be overtly expressed and that they must enter to sustain visibility, relevancy, and value. They also understand that intimacy is required to enter this conversation: intimate knowledge of the information users' work context and challenges, which is the basis for trust. They then put these principles into action using these tactics:

- Engage in multiple, reciprocal interactions with information users at all levels of their communities, informally (for example, through hallway chats, shared meals, and volunteering for committees) and formally, sometimes relocating to their workspace, but more often showing up where they work (at meetings, in projects, seminars, and classrooms).
- Broaden reach, effectiveness, and sustainability by focusing on communities rather than individuals.
- Cultivate management support: to sanction and fund the embedded position; to integrate the librarian into the community and its work; to manage information needs at the community level; to get feedback and direction on work priorities.
- Stay alert for unspoken information challenges and offer solutions using the opportunities inherent in multiple and varied interactions.
- Increase subject knowledge and capabilities by attending the same in-house classes, external educational programs, professional meetings, and conferences.
- Leverage increased knowledge and capabilities to take on more complex work.
- Disregard typecasting by function and break boundaries by taking on non-traditional assignments and roles. Ask big questions: "What gets in the way of your work?" not "What information do you need?"
- Earn trust: Trust is required to enter the conversation; knowledge, credibility, and accountability are the foundations of trust.

The characteristics and practices of collaborative librarians are illustrated in this chapter's sections on embedded roles in law firm and academic libraries.

Collaborative Roles in Law Libraries

There is little in the professional literature on the adoption of embedded roles in law libraries, although it's not a new concept to either law or general librarianship Medical libraries were early adopters of the embedded concept. In the 1960s, librarians began accompanying doctors on their rounds and providing specialized research in the context of a patient's situation. This new arrangement changed the relationship between librarian and information user from service provider to collaborator. Law librarians, in their own right, have had a long tradition of developing subject expertise and providing targeted services to practice groups. Academic law libraries have been putting librarians in classrooms to teach legal research for many years. In addition, there is a general indication from recent research and other surveys that collaborative librarianship is a growing trend in legal librarianship. A 2009 survey of SLA members—including a substantial number of law librarians—conducted for the Models of Embedded Librarianship research project, showed that embedded librarians were present in similar proportions across seventeen different industries, including legal services. In fact, the percentage of embedded librarians vs. non-embedded in law libraries was second only to that of academic libraries.[6] In a 2012 survey[7] of embedded practices in 244 law firms, conducted by Ark Group, 57 percent of law firm libraries either had some type of embedded role (26.2 percent) or were considering establishing one (20.5 percent). According to this survey, embedded librarianship is in the midst of a growth spurt: 63 percent of the embedded librarian programs represented in the survey were established within the past five years. Unfortunately, similar data are not available for academic law libraries to determine prevalence or growth.

Collaborative Roles in Law Firm Libraries

Law firm librarians have seen their libraries empty out as legal materials migrate to e-resources available at the desktop while, frequently, seeing information requests increase. Still, law firm librarians face multiple challenges to their visibility, relevance, and value. Cindy Adams asked, in her post "Law Librarians, Out of Sight, Out of Mind?" on the *3 Geeks and a Law* blog, "If no one comes to visit the library, is the facility, and its staff, still relevant?"[8] Another challenge to visibility and relevance, underlying the question, is the perception of librarians who remain located with the least relevant of resources: print materials. The performance-based environment of the law firm, in which value is determined by clients and billable hours, challenges the value of the law firm librarians, who are often gathering numbers and statistics to prove that they are of value. Some librarians have effectively met these challenges by assuming collaborative roles.

The programs and relationships described in the following profiles illustrate how actual law firm libraries are doing it and why it is working.

Profile #1: Collaborative Relationships with Peer Departments and in a Firm-Wide Initiative; Successful Relationship-Building Strategies

Jill Strand, the director of the Information Resources Library for a mid-size law firm in Minneapolis, manages library operations, technology deployment, and e-resource acquisition, and has primary responsibility for reference and research requests—all with the staff of one.[9] Partnering with an individual practice group and locating to their workspace to engage in complex, customized work is not possible in such a small operation—but building collaborative relationships is. From the start, she set out to personally meet with every attorney, until as she put it, she "realized this is what I should be doing every day, not just once." Relationship building is part of Jill's day-to-day work life and her management style, which she calls "management by movement": engineering frequent, face-to-face interactions with attorneys, practice group leaders, and firm management. She engages with them on both work and personal levels (remembering birthdays and other special events in their lives), building her presence into attorneys' day-to-day work life. Relationship building extends to peer department heads with whom she meets frequently for lunch to establish common ground around firm issues.

A number of effective collaborations have come out of these relationships, enriching the firm and Jill's work and adding to her capabilities. On one, she partnered with the IT director on a library orientation program to familiarize IT staff with the library's e-resources. On another project, she provided a major component in the evaluation, negotiation, and deployment of cite-checking software for firm secretaries, which she managed after the software's deployment. Jill's expertise in these areas is recognized by the firm and has increased her credibility.

Recently, Jill was tapped by the IT director to serve on the firm's Technology Committee and invited to participate by the firm's managing partner. She now partners with attorney and department heads to develop firm-wide strategies that improve work processes, has assumed a major role in the evaluation and selection of a technology consultant, and has discovered an internal IT technology that she is adapting to improve workflow processes. Insight into "what keeps her bosses up at night" and a broader understanding of firm priorities have enabled Jill to focus the library's priorities to better match those of the firm. Taking on nontraditional roles has given her considerable capabilities, a wider audience, and more recognition in the firm. Jill considers part of her function is to improve the firm efficiency. Now, as part of the Technology Committee, she shares its purpose of improving core firm processes and visibly contributes valued solutions, increasing her knowledge and credibility.

Profile #2: Embedded Librarians Colocated with Practice Groups

The experience of Cindy Adams, codirector of library services for McKenna, Long and Aldridge, is an example of how a creative approach can overcome barriers to developing collaborative roles.[10] In addition to managing firm-wide services, she also directs four research librarians in the Atlanta office. The library there is inconveniently located on many floors and two banks of elevators away from the attorneys. Opportunities for face-to-face interactions—and for relationship building—were rare, as the attorneys had less and less need to come to the library for information. The librarians felt isolated and concerns arose about the library's visibility and relevance. Efforts to create paths for face-to-face interactions between the librarians and the attorneys and peer department heads were stymied at every turn. The librarians needed to get out of the library, but their workload—which had never decreased—kept them at their computers. Efforts to participate in team meetings had only weak results as tensions persisted between the library and the lead administrative department. Pairing a librarian with a practice group met with other obstacles, including the need to choose one librarian over others and pushback from practice group leaders.

Cindy's solution to these roadblocks was to locate each of the librarians with a practice group. To make the plan work, she sought buy-in, first, from her department head, the director of IT, and then from executive management, who endorsed it. In fact, they saw the codirector as forward-thinking and the plan as an innovative solution to increasing the librarians' value to the firm. Buy-in and support from firm decision makers was essential: Sanctioning the embedded program negated the need for individual practice group approval and partnering the librarians with the practice groups increased their standing, which facilitated the librarians' inclusion in team meetings and contributed to the decrease in interdepartmental tensions. Management support also paved the way for the allocation of office space—although it took a year to relocate one of the four. Finally, involving all four librarians ensured that no one was left behind and that practice groups received equal coverage.

Today, the librarians are taking initial steps toward collaborative work with the practice groups. Proximity has led to frequent conversations at the point of need, allowing them to proactively identify the need for targeted news alerts and to introduce pertinent e-resources attorneys were previously too busy to learn. Accessibility raises visibility and relevance: Attorneys have noted that they are going to the librarians for questions they wouldn't have otherwise thought of asking. Cindy reports uniformly positive feedback on the librarians' presence in the practice groups and an increase in questions from the attorneys as well. Being close to the practice groups has given the librarians the opportunity to interact and develop relationships with partner-level attorneys, who, in turn, are seeing the librarians' capabilities in action. Positive interactions have also been facilitated with administrative department heads. As a result, the librarians are experiencing a stronger sense of inclusion in the teams and in the life of the firm.

Profile #3: Multiple Collaborative and Embedded Relationships for Targeted Areas of the Firm, Including Executive Management, Business Development, and Knowledge Management Initiatives

Collaborative and embedded roles have been an integral part of this large, international law firm's library for the past ten to twelve years.[11] Recently, these roles have gained in importance and demand that all librarians in the firm have the ability to build relationships and work collaboratively. In a recent interview, the firm's chief library and records officer stated, "Reactive service is not working in the modern world. To ensure longevity, we have to get out of the library."[12] The firm's culture reinforces this perspective: The chief has described it as collaborative with decentralized practices and management.

Under the encouragement of the chief, collaborative roles and programs initially grew out of existing, informal work relationships or in response to needs articulated by the practice groups. These programs included knowledge management analysts embedded with specific practice groups; research analysts embedded in administrative departments, such as business development; and other researchers collaborating on cross-functional teams in such areas as mergers and acquisitions.[13] In the past two years, the chief has strategically expanded collaborative programs to serve high-value segments of the firm and to extend collaborative roles to include all library staff.

In 2012, the chief expanded the small, decentralized research analyst program into a coordinated group of five librarians dedicated to the information needs of senior executive management. This new arrangement corrects several issues: uneven levels of support, favoring some executive managers over others that arose from a system of decentralized, embedded analysts; and the research and analysis work contributed to executive management reports by non-embedded librarians that had gone uncredited. Now, the original model of embedding a librarian in a single department or practice group and separating them from the library has been reversed. Individual embedded analysts now operate as part of a global research team, collaborating with all executive managers. A manager heads the team, coordinates assignments, and provides relief from assignment overload. The chief reports that practice chairs and other executive managers have enthusiastically embraced the new global research team concept and that assignments are on the increase. Today, analysts receive full credit for their research contributions; their names appear on the covers of the high-level reports they help produce, improving both visibility and value.

In this firm, building capacity for embedded roles was formerly an evolutionary process. Embedded librarians were self-selecting, developing collaborative roles after first learning the firm's culture and practice groups and building credibility and trust with the attorneys. The library has several collaborative programs that reference librarians can participate in, such as one in which librarians partner work with attorneys on developing content for a regulatory tracking site. This self-selecting approach has changed in the past two years: Collaborative capabilities are now directly encouraged in all of the reference and research librarians. Each librarian has been tasked with devising a strategy to partner with a practice group either as a liaison or specialist. Library management provides training and specifies the plan's components, such as a justification for the selection of the practice groups, identification of the appropriate person to approach within the practice group and how, and a description of the skills required and potential services or contributions. Library managers meet with the librarians periodically to evaluate progress and provide further assistance. This program is still a work in progress, but early evidence shows that the reference and research librarians are well on their way to becoming a more collaborative workforce.

Here, a law firm library with a well-developed and successful embedded program continuously expands, restructures, and innovates collaborative relationships. The chief library and records officer strategically drives the programs and builds capacity in the firm librarians. The chief reaffirms the importance of collaborative relationships by nurturing his own with practice group and firm management, and in return finds new opportunities to build collaborative relationships between the library and the firm's attorneys and management.

EMBEDDED ROLES IN ACADEMIC LAW LIBRARIES

The role of the embedded librarian continues to evolve within academic law libraries. This new model has increased collaboration between students and faculty as the librarians have started to actively push their services beyond the walls of the library. Although the level of collaboration will continue to expand in the future, academic law libraries have embraced the embedded librarian model. This section of the chapter will focus on specific examples in three different areas that the embedded librarian model is currently being utilized to more effectively assist patrons: embedding to advance legal research instruction, physical embedding outside of the law library, and embedding within classes.

Embedding to Advance Legal Research Instruction

Law students obtain basic research skills through their first year legal research and writing classes and some students take advanced research classes or topical research classes to expand upon these skills. However, law students' legal research skills are still lacking when they enter the workforce.[14] Librarians have joined together with faculty members to use librarians' expertise in research to instruct students on this topic in other environments. This increased interaction between librarians, students, and faculty allows students to obtain necessary research skills during law school and has brought about a new partnership with faculty members.

One-Shot Instruction

On the sliding scale of embedded librarianship, "one-shot instruction" falls on the lower end. In this arrangement, librarians are typically given one class session or a portion of a class to provide legal research instruction on the applicable legal topic. This setup offers the librarians limited access to teaching research on the legal subject, but librarians can use this opening to work with faculty to offer research training to students in a combination of ways.

Course Management Software

One way that law libraries have expanded legal instruction is through Westlaw's TWEN, Lexis Web Courses, or other university platforms (Blackboard, Desire-2Learn, Moodle). These course sites allow law librarians to go directly where the students will be online. Students' research no longer centers on the physical space of the library or the library website.[15] These course websites offer librarians an opportunity to go where students are and provide legal instruction to a captive audience. There are a variety of ways that law librarians can use these class sites to provide training:

- *Library Links:* The easiest way to let students know about the resources on a legal topic is to create links to the resources at the law library. These links can bring the students directly to the books, databases, journals, websites, tutorials, and research guides on the relevant legal subject. This is one of the easiest and least time-consuming methods to make students familiar with the various resources available to them.
- *Posting:* Law librarians can also use the discussion boards of course management software to increase legal research instruction. Law librarians can start a discussion topic on legal research, allow students to ask questions directly to the law librarian, provide daily or weekly research tips to the students, provide information on citation, or deliver information on how to find a research topic.[16] Law librarians can meet with faculty members to come up with different topics to post throughout the semester and also to brainstorm additional learning opportunities for students on the class site.

- Other ideas include having live chat sessions between the librarians and students throughout the semester, posting research exercises for the students, having quizzes, or creating online tutorials specifically for the class.

Below are some keys to successfully using course management software for legal research instruction:[17]

- Law librarians should get to know the platform and what options are available to users. Does it allow you to directly link to certain databases or articles? Are there online quizzes available? Is a wiki feature present? Does it permit instant polling? Work with your Westlaw, Lexis, or other representative to become knowledgeable about the product.
- Law librarians need to provide more than links to their patrons. Although links to helpful sources permit students to learn about the resources available to them, there needs to be interaction between the students and the librarians. This interaction allows students to ask questions directly to the librarian and permits the librarian to better understand what the student is looking for or with which part of the legal research process he is having trouble.
- Law librarians must devote sufficient time to review posts, locate relevant resources, and come up with interesting topic discussions or exercises for students. In order to have a successful online embedded program, the library needs to fully commit to the project.
- Law librarians need to dedicate themselves to marketing this service. Go further than the simple e-mail to faculty letting them know about the availability of librarians to embed in their class sites. Provide a free lunch presentation, letting them know how it benefits their students, design a short brochure, create a webpage about the service, or stop by and talk directly to faculty members in their office. Also, librarians should introduce themselves to the class at the beginning of a semester so that the students will know about the embedded librarian program.
- At the conclusion of each semester, law librarians should immediately assess what worked and what did not work for each class site that had an embedded librarian. Gather feedback from the students and faculty on what methods you used that helped them or were not helpful; see if they have any suggestions for other things that would be beneficial. This assessment allows law librarians to better assist the students and faculty in future class sites.

Creation of Websites or LibGuides

Another option for furthering legal instruction outside of the classroom or course management system is to use a website or a LibGuide that is specifically tailored to a class. Either of these options permits librarians to create links, embed images, video, RSS feeds, utilize Google Gadgets, and create a functional, intuitive platform for students.[18] There are many examples of law libraries using LibGuides or webpages to expand legal research instruction. William Mitchell College of Law Library produced a LibGuide for a Comparative and International Patent Law class (http://libguides.wmitchell.edu/IntPatentLaw), Washington University Law Library created a website for a course on Bioethics Research Law and Policy (http://law.wustl.edu/library/pages.aspx?id=4907), Southern Methodist University School of Law Library has created multiple LibGuides for courses offered at the law school (http://libguides.law.smu.edu/cat.php?cid=41603), and University of Maryland Law Library lists the research websites that it created specifically for its law school's classes (https://www.law.umaryland.edu/marshall/researchguides/coursepages/index.html).

Online Tutorials/Webinars

Online tutorials and webinars are another tool law librarians can use to push legal instruction to students at a point of need.[19] Once again, librarians can team up with faculty members to develop relevant tutorials based on the desires of faculty and students. The Stetson University College of Law Library began to use webinars and tutorials to reach students more effectively because of the scarcity of class space and scheduling problems with students.[20] The law library coordinated this project with the legal research and writing instructors to develop research topics on basic legal research tasks as well as advanced legal research skills. This embedded project was a success at Stetson with the webinars being attended by 403 students over three semesters and the tutorials being used over seventy times each month.

Legal research instruction is a major responsibility for academic law libraries. Law libraries can expand this instruction beyond the standard legal research classes into other places where students are located. The embedded model for legal research increases the exposure students have to research instruction and allows them to enhance their research skills prior to starting their first job after law school. In addition, all of these examples provide a way for librarians to better serve patrons by increasing the interaction with them.

Physical Embedding Outside of the Law Library

Physical embedding has recently started to gain momentum within academic law libraries. Almost all law libraries have a liaison program between the law library and faculty to support collaboration. Rather than wait for faculty members to contact them, law librarians are now starting to move out of the library and embed directly with both faculty and students to strengthen the relationship between the law library and these individuals.

Office Hours

A first easy step to establish the "embedded roles" with patrons is to establish office hours outside of the library. Although law librarians are always accessible through e-mail or phone, having office hours where patrons congregate allows the law library to become more familiar with what patrons are working on and create more connections between faculty and students. This is a low-cost embedding method that can permit law libraries to wade into the embedded librarian model without using a lot of library resources.

Law librarians can set up remote reference tables on a weekly or biweekly basis where students or faculty members typically are present. The "office" needs to be in a location where it will be noticed.[21] Places include student lounges, faculty libraries, faculty dining areas, law review offices, open spaces in front of elevators, student group meetings, or any space that is frequented by the students or faculty. Once the office hours are established, the librarian needs to be assertive in promoting the library's resources and services to the patrons that pass by. Always come prepared for the office hour session by having information about relevant databases, books, a recent court decision, or a new article that will interest students or faculty members. However, it is also just as useful to discuss what projects faculty members or students are currently working on and then market how the library can assist them in their research. It also never hurts to offer patrons something to initiate the conversation whether it is food, bookmarks, or a new brochure.

Roving

Another option is to simply have the librarians leave their office or reference desk and roam where faculty or students tend to be to make the law library available at the point of need. At Texas Tech University Law School, law librarians have established an informal program where the librarians will grab an iPad and visit faculty members' offices to see if there are any informational services that they can assist them with.

Stetson University College of Law Library started a more formal program named iLibrarian to connect with its patrons.[22] In this program, librarians roam with an iPad at established times and locations where students and faculty congregate. After patrons became used to the new locations, they began to approach the librarians (rather than the librarians starting the conversation).

Physical embedding offers librarians the ability to strengthen relationships among the law school community so that they better understand their needs.

This allows law libraries to provide services at a point of need and users end up with a better final product by employing all of the applicable resources available at the law library.

Embedding within Classes

One of the best collaborative models is to have law librarians embed physically within classes. This can be done easily as becoming established as the "personal librarian" for the class.[23] This has also been used in classes that have a writing component. In these classes, law librarians have worked with faculty members to require each student to meet with a reference librarian throughout the semester in order to explore different resources that will assist students in finding sources and topics for their papers.

Fully embedding within law school classes presents a wonderful opportunity for law libraries to assist both faculty and students. During 2010–2011, law librarians at the University of the District of Columbia, David A. Clarke School of Law, began to fully embed within three of the law school's eight clinics.[24] In their model, librarians would attend the clinic classes so that they could better understand the issues that the students were encountering during their work with clients. Faculty members and librarians worked together in designing the syllabus so that the class required students to meet with the librarians for their research questions and incorporated a research workshop at the beginning of the class. Throughout the semester the librarians would meet with all of the clinical students to know what issues they were encountering in research. This proved to be a successful program for the librarians, students, and faculty. Librarians were able to improve the way they teach legal research, students were able to improve their legal research skills, and faculty members were provided with a better work product from the students.

Physically embedding with classes offers law librarians a perfect opening to collaborate with faculty and students. This model allows librarians to understand firsthand the problems students are having and provide immediate solutions. They are also able to foster the relationship with faculty and can expand this into helping faculty with their other responsibilities.

Future of Embedded Librarians in Academic Libraries

The ultimate goal of the embedded librarianship model is to change the relationship between law librarians and patrons so that a partnership forms between them. Law libraries can look to how other academic libraries and law firm libraries are using the embedded model in order to increase their own embedded activities and reach the goal of having a true partnership.

Possibilities include having a librarian permanently relocate his or her office within faculty members' or clinic offices. Librarians could also work in partnership with faculty members in course design to directly incorporate legal research instruction throughout the course.[25] Law libraries can adopt a model like law firm libraries and have a librarian located with each "practice group" (i.e., tax law, civil procedure, constitutional law, etc.), or have a librarian embed within a legal institute or center at the law school. Libraries also could establish a librarian that focuses specifically on collaboration for faculty research and promotion of the final product.[26] The examples set forth in this section are just a starting point, academic law libraries need to continue to adopt strategies that will further permit librarians to become partners with their patrons.

CONCLUSION

Law librarians from academic and private libraries are transcending the limitations of traditional service models and functions and collaborative and embedded roles are providing the way. They are moving beyond the confines of their libraries, focusing outward on their communities' success, not inward on library functions and participating in the life of their communities. They are surviving and thriving in new, collaborative roles.

FURTHER READING

Kolonay, Brittany, and Gail Mathapo. June 2012. "Experimenting with embedding." *AALL Spectrum* 16(8): 18–20

Kvenild, Cassandra, and Kaijsa Calkins. 2011. *Embedded librarians moving beyond one-shot instruction.* Chicago: Association of College and Research Libraries.

Schulte, S. J. 2012. "Embedded academic librarianship: A review of the Literature." *Evidence Based Library and Information Practice* 7(4): 122–138.

Shumaker, David. 2012. *Bibliography: The embedded librarian: Innovative strategies for taking knowledge where it's needed.* Medford, N.J.: Information Today. http://embeddedlibrarian.com/2012/07/13/a-bibliography-of-embedded-librarianship/embeddedlibrarianbibliographyversion1/.

Shumaker, David, and Mary Talley. 2010. "Models of embedded librarianship: A research summary." *Information Outlook: The Monthly Magazine of the Special Libraries Association* 14(1): 26. http://www.sla.org/io/2010/01/802.cfm.

Talley, M. 2011. "Success and the embedded librarian." *Information Outlook* 15(3): 25–28. http://www.sla.org/io/2011/04/995.cfm.

———. 2010. "Rethinking value: How embedded roles make us valuable." In *The New Face of Value: 2010 Best Practices for Government Libraries*, edited by Marie Kaddel, 114–116. http://www.lexisnexis.com/tsg/gov/Best_Practices_2010.pdf.

NOTES

1. R. David Lankes, "A new librarianship for a new age," video presentation, http://vimeo.com/32400532.

2. David Shumaker and Mary Talley, "Models of embedded librarianship: A final report" (Alexandria, Va.: Special Libraries Association, 2009).

3. See, for example, articles in *Information Outlook*, such as the January/February 2010 issue devoted to embedded librarianship and articles in the May and June 2011 issues of same.

4. See David Shumaker, *The embedded librarian: Innovative strategies for taking knowledge where it's needed* (Medford, N.J.: Information Today, 2012); Cassandra Kvenild and Kaijsa Calkins, *Embedded librarians moving beyond one-shot instruction* (Chicago: Association of College and Research Libraries, 2011).

5. The terms "community" and "communities" are used to stand for the multiple types of internal groups within organizations, such as teams, groups, units, division, departments, and so forth.

6. Shumaker and Talley, "Models of embedded librarianship: A final report," 36.

7. Ark Group, Embedded Librarian Statistics.pdf, https://docs.google.com/file/d/0B8BSC1okpVKCYm1lYVlzZWlSQlNlWXZDM1IwVHJldw/edit.

8. Cindy Adams, "Law firm librarians: Out of sight, out of mind," *3 Geeks and a Law Blog*, January 10, 2012, http://www.geeklawblog.com/2012/01/law-firm-librarians-out-of-sight-out-of.html.

9. Interview with Jill Strand, director of information resources, conducted by Mary Talley, February 5, 2013.

10. Interview with Cindy Adams, codirector of library services, McKenna, Long and Aldridge, conducted by Mary Talley, January 28, 2013.

11. Interview with the chief conducted by Mary Talley, February 4, 2013. This law firm's embedded program was initially described in Shumaker and Talley, *Models of embedded librarianship: A final report*, Appendix B: Site Visits, pp. B1–8; and in D. Shumaker, "Site visit 4," in "Models of Embedded Librarianship: Addendum 2011b" (Alexandria, Va.: Special Libraries Association, 2009), pp. 22–27. This interview is an update to the two previous reports.

12. From an interview conducted by Mary Talley, February 4, 2013.

13. For a description of the Knowledge Management Analyst and the Global Research Analyst's positions and work contributions, see Shumaker and Talley, *Models of embedded librarianship: A final report*, Appendix B: Site Visits, pp. B1–8; and Shumaker, "Site visit 4," in "Models of embedded librarianship: Addendum 2011b."

14. Patrick Meyer, "2012 law firm legal research requirements for new attorneys," working paper, SSRN, September 26, 2011, http://ssrn.com/abstract=1953437.

15. B. E. Tumbleson and J. J. Burke, "When life hands you lemons: Overcoming obstacles to expand services in an embedded librarian program," *Journal of Library Administration* 50, 7 (2010): 972–988.

16. Ann Schroeder, "Replacing face-to-face information literacy instruction: Offering the embedded librarian program to all courses," in *Embedded librarians moving beyond one-shot instruction*, edited by Cassandra Kvenild and Kaijsa Calkins, 63–78 (Chicago: Association of College and Research Libraries, 2011).

17. Amy C. York and Jason M. Vance, "Taking library instruction into the online classroom: Best practices for embedded librarians," *Journal of Library Administration* 49, 1–2 (2009): 197–209.

18. Emily Daly, "Instruction where and when students need it: Embedding library resources into learning management systems," in *Embedded librarians moving beyond one-shot instruction*, edited by Cassandra Kvenild and Kaijsa Calkins, 79–91 (Chicago: Association of College and Research Libraries, 2011).

19. For a discussion on creating online tutorials in a law library, see Tawnya K. Plumb, "Creating electronic tutorials: On your mark, get set, go!" *Journal of Electronic Resources Librarianship* 22, 1–2 (2010): 49–64.

20. Whitney A. Curtis and Robert Brammer, "We go out looking for trouble: Taking library services to the patron's point of need," *Perspectives: Teaching Legal Research & Writing* 21 (2012): 13–17.

21. Jennifer Davitt, "10 relatively effortless high-impact ideas for faculty services librarians: Impress your patrons without breaking the bank," *AALL Spectrum* 15, 6 (April 2011): 12–14.

22. Curtis and Brammer, "We go out looking for trouble."

23. John B. Nann, "Personal librarians," *AALL Spectrum* 14, 8 (June 2010): 20–23.

24. V. Feliu and H. Frazer, "Embedded librarians: Teaching legal research as a lawyering skill," *Journal of Legal Education* 61, 4 (2012): 540–559.

25. Lisa Coats and Bojana Beric, "More than a one-shot: Innovative faculty-librarian collaboration," in *Embedded librarians moving beyond one-shot instruction*, edited by Cassandra Kvenild and Kaijsa Calkins, 165–184 (Chicago: Association of College and Research Libraries, 2011).

26. Shumaker, *The embedded librarian*.

3

Copyright in the Digital Age

Kyle K. Courtney
HARVARD LAW SCHOOL

Today's law librarian faces an excess of copyright-related questions dealing with every facet of library services: digital archives, virtual reference, interlibrary loan, access to law, e-reserves, licensing, and so forth. Many librarians do not have the formal instruction necessary to answer all of these questions, but many find themselves becoming de facto library copyright experts by reading, writing, collaborating, attending conferences, and working with informed colleagues. In the digital, on-demand environment in which law libraries exist, it is almost a requirement that librarians have a basic understanding of copyright to successfully assist students, faculty, and staff competently.

The goal of this chapter is to provide the fundamental instruction for some of the most common copyright topics facing law libraries in the digital age. Whether it is scanning chapters for e-reserves or accessing databases online, knowledge of copyright law can help mitigate risk and enhance our patrons' services. As librarians, we want to provide whatever our patrons desire. But we also must balance the law versus the patrons' needs. Fortunately, copyright law does not always restrict a patron's uses. In many cases, a solid understanding of copyright can help ease a patron's fears, provide legal alternatives to a patron's request, or help educate the community at large.

COPYRIGHT LAW

Copyright is a form of intellectual property protection provided by the laws of the United States. The law protects original works of authorship, including literary, dramatic, musical, and artistic works such as novels, music, movies, art, architecture, and computer programs. This property right gives the holder the exclusive right to reproduce, distribute, rework, perform, and display the work.

Origins and the U.S. Constitution

All of the copyright laws passed by Congress exist pursuant to Article 1, Section 8, Clause 8, of the Constitution, which gives Congress the power "[t]o promote the Progress of Science and useful Arts, by securing for limited Times to Authors and Inventors the exclusive Right to their respective Writings and Discoveries."[1] This section of the Constitution is known as the "Copyright Clause." The current copyright law in the United States is the Copyright Act of 1976 ("Copyright Act").[2]

The Copyright Act made some significant changes to the previous U.S. copyright laws. Under the current law, copyright protection begins immediately at the creation of an original "work of authorship fixed in any tangible medium of expression."[3] Typically, a copyright lasts, under the current Copyright Act, for the life of an individual author or authors, plus 70 years. Alternatively, if a corporation is the author, the term is 95 years from creation or 120 years from publication, whichever is sooner.[4]

On March 1, 1989, the United States agreed to the Berne Convention of International Copyright. This eliminated the formalities of copyright registration. Authors no longer needed to put a copyright notice on their work. Additionally, it no longer required registration with the U.S. Copyright Office. However, although registration is optional, the Copyright Office offers advantages for registration. For example, registration is a prerequisite to filing a copyright infringement suit in federal court.

What Is Copyrightable?

The Copyright Act lists the following types of works of authorship as those that can be protected by copyright: literary works; musical works, including any accompanying words; dramatic works, including any accompanying music; pantomimes and choreographic works; pictorial, graphic, and sculptural works; motion pictures and other audiovisual works; sound recordings; and architectural works.[5] Typically, these categories are viewed quite broadly. For example, since a computer program is written in computer code, it is considered a "literary work."

Rights of a Copyright Holder

Let's say you do create an original work of authorship. What rights do you have as the owner of that work? Copyright owners have exclusive rights to: reproduce the work (make copies); prepare derivative works based upon the copyrighted work; distribute the work; publically perform the work; and publically display the work. There are also other specific rights enumerated for particular kinds of works.[6]

These are all considered the "bundle of rights" of the copyright owner. Any of these individual rights may be given away, licensed, sold, or any combination thereof. They are like pieces of property that can be allocated by the owner. For example, many musicians license their rights of reproduction and distribution to music distributors, so the distributer can legally make, ship, and sell CDs, mp3s, and other copies of the music. Or alternatively it is possible to transfer one right, while retaining all other rights.

If a person utilizes any of these rights without permission of the copyright owner, they may be liable for infringement. From this context, the Copyright Act not only defines the rights of the owner, but also all the potential methods of infringement. To succeed on claim for direct infringement under Copyright Act, plaintiff must show: (1) ownership of valid copyright, and (2) unauthorized copying or violation of one of other exclusive rights afforded copyright owners pursuant to Copyright Act.[7]

Beyond direct infringement is the doctrine of secondary liability. This arises when one party is held legally responsible for the actions of another party. There are generally two kinds of copyright secondary liability: vicarious liability and contributory liability.

However, if the infringer is a library or archive, or the employee of a library or archive, then the court can lower or eliminate damages altogether if the infringer "believed and had reasonable grounds for believing that his or her use of the copyrighted work was a fair use under § 107."[8] To help support this in practice, a library should try to retain any documentation that determined that the alleged use was a fair use. We will learn more about fair use below.

PUBLIC DOMAIN

Works that are not protected by copyright law belong to the public domain. There is no need to get permission to use a public domain work. It is interesting to note that the Copyright Act does not define the phrase "public domain."

There are two kinds of works that belong to the public domain. First are works whose copyright protection has expired. For example, *The Adventures of Tom Sawyer*, first published in the United States in 1876, is in the public domain because its copyright term has expired. The length of copyright protection varies depending on when a work was created and if, when, and where it was published.

Second are works that were never protected by U.S. copyright law. This category includes works created by an employee of the U.S. federal government in the course of her employment, works that did not comply with the formalities by former U.S. copyright regimes, and works published in certain countries that have not joined international copyright treaties. For example, decisions of the U.S. Supreme Court are in the public domain because U.S. federal government employees create them during the course of their employment.[9] There are several resources that may help you determine if the work you wish to use is in the public domain.[10]

EXCEPTIONS TO THE RIGHTS OF COPYRIGHT OWNERS

Fair Use (§107)

One of the most famous, yet the least clear, of all the copyright limitations in the Copyright Act is the doctrine of fair use.[11] Under fair use, you may use copyrighted material without permission from the copyright owner. The doctrine itself was rooted in both English and U.S. case law, but was eventually codified in the Copyright Act.[12]

The source of fair use law is currently statutory: Section 107 of the Copyright Act provides that fair use of a work "for purposes such as criticism, comment, news reporting, teaching (including multiple copies for classroom use), scholarship, or research"[13] is not copyright infringement. This list is not exhaustive; other uses of copyrighted work without permission may also be considered fair use.

Section 107 of the Copyright Act provides:

> Notwithstanding the provisions of sections 106 and 106A, the fair use of a copyrighted work, including such use by reproduction in copies or phonorecords or by any other means specified by that section, for purposes such as criticism, comment, news reporting, teaching (including multiple copies for classroom use), scholarship, or research, is not an infringement of copyright. In determining whether the use made of a work in any particular case is a fair use the factors to be considered shall include—
>
> 1. the purpose and character of the use, including whether such use is of a commercial nature or is for nonprofit educational purposes;
> 2. the nature of the copyrighted work;
> 3. the amount and substantiality of the portion used in relation to the copyrighted work as a whole; and
> 4. the effect of the use upon the potential market for or value of the copyrighted work.
>
> The fact that a work is unpublished shall not itself bar a finding of fair use if such finding is made upon consideration of all the above factors.[14]

When the fair use provision was being proposed, Congress took the position that "since the doctrine is an equitable rule of reason, no generally applicable definition is possible, and each case raising the question must be decided on its own facts."[15] Therefore, when courts are examining a fair use claim, they analyze each factor using the specific facts of the case to make a fair use determination. This examination of the four factors determines whether the use is "fair" or constitutes "copyright infringement." Courts weigh each factor and make a decision based on the overview of all four factors.

Frequently, this four-factor test is also used for risk mitigation. For example, law libraries frequently use this test to determine whether or not they can perform a certain activity or function involving copying or scanning. By reviewing the four factors, as a court might, a law librarian can determine whether or not the action he or she is taking might risk infringement or fall squarely within the realm of fair use. Because the four factors can act as a guideline for liability, the following section examines each factor closely, and also through the lens one of the most recent digital fair use cases, *Authors Guild v. HathiTrust et al.*

Authors Guild v. HathiTrust et al.

The HathiTrust partnership is in the process of creating a shared digital repository which, to date, has almost 10 million scanned digital volumes from various university libraries around the United States. The process started when Google created digital copies of works in the universities' libraries in exchange for which Google provided digital copies to the universities. The universities then contributed these digital copies to the HathiTrust Digital Library.[16] In 2011, the Authors Guild, the Australian Society of Authors, the Union des Écrivaines et des Écrivains Québécois, and eight individual authors filed a lawsuit against HathiTrust, five major universities, and many university officials for copyright infringement. Specifically, the Authors Guild and others alleged that HathiTrust's storage of its repository of millions of full-text digital book scans was copyright infringement. In 2012, U.S. District Court Judge Harold Baer summarily dismissed the Authors Guild's lawsuit against the HathiTrust library collaborative. Judge Baer employed the fair use test and found that the repository was not only providing new avenues for scholarship but also enhancing access for the print and visual disabled. He stated, dismissing the motion, "I cannot imagine a definition of fair use that would not encompass the transformative uses made by the defendants and would require that I terminate this invaluable contribution to the progress of science and the cultivation of the arts that at the same time effectuates the ideals of the ADA."[17] Let's examine each factor and how they were utilized to analysis a massive digitization project.

Fair Use Factor One: Purpose and Character of the Use

The first factor plays a major role in the fair use analysis. This factor distinguishes "nonprofit educational" uses with "commercial" ones. Obviously, nonprofit educational uses are considered more of a fair use, while commercial uses may be a sign of infringement. Law libraries that are part of a for-profit firm, private law libraries, and membership-based law libraries would not always receive a favorable evaluation on the first factor because of their inherent "commercial" nature. Questions that arise during this inquiry ask: Are you making a profit using the copyrighted work in some commercial way, such as selling copies, charging clients, or are you using it for teaching purposes? Typically a court will also enquire about making a transformative use of the work.

Transformative use is the use of copyrighted material in a manner, or for a purpose, that differs from the original use.[18] This use develops in such a way that the expression or meaning is essentially new. For example, when copyrighted concert posters from a Grateful Dead archive were reproduced without permission in a Grateful Dead visual biography, the use of the posters was found to be transformative, because they were a new and different use than the poster's original intent—namely, concert promotion.[19]

In *Authors Guild v. HathiTrust et al.*, the court closely examined the first factor. Here, the character of the use was clearly a nonprofit, educational use. The university libraries had entered into agreements to convert the books in their libraries into a digital format, and then deposit them into the collective HathiTrust Digital Library. When a user searches a term in the Digital Library, it reveals only the pages on which the term is found and the number of times the term is found on each page. Accordingly, no actual text from the book is ever shown, except to those students who are visually impaired. This clearly falls within the context of nonprofit, educational use.

Judge Baer also examined the transformative use. Again, a transformative use occurs when a change in the content has been made into something new or different. Here, Judge Baer stated that the use was transformative because it serves a different purpose than the original work: namely, providing a searchable database of millions of books and making full text only available to the visually impaired.

Fair Use Factor Two: Nature of Copyrighted Work

Courts often begin the second factor work by examining whether the work is informational or fact-based versus creative or fictional. The court also may examine whether copyrighted work is published or unpublished. The law is more protective of unpublished works because authors have the right to decide whether and when they will make their work public. Using a previously published work that has fallen out of print may tilt toward fair use since the work is not otherwise available. The law is also more protective of highly creative and imaginative works, so using an excerpt from a novel about a beloved boy wizard may lean further away from fair use than a snippet from a heavily fact-based nonfiction book. Copyright protects expression, not ideas or facts, and copyright on factual works is considered to be "thinner" than copyright on highly creative works. It is also possible that a court may take into consideration whether a digital work should be treated differently from one in print, because of the technological ease in which it can be copied.

In *Authors Guild v. HathiTrust*, the Court went directly to the creative examination, since the majority of the works inside HathiTrust were already published. Judge Baer acknowledged that there is more room for fair use claims of more factual materials, but he did not find it relevant to this case. He held that in transformative uses, like HathiTrust, the second factor was "not dispositive."[20] In other words, it did not help bring about any settlement on the fair use issue.

Fair Use Factor Three: Amount and Substantiality Used

This factor is not only about quantity. In many previous fair use cases, both quantity and quality are taken into account. For example, in *Harper and Row v. Nation Enterprises*, a magazine, without permission, published roughly 300 words taken from former president Gerald Ford's book-length memoir. Despite the small percentage of the overall work that the magazine used, the Supreme Court held that the third factor weighed especially against fair use because the excerpts were all about Ford's pardon of Richard Nixon and these pieces made up the "heart" of the book.[21]

Therefore, a court may look to determine whether there is a small piece of a copyrighted work being used, like a paragraph from a 300-page book or a minute-long clip from a feature film, or if there is a use of the entire work. However, it can be fair to use an entire work; the important thing to remember is that the law favors using just the amount you need for your purpose. Always remember the *Harper and Row* holding, however. It still can be a violation to use even a small portion of a copyrighted work if what you're taking is the "heart of the work." Generally, there are no bright lines when it comes to what percentage of the copyrighted work is too much to use, but the higher the percentage, the more likely this factor is to count against fair use.

In *Authors Guild v. HathiTrust*, HathiTrust indisputably had made copies of entire books and was showing copies of entire books to users with print disabilities. Judge Baer simply stated that the law favors using just the amount you need for your purpose. He concluded that for HathiTrust to facilitate word searches and provide access for print-disabled individuals "it [was] necessary to copy entire works."[22] The digitally scanned copies of entire books were necessary, and therefore, a legitimate fair use to fulfill the purpose of the HathiTrust.

Fair Use Factor Four: Market Effect

The purpose of copyright is to encourage creation. Using something in a way that means lost sales to the copyright owner will weigh against fair use. The reasoning is logical: If authors or filmmakers thought that they would never be able to receive value for their work, they might be less likely to share that work with the public, which would be a loss for everyone. In considering market effect, the law tries to protect creators. Courts, then, often enquire as to whether the use of this copyrighted work take away a sale of the work. They ascertain whether a particular use will cut into the potential market or value of the work.

In *Authors Guild v. HathiTrust* there wasn't much of an argument about the current market, as there was for a potential market. The Authors Guild claimed that HathiTrust was precluding opportunities for the development of future licensing opportunities. In a well-reasoned analysis, Judge Baer stated that where a use is transformative, copyright holders cannot undermine fair use just by claiming that they intend to license that use in the future. "A copyright holder cannot preempt a transformative market."[23] Plaintiffs actually had to prove, and failed to prove, that there was some market harm, or a meaningful likelihood of future harm exists. Without evidence of any actual harm, the court found the assertions of potential future licensing revenue to be mere "conjecture."[24]

Case Conclusion

Judge Baer relied heavily on the transformative nature of creating a digital repository of books. Transformativeness seemed to drive the analysis and received a great deal of attention in every fair use factor. Judge Baer found that the repository was not only providing new avenues for scholarship but also enhancing access for the print and visual disabled. He concluded that this was definitely a transformative use and stated, "I cannot imagine a definition of fair use that would not encompass the transformative uses made by the defendants and would require that I terminate this invaluable contribution to the progress of science and the cultivation of the arts that at the same time effectuates the ideals of the ADA."[25]

For Private Law Libraries

What does the *HathiTrust* decision mean for private law libraries that are not involved in nonprofit, educational uses? Commercial versus noncommercial work does not have as much of a bearing on fair use analysis as one would think. In a seminal article published in 2008 titled "An Empirical Study of U.S. Copyright Fair Use Opinions, 1978–2005," data suggested that a finding that the defendant's use was for a commercial (rather than noncommercial) purpose "had no significant influence on the outcome of the [fair use] test."[26] This was after examining 306 reported opinions from 215 cases that made substantial use of the fair use test. That means that private libraries *could* consider making digital transformative uses of materials, provided that they continued to fall within the realm of fair use for the other factors. Commercial uses were far from dispositive in deciding for fair use.

Library Exceptions (§108)

Section 108 of the Copyright Act provides under certain conditions it is not infringement for a library or archives, or its employees acting within the scope of their employment, to reproduce or distribute one copy of a work (or in particular cases, three copies). There are very specific conditions in Section 108(a) that must be met to qualify for this exception.

First, copies cannot be made for any direct or indirect commercial advantage. There is conflicting analysis about whether libraries associated with for-profit organizations (law firm libraries, corporate libraries, etc.) may rely on this law. A House of Representatives report, in the legislative history of the Copyright Act, indicates that a for-profit library can make copies as long as the copying itself is not the overall goal of the institution where the copying takes place. A Senate report, found in the same legislative history, has a different interpretation.[27]

Second, the collections of the library must be open to the public, available to researchers affiliated with the library, or other persons doing research in a specialized field. Again, for-profit libraries might have some trouble meeting this condition. Very few private law libraries or corporate libraries would open their doors to other lawyers, competitors, or business rivals in a particular field.[28]

Lastly, the reproduction or distribution must include a notice of copyright or a statement that the work may be protected by copyright if no notice is found on the original.

The exceptions granted under Section 108 extend only "to the isolated and unrelated reproduction or distribution of a single copy . . . of the same material on separate occasions"[29] and do not apply to "a musical work, a pictorial, graphic or sculptural work, or a motion picture or other audiovisual work other than an audiovisual work dealing with news."[30]

Once a library has met the requirements above, what type of copies and in what situations can you copy, scan, or distribute? The following is a selected samples of various §108 rights.

Replacement Copies (§108(c))

Up to three copies of a published work may be reproduced solely for the purpose of replacing a copy that is damaged, deteriorated, lost, stolen, or in an obsolete format. This reproduction is contingent on the library making reasonable efforts to determine that an unused replacement cannot be acquired at a fair price, and any such copy that is reproduced digitally is not made available to the public outside of the library.

What exactly is an obsolete format? Are VHSs, cassette tapes, or record players considered "obsolete" under the statute? The definition states that "a format shall be considered obsolete if the machine or device necessary to render perceptible a work stored in that format is no longer manufactured or is no longer reasonably available in the commercial marketplace."[31] In today's global economy, it is easy to see how Betamax, VHS, and Laserdisc might not be considered obsolete, considering you can easily find these players on eBay and other secondary markets.

Archival Copies (§108(b))

If a work is unpublished and currently resides in the collection of the library making a copy, the library may make three copies. These copies must be made solely for preservation and security or for the deposit for research use in another library. In the legislative history of the Copyright Act, the House Report noted that this right "would extend to any type of work, including photographs, motion pictures, and sound recordings."[32] Although this seems to be a fairly encouraging section for digital reproduction of many types of works, any of these archival copies made in a digital format under this section must not be made available to the public outside of the library.

Out of Print Works (§108(e))

A copy of an entire work may be made if the library has determined a copy cannot be obtained at a fair price, subject to the condition that the copy becomes the property of the user, the library has no notice that the copy is going to be used for any purpose other than private study, research, or scholarship, and the library must prominently display a copyright warning notice on any request form, including print or Web forms, at the location where the copy orders are accepted.

Library Copiers, Scanners, and other Reproduction Devices (§108)(f))

One of the most common questions about copiers in the law library is "Is the law library liable for the user's violation of copyright on the copy machines?" Thanks to Section 108(f), libraries are protected from liability resulting from a user's potential copyright infringements when using copy machines in the library. Like most §108 sections, in order to claim the benefit of this limited liability, there are two requirements. First, the copying must be unsupervised by any library staff. Second, the copier must "display a notice that the making of a copy may be subject to the copyright law."[33] You will find the some sample copyright notices at the end of this chapter.

By complying with these simple provisions, a law library can effectively remove any potential liability for copyright infringement from patron copier use. The liability then shifts directly to the user making the copies. In a modern law library, this notice might appear on copy machines, scanners, microfilm readers, printers, or any other equipment capable of reproduction.

Digital Reproduction during the Last Twenty Years of the Term (§108(h))

As we have learned above, digital archives do not fall under the various §108 library exceptions. First, it does not apply to material the library or archive does not own. Second, a library cannot distribute digital copies or make them available to patrons outside the library premises. However, §108(h) allows libraries and archives to reproduce, distribute, display, or perform, in digital form, a copy of a published copyrighted work *during the last twenty years of any term of copyright* for purposes of preservation, scholarship, or research. This exception will apply as long the work is: (1) not still being commercially exploited, (2) the work cannot be obtained at a reasonable price, and (3) the copyright holder does not provide notice that one of the above conditions applies. There is no definition to the term "reasonable price," nor has any court specifically spoken to that portion of the statute. This section was, at the time period, a concession to libraries for public domain problems that arose because of the passing of the Sonny Bono Copyright Term Extension Act.[34]

RISE OF DIGITAL TECHNOLOGY IN LAW LIBRARIES

In the past decade, digital technology has altered the traditional concept of the law library. An attorney may no longer have a treatise or statute in her office. A student may never have to access the Federal Reporter in print during their time in law school. As a result of laptops, desktops, mobile phone, and tablets, patrons have a virtual law library "at their fingertips." Mobile technology usage is rapidly increasing, and patrons may not be anywhere near the law library but require its services. Using technology to modernize processes and collections, in light of these changes, has become a priority for law libraries to show their real value, in the face of shrinking budgets and instability in the legal job market.

Law libraries make extensive use of technology to catalog, store, and disseminate information. Some are digitizing print materials, creating searchable metadata, and developing websites. All of these functions involve copyright in some way. However, even though these new technological-focused projects are vital, access to information and its delivery has emerged as the focus of copyright law with regard to law libraries and their patrons. Over the next few sections we examine some of the top delivery and access questions involving copyright, including e-reserves and modern document delivery programs.

e-Reserves and the Law Library

The appeal of e-reserves, especially in law schools, is apparent. The study of law is frequently filled with heavy casebooks, enormous hornbooks, or unmanageable treatises. Regardless of the class, typically a professor assigns some reading, and many times these readings end up in a law library reserve collection. Providing digital access to the reserve readings would not only alleviate the "heavy bag" dilemma but also allow students to have equivalent access to the materials without having to wait for a fellow student to return the material from the standard two-hour loan. Regardless of the advantages, copyright law is a definite factor to consider when scanning a copyrighted work into an e-reserve system.

In 1993, President Clinton formed the Information Infrastructure Task Force and specialized subgroup called the Working Group on Intellectual Property Rights. This group convened a Conference on Fair Use (CONFU) to bring together copyright owner and user interests to discuss fair use issues and, if possible, develop guidelines for fair uses of copyrighted works by librarians. One of the most hotly contested areas covered by CONFU was e-reserves. Over ninety organizations representing for-profit and nonprofit publishers, the software industry, government, scholars, authors, photographers, the movie industry, public television, libraries, museums, universities, and colleges spent more than two and a half years to find agreement on the scope of e-reserves.

In March 1996, the Draft Guidelines for Electronic Reserves were completed. The guidelines were rejected by the majority of CONFU participants because there was no general consensus that the document represented an understanding of fair use by all participants. Some thought the guidelines were overly restrictive, and the content owners thought they went too far. It was decided at the CONFU session in November 1996 that the guidelines would not be distributed as part of the final CONFU report.

In November 2003, the American Library Association (ALA), with assistance from Georgia Harper, from the University of Texas System, and Peggy Hoon, from North Carolina State University, released a statement titled "Applying Fair Use in the Development of Electronic Reserves Systems." Instead of trying to approach the e-reserve problem with a new set of guidelines, this statement expressed reliance on the already well-established, and statutorily created, fair use doctrine. The document was endorsed by American Association of Law Libraries, American Library Association, Association of Research Libraries, Medical Library Association, and Special Libraries Association.[35] Many libraries adopted similar recommendations, and based their e-reserve systems on fair use.

In April 2008, three textbook publishers, Oxford University Press, Cambridge University Press, and Sage Publications, filed a copyright infringement law suit against Georgia State University (GSU).[36] The suit claimed ninety-nine specific instances of copyright infringement through the distribution of copyright-protected material via GSU's online course management systems. Specifically, the publishers alleged that the library allowed the faculty to use the university library e-reserves systems to copy and distribute book excerpts to students without paying any licensing fees. In May 2012, U.S. District Court Judge Orinda D. Evans ruled in favor of the university in 95 percent of the excerpts under the fair use doctrine. This is one of the most important digital fair use decisions to be released, and more importantly, for our discussion, specifically based on a library e-reserve policy. Let's examine this case closely, and look at how the court came to the conclusions it did for each of the four factors of fair use for digital e-reserves.

First Factor

What was unique about this case was that it was the first e-reserves case to come from a real educational system, GSU. The court itself states "there is no precedent . . . for how the [fair use] factors should be applied where excerpts of copyrighted works are copied by a nonprofit college or university for a nonprofit educational purpose." Previous reserves and coursepack cases were examined in the commercial, for-profit context. In *Basic Books, Inc. v. Kinko's Graphics Corp.*, the court found that Kinko's, a commercial enterprise, was infringing copyright when it photocopied book chapters for sale to students as coursepacks for local university classes.[37] In *Princeton University Press v. Michigan Document Services, Inc.*, in facts very similar to the *Kinko's* case, a copy shop created and sold coursepacks for a local university. The court found that the copy shop was infringing copyright by creating and selling the coursepacks.[38]

Judge Evans distinguished this case from *Kinko's* and other commercial defendants, noting that Georgia State University "is a purely nonprofit, educational institution and the excerpts at issue were used for purely nonprofit, educational purposes."[39] As a result, the court held that the purpose factor was squarely in favor of fair use. Judge Evans stated, "the language of § 107 itself . . . compel the decision that the first fair use factor favors Defendants. This case involves making copies of excerpts of copyrighted works for teaching students and for scholarship, as specified in the preamble of § 107. . . . The fact that the copying is done by a nonprofit educational institution leaves no doubt on this point."[40]

The Second Factor

Here Judge Evans, although considering the argument from the publishers that compiling scholarly works can be difficult, nevertheless found most of the works at issue were scholarly nonfiction, and were therefore less creative in nature. Because the works are less creative, and more information-heavy, the scope for finding fair use is much wider, even if they are digital copies of those works. Analyzing the most helpful case law, the judge found that factor two favored GSU, specifically because they were using e-reserves of scholarly, nonfiction works.[41]

The Third Factor

As stated in the previous fair use section, this factor looks at the amount and substantiality of the portion copied or used. Although the court stated a general conflict between the flexible fair use standard and a strict quantity measure, Judge Evans stated that e-reserves "must fill a demonstrated, legitimate purpose in the course curriculum and must be narrowly tailored to accomplish that purpose."[42]

With that statement, a bright-line test was developed, based on the safe harbor of a "decidedly small" portion of the work. The court stated the following formula:

> Where a book is not divided into chapters or contains fewer than ten chapters, unpaid copying of no more than 10% of the pages in the book is permissible under factor three. . . . In practical effect, this will allow copying of about one chapter or its equivalent. Where a book contains ten or more chapters, the unpaid copying of up to but no more than one chapter (or its equivalent) will be permissible under fair use factor three. Excerpts which fall within these limits are *decidedly small*, and allowable as such under factor three.[43]

This was a surprising part of the decision for many advocates of fair use. Fair use was created to be a flexible standard, not set with ridged rules. However, it is important to stress that this is just one factor in a four-factor test. Just because there is a use over 10 percent or one chapter does not necessarily mean it is automatically infringement, nor does it trump any of the other factors. In fact, as we show below, even when there is use over the 10 percent or one chapter standard, there can still be an overall fair use.

Two real positive results from the examination of the third factor in this case were the wholesale rejection of the "1976 Classroom Guidelines" and the "one-time use limit." The 1976 Classroom Guidelines were leftover products from the legislative history of the Copyright Act. They were negotiated on behalf of private-interest groups and attempted to define the maximum amount of copying allowable for "safe harbor" from litigation. They set horrendously small and useless portions of articles, newspapers, prose, and other material for use in the classroom.[44]

The court rejected the guidelines and stated they certainly did not represent the legal definition of fair use. Judge Evans said, "Plaintiffs do not explain their decision to seek acceptance of the minimum standards as the maximum standard."[45]

The "one-time use limit" had previously haunted e-reserves specialists because it was thought that repeated use of the same excerpt across multiple semesters would count against fair use. The court said, "The idea that professors be prohibited from unlicensed use of the same chapter from one academic term to the next is an impractical, unnecessary limitation. The right approach is to select a percentage of pages which reasonably limits copying and to couple that with a reasonable limit on the number of chapters which may be copied."[46]

The Fourth Factor

Fundamentally, it is understood how the fourth factor truly expresses the balance of fair use against a rightsholder's work. Whether or not the market is harmed by the use is an important part of the economic monopoly given to authors for limited times, as an incentive to continue to create new works.

In the GSU decision, the judge, once again, broke into new territory. "For loss of potential license revenue to cut against fair use, the evidence must show that licenses for excerpts of the works at issue are easily accessible, reasonably priced, and that they offer excerpts in a format which is reasonably convenient for users."[47] The rule that develops from this decision is that when digital permissions are available and the library did not seek the digital permission, this tilts the fourth factor strongly in favor of the rightsholder. However, the language of the court is clear: It is important that the work for e-reserves must be readily available and reasonably priced specifically in the context of *digital excerpts*. By analogy, if the there was only a license to access the entire work, or only a license to make photocopies, this might favor the factor toward fair use, since there was no specific e-reserve excerpt license.

Case Conclusion

The ruling, which took a year to complete, and comprises 350 pages, held that the university library's e-reserves fair use policy was a good faith effort to comply with the copyright law and, for the purpose of awarding costs for the case, determined that GSU was the prevailing party. Although the judgment is limited to the jurisdiction and the parties involved, this case will continue to influence future fair use decisions in the digital e-reserves arena for years to come. Law librarians would be best served by reading, discussing, and considering some of the recommendations and analysis offered by this decision for their own e-reserve programs.

Document Delivery in Law Libraries

Law libraries across the United States are finding new ways to take advantage of their special role in the Copyright Act and provide enhanced "at the fingertip" research capabilities for their faculty, students, and staff. As an example, we briefly discuss the Scan and Deliver services implemented at Harvard, University of Chicago, and Yale.

Harvard Library launched the Scan and Deliver service in May 2009.[48] The intention was to provide Harvard students and researchers, on-campus, off-campus, across the country, or on another continent, access to the 10 million items collected in all the Harvard libraries. Prior to that service being launched, interlibrary loan (ILL) units utilized OCLC as the mechanism for requesting and fulfilling article requests through the local Harvard system, but this system wasn't ideal. It required a lot of work around and frequently impacted standard ILL operations. Also, there was patron confusion about what items/collections were eligible under the ILL system. The Harvard Scan and Deliver system changed all of that by allowing Harvard affiliated students, faculty, and staff to request copies of book chapters or articles directly through the Harvard Library catalog, regardless of which library the work was held.

University of Chicago and Yale University launched similar Scan and Deliver services in 2012.[49] Similar to the Harvard model, they allow members of the university community to request a copy of an article or chapter to be delivered to them via e-mail in a PDF format. Users can simply find the book or journal they are looking for in the library catalog and click on the link, and within four days, they receive a copy of the requested material, regardless of the location. One main difference is that Yale also offers this Scan and Deliver service to its alumni.

Looking at the rules and guidelines, you can see that each of these programs worked with university counsel and the library staff to determine what limits should be set, so that there was little possibility of copyright infringement. As we reviewed in the sections above, libraries have some special exceptions under the Copyright Act. In particular, §108(d) and (e) deal with copies for private study and relate closely to the Scan and Deliver services. Specifically, §108(d) is utilized for copies of articles, book chapters, or portions of other works. The section requires that the copy become property of the user, that the library has no notice that the copy is for any other purpose than scholarship or research, and that a copyright warning notice is prominently displayed. All the Scan and Deliver systems at all three universities meet the criteria for §108(d).

Additionally, each library doubles down on the risk mitigation by creating its own "built-in" fair use guidelines. All the university libraries set a maximum amount of requests and pages that a user can order from the Scan and Deliver service. A user may make two requests a day. They may not request from the same source in a single semester. Depending on the university, the users are also limited to approximately thirty pages, or one article, or one chapter per request. Again, this is to absolutely ensure that the libraries are well within fair use ranges, and compliance with §108(d).

The service has come to enhance the libraries' offering of "just in time" research methods. Harvard reported that one graduate student in China was able to complete her dissertation without having to return to the United States to access the library collection. Other improvements in the future may include more microform eligibility and OCR-ready PDFs to aid in the searchability of a document.

Private law firms have offered "desktop delivery" of their services over intranet systems for years. However, they frequently do not rely on fair use or §108(d) to make copies available to their patrons. Private law firms often license with vendors and define their uses very specifically. The license is what governs the ability to share articles, chapters, and other material over a law firm library network.

If you are working with a vendor and want to include some digital content in your law firm's document delivery program, or start a document delivery program, make sure to spell it out clearly in the license. The terms for "authorized users" and "authorized sites" are critical to a license, and you may negotiate these terms with the vendor. Do you want access for partners, in-house counsel, consultants, or independent contractors? Additionally, *where* these people might access the materials is typically found under the "authorized sites" in the license. Normally these are either "in office" or "remote access." Make sure you clearly define where the majority of the documents or content will be delivered. For example, if some of your attorneys are constantly in court, make sure that remote access, set for a mobile device, is part of the license.

Other factors to consider are whether or not you have access to digital content from a single computer, from a network of specified users, or an unlimited number of users. Again, these may be options for your negotiation with the vendor to provide the best service possible to the employees in the law firm.

CONCLUSION

A basic understanding of copyright is essential to navigate all the potential interactions we face as law librarians when we assist students, faculty, staff, patrons, clients, and others. We have reviewed some of the Copyright Act, especially as it pertains to the special provisions libraries have inside the law. We have reviewed some of the newest library and technology-focused decisions affecting our day-to-day operations in law libraries.

However, we should not stop here. An interest in copyright law requires frequent reading and reviewing. By keeping up with current awareness for these and other copyright issues, law librarians can take an active role in becoming advocates in areas such e-book rights, licensing, contracts, and fair use. The knowledge of copyright law can help mitigate risk in our own organizations, and it can help us increase awareness in our community at large.

FURTHER READING

Association of Research Libraries et al. 2012. *The code of best practices for fair use in academic and research libraries.* Washington, D.C.: Association of Research Libraries. http://www.arl.org/pp/ppcopyright/codefairuse/code/.

Aufderheide, Patricia, and Peter Jaszi. 2011. *Reclaiming fair use: How to put balance back in copyright.* Chicago: University of Chicago Press.

Columbia Copyright Advisory Office. *"Fair use checklist" Copyright Advisory Office.* Revised November 11, 2011. http://copyright.columbia.edu/copyright/fair-use/fair-use-checklist.

Crews, Kenneth D. 2012. *Copyright law for librarians and educator: Creative strategies and practical solutions*, 3rd ed. Chicago: American Library Association.

Gasaway, Laura N. 2009–2010. "Copyright basics: From earliest times to the digital age." *Wake Forest Intellectual Property Law Journal* 10(3): 241.

NOTES

1. U.S. Const., Article I, § 8, cl. 8 (1789).
2. 17 U.S.C. §101 et seq. (2006).
3. 17 U.S.C. §102 (2006).
4. 17 U.S.C. §302 (2006).
5. 17 U.S.C. §102 (2006).
6. For example, in the case of sound recordings, copyright owners may perform the work publicly by means of a digital audio transmission. See 17 U.S.C. §106 (2006).
7. 17 U.S.C. § 101 et seq. (2006).
8. 17 U.S.C.§ 504(c)(2) (2006).
9. 17 U.S.C. §105 (2006). However, much of the law library community is aware that the editorial enhancements added to cases, such as Westlaw or Lexis headnotes, key numbers, and summaries, may be copyrighted material not in the public domain. There-

fore, the opinion of the court itself is in the public domain, and the vendor-added enhancements are covered by copyright. See *Little v. Gould*, 15 F. Cas. 612 (C.C.N.D.N.Y. 1852) (No. 8,395) (noting that "the written opinions of the judges . . . were regarded as having become the property of the public, and, therefore, as not the subject of a copyright."); *West Publ'g Co. v. Lawyers' Coop. Publ'g Co.*, 79 F. 756, 761 (2d Cir. 1897) (noting that a reporter of court opinions "may acquire a valid copyright for the headnotes, footnotes, tables of cases, indexes, statements of facts, and abstracts of the arguments of counsel, where these are prepared by him and are the result of his labor and research").

10. *The Copyright Slider* created by the American Libraries Association Office for Information Technology Policy (http://librarycopyright.net/resources/digitalslider/); *Copyright Term and the Public Domain in the United States* created by Peter Hirtle of the Cornell University Library (http://copyright.cornell.edu/resources/publicdomain.cfm); *The Durationator created by* Professor Townsend Gard of the Tulane Center for Intellectual Property Law and Culture (http://www.durationator.com); *U.S. Copyright Office Records* (http://www.copyright.gov/records/). The Copyright Office has digitized its copyright registration records from January 1, 1978, to the present. These records list publication and creation dates.

11. 17 U.S.C. §107 (2006)

12. See *Folsom v. Marsh*, 9 F. Cas. 342 (C.C.D. Mass. 1841), where Justice Story laid out the initial statements that were to become the four-factor test used for fair use analysis today.

13. 17 U.S.C. §107 (2006).

14. Ibid.

15. H.R. Rep. No. 1476, 94th Cong., 2d Sess. 65 (1976); S. Rep. No. 473, 94th Cong., 1st Sess. 62 (1975), quoted in *American Geophysical Union v. Texaco Inc.*, 37 F.3d 881, 884 (2d Cir. 1994).

16. For a more detailed description of HathiTrust and its partners, please see their official website at http://www.hathitrust.org/.

17. *Authors Guild v. HaithiTrust*, 2012 WL 4808939 (2012) *Authors Guild v. HaithiTrust*, 2012 WL 4808939 (2012) *Authors Guild, Inc. v. HathiTrust*, 2012 WL 4808939 (S.D.N.Y. Oct. 10, 2012).

18. "Transformative use" was first used in the Supreme Court fair use case *Campbell v. Acuff-Rose Music, Inc.*, 510 U.S. 569 (1994). ("The central purpose of this investigation is to see, in Justice Story's words, whether the new work merely 'supersede[s] the objects' of the original creation, or instead adds something new, with a further purpose or different character, altering the first with new expression, meaning, or message; it asks, in other words, whether and to what extent the new work is 'transformative.'") Ibid., 579.

19. *Bill Graham Archives v. Dorling Kindersley Ltd.*, 448 F.3d 605 (2d Cir. 2006).

20. *Authors Guild*, 12.

21. *Harper and Row v. Nation Enterprises*, 471 U.S. 539 (1985).

22. *Authors Guild, 12.*

23. Ibid., 13.

24. Ibid.

25. Ibid., *14.*

26. Barton Beebe, 2008, *An Empirical Study of U.S. Copyright Fair Use Opinions, 1978–2005*, 156 U. Pa. L. Rev. 549, 556.

27. H.R. Rep. No. 1476, 94th Cong., 2d Sess. 47, 75 (1976) (hereinafter *House Report*). Cf. S. Rep. No. 473, 94th Cong., 1st Sess. 60, 67 (hereinafter *Senate Report*). ("108 . . . is intended to preclude a library or archives in a profit-making organization from providing photocopies of copyrighted materials to employees engaged in furtherance of the organization's commercial enterprise, unless such copying qualifies as a fair use, or the organization has obtained the necessary copyright licenses. A commercial organization should purchase the number of copies of a work that it requires, or obtain the consent of the copyright owner to the making of the photocopies.") To date, no court has adjudicated the difference between the House Report and the Senate Report.

28. Register of Copyrights, *Library Reproduction of Copyrighted Works (17 U.S.C. §108)* (Washington, D.C.: U.S. Copyright Office, 1983), 78–79.

29. 17 U.S.C. §108(g) (2006).

30. 17 U.S.C. §108(i) (2006).

31. 17 U.S.C. §108(c) (2006).

32. House Report, 75.

33. 17 U.S.C. § 108(f) (2006).

34. Sonny Bono Copyright Term Extension Act, Pub. L. No. 105-298 (1998) (codified as amended in various sections of title 17 of the United States Code (U.S.C.)). In 1998, the Sonny Bono Copyright Term Extension Act extended the term of copyright for twenty years, for both existing and future works.

35. American Library Association, "Applying fair use in the development of electronic reserves systems," accessed December 4, 2012, http://www.ala.org/advocacy/copyright/fairuse/fairuseandelectronicreserves.

36. *Cambridge Univ. Press v. Becker*, 863 F. Supp. 2d 1190, 1224 (N.D. Ga. 2012).

37. *Basic Books, Inc. v. Kinko's Graphics Corp.*, 758 F.Supp. 1522 (S.D.N.Y. 1991).

38. *Princeton University Press v. Michigan Document Services, Inc.*, 99 F.3d 1381 (6th Cir. 1996).

39. *Cambridge Univ. Press*, 1224.

40. Ibid.

41. Ibid., 1225–1227.

42. Ibid., 1233.

43. Ibid., 1243.

44. The Classroom Guidelines are known officially as "Agreement on Guidelines for Classroom Copying in Not-for-Profit Educational Institutions," House Report, 68. Specifically, the parties who agreed to the Classroom Guidelines were the Ad Hoc Committee on Copyright Law Revision, the Author–Publisher Group, the Authors League of America, and the Association of American Publishers; ibid., 70. The American Association of University Professors and the Association of American Law Schools strongly criticized the Guidelines as being too restrictive; ibid., 72.

45. *Cambridge Univ. Press*, 1228.

46. Ibid., 1234–1235.

47. Ibid., 1237.

48. "Harvard Libraries launch Scan and Deliver service," *Harvard College Library News*, accessed December 30, 2012, http://hcl.harvard.edu/news/articles/2009/scan_and_deliver.cfm.

49. Aspara Iyer and Julia Zorthian, "Library launches 'Scan and Deliver' service," *Yale Daily News* (online), last modified September 18, 2012, http://yaledailynews.com/blog/2012/09/06/library-launches-scan-and-deliver-service; "Scan & Deliver service brings the library to you," *University of Chicago Library News*, last modified January 18, 2012, http://news.lib.uchicago.edu/blog/2012/01/18/scan-deliver-service-brings-the-library-to-you-2/.

4

Open Access to Legal Scholarship

Cheryl Kelly Fischer
UCLA LAW LIBRARY

Vicki Steiner
UCLA LAW LIBRARY

The majority of law reviews and law journals in the United States are managed by law student editors who select, edit, and publish legal scholarship written by academics, lawyers, and students. Thus, it is law students who, with the permission of authors and the financial support of their institutions, decide the means available for accessing this legal literature. Historically, law reviews and journals have been published primarily in print issues and distributed to subscribers through the mail. Electronic access to this content has been delivered mainly through costly commercial legal databases such as Westlaw, LexisNexis, and HeinOnline. In the past decade, however, there has been progress—albeit slow progress—toward making legal scholarship freely available electronically through open access, and academic law libraries are increasingly playing pivotal roles in working with law students and law faculty to facilitate such progress. In this chapter, we provide an introduction to the open access movement, discuss key milestones in the development of open access to legal scholarship, consider the benefits and challenges to moving to open access, and examine some of the technologies available to assist in the process. We conclude with a look ahead to potential future developments in open access.

WHAT IS OPEN ACCESS?

There are varied definitions of open access (OA),[1] but most track closely with the definition set forth in the Budapest Open Access Initiative, which states:

> By "open access" [to peer-reviewed research literature] we mean its free availability on the public internet, permitting any users to read, download, copy, distribute, print, search, or link to the full texts of these articles, crawl them for indexing, pass them as data to software, or use them for any other lawful purpose, without financial, legal, or technical barriers other than those inseparable from gaining access to the internet itself. The only constraint on reproduction and distribution, and the only role for copyright in this domain, should be to give authors control over the integrity of their work and the right to be properly acknowledged and cited.[2]

Additionally, literature in the OA movement often refers to "Gold" OA and "Green" OA. The distinction between the two is the vehicle used to deliver content.[3] Gold OA refers to publishing, thus journals are the means of delivery, while Green OA refers to archiving, and so institutional or hosted disciplinary repositories are the vehicles used to deliver the content.[4]

The primary focus of this chapter is on Gold OA, the publication of legal scholarship in open access law journals.[5] For Green OA, the deposit of legal scholarship into OA repositories has long been common practice for legal scholars. For instance, law faculty members regularly deposit their work (typically preprints, versions of works not yet reviewed, copy-edited, or published) in SSRN, a hosted disciplinary repository that affords users free access to author-submitted abstracts and full-text content (again, usually preprints) and free downloading of said content in most cases.[6] Given this widespread practice, it should logically follow that legal scholars would similarly embrace the deposit of their works into institutional repositories. Unfortunately, this is not always the case.

HOW "OPEN" IS OPEN ACCESS?

The question of openness is equally varied and is informed by issues such as readers' rights, authors' rights, availability for deposit into a repository, and machine-readability.[7] Although the extent of "openness" is varied in all disciplines, discussions of OA often focus on the traditional publication model (closed, subscription, or pay-per-view required to read) versus a fully open and digital-only publication model. Within the context of legal scholarship, while there are some fully open, digital-only law journals[8] (typically specialty journals, such as law and technology journals, or online companions to law reviews), most law journals that have transitioned to OA do so using a hybrid or "OA Plus" publication model, in which a journal's content is freely available online and made available in traditional subscription-based databases, and print subscriptions are maintained by the journal or fulfilled via print-on-demand.[9]

KEY MILESTONES IN THE ONGOING DEVELOPMENT OF OPEN ACCESS TO LEGAL SCHOLARSHIP

Compared to the overall movement for OA to scholarly literature, which dates back to the 1960s,[10] the development of OA to legal scholarship is still in its nascent stage. Key milestones in the legal academy's adoption of OA principles for legal scholarship include the following:

- 1998: Professor Richard A. Danner leads Duke Law to become the first law school in the United States to make the full runs of its law journals freely available online. The journals grant authors the right to repost their works without restriction in other freely accessible repositories or websites.[11]
- 2005: Professor Dan Hunter publishes his self-proclaimed "manifesto in favor of open access to legal scholarship and a polemic against the walled gardens of law reviews."[12] In reporting the findings of his 2004 survey of the publication policies of law journals at all ABA-accredited law schools in the United States, Hunter concludes that while law journals are "a long way behind other disciplines in adopting open access,"[13] they are "obvious candidates to benefit from open-access models" since they are inexpensive to produce and the institutions with which they are affiliated have values that are aligned well with those of the OA movement.[14]
- 2005: The Open Access Law Program (OALP), a project intended to support OA to legal scholarship, is launched.[15] OALP resources include an author pledge,[16] OA principles[17] and a list of law journals adopting OA principles,[18] and a model publication agreement.[19] Early adopters of the author pledge include legal scholars James Boyle, Mark A. Lemley, Lawrence Lessig, and Margaret Jane Radin.[20]
- 2006: *Lewis & Clark Law Review* sponsors a daylong symposium on OA and the future of legal scholarship, which leads to publication of articles on topics such as the economics of OA legal publishing, alternative forms of electronic publishing, and the impact of OA on prestige.[21]
- 2009: Directors from several academic law libraries release the "Durham Statement on Open Access to Legal Scholarship," which calls upon law schools to forgo print publication of law journals in favor of publication in stable, open, digital formats.[22]
- 2012: Professor Ross E. Davies's historical analysis of total paid subscriptions to flagship law reviews reveals sharp declines in paid subscriptions.[23] Davies notes that in 2011, and for the first time since the U.S. Postal Service began requiring law reviews to report circulation data, "no major law review had more than 2,000 paying subscribers. The *Harvard Law Review* remains the top journal, but its paid circulation has declined from more than 10,000 during . . . 1960s and '70s to . . . 1,896."[24]

BY THE NUMBERS: OA LEGAL SCHOLARSHIP TODAY

There has been slow, but steady, progress toward increased OA to legal scholarship since the aforementioned pronouncements of OA principles and the trend of rapid decreases in the number of total paid subscriptions has persisted in recent years. As of January 31, 2013:

- More than 400 law journals have full-text articles that are searchable and freely accessible online.[25] Some, but not all, of these journals permit use of said content—such as reproduction and distribution—without copyright restrictions.
- Thirty-five U.S. law schools use the Digital Commons as their institutional repositories.[26] Many others prefer to use free, open-source software platforms such as EPrints (http://www.eprints.org/) and DSpace (http://www.dspace.org/) to establish their institutional repositories.
- More than 156 worldwide law journals have met the selection criteria[27] required for registration with the Directory of Open Access Journals, 30 of which are U.S. law journals.[28]

MAKING THE CASE FOR OPEN ACCESS

Making the case for OA to legal scholarship is important because the traditional print model of law journal publication is economically, academically, and culturally untenable. We now turn our attention to some of the compelling arguments that support adoption of OA principles and some of the counterarguments to common concerns expressed by legal scholars about OA to legal scholarship.

Open Access Is Economically Sustainable

The traditional print publication model causes many law journals to lose money each year, due to increasing costs of production and distribution and inefficiencies in the traditional print publication model. Exacerbating the problem is the significant loss of total paid subscriptions for many journals each year.[29] OA publication can defray many of the costs of print production and distribution without diminishing existing royalties received from vendor agreements with Westlaw, LexisNexis, and HeinOnline. OA publications that incorporate print-on-demand (what we term OA Plus)[30] can also help law journals to stop operating at a loss. To illustrate this point, we summarize an actual example UCLA Graduate Student Association's (GSA)[31] director of publishing Stacey Meeker provided via e-mail[32] in the course of our consultations with GSA to facilitate our law journals' transitions to OA. The example compares the traditional print publication of a law journal with the publication of a nonlaw journal using OA Plus, and the journals' respective uses of fund allocations in 2012 to publish issues comparable in size.

- Traditional Print: UCLA School of Law's *Journal X* used an allocation of more than $5,800 for its publication of its latest issue, which paid for costs of compositing (with corrections), production of about 300 copies, offprints, and shipping.
- OA Plus: *Journal Y* used an allocation of $3,000, which paid for "layout (corrections and prepress included) for two years, one issue each (comparable in size to [*Journal X*]), global distribution through Lulu, 115 printed copies for distribution to authors and [*Journal Y*'s] department, the full item-level processing of 37 years' worth of back issues, including upload to eSchol[arship]."[33]

As is clear in the foregoing example, even if law *Journal X* in the above example "were to recoups its costs via subscriptions, there is a substantial outlay of cash to produce each issue"[34] that is not incurred with the OA Plus publication model *Journal Y* used, wherein many of the costs of production become one-time, rather than recurring, expenses and additional savings are realized through a simplified, more centralized publication process.[35]

Open Access Is Academically Sustainable

In addition to the aforementioned significant costs incurred in the traditional production and distribution of law journals in print form, there are potentially substantial academic costs that law schools and legal scholars bear when law journals strictly follow the traditional law journal publication model. We are now in an era in which technology makes it possible for law schools and legal scholars to extend the reach of their work in unprecedented ways. Despite innovations in technology, most of the legal academy relies on commercial legal vendors such as Westlaw, LexisNexis, HeinOnline, and the Index to Legal Periodicals and Books and print volumes collected by libraries to make its scholarly output available to others. However, such reliance creates substantial cost and physical barriers to accessing legal scholarship.

First, commercial legal databases limit the audience for legal scholarship since, for the most part, only researchers affiliated with academic institutions in the U.S. legal academy (and perhaps a number of large law firms and courts) can afford subscriptions to such databases. Second, reliance on print volumes as the sole supplement to commercial legal databases also limits the audience for legal scholarship, by extending access to include those who have the means to visit or borrow from law libraries. Overall, OA increases the visibility, and thus the potential scholarly impact, of legal scholarship by removing these barriers to legal scholarship and thereby making such content freely available to all those who seek it, regardless of who they are or where they live.

Perceptions of Prestige

The leading obstacle OA advocates confront in making the case for OA to legal scholarship, and the increased access it affords, relates to the perception of prestige. Legal scholars may believe that restricted access and the resultant perceived "aura of scholarly exclusivity"[36] creates prestige. If—as the argument goes—unfettered access to legal scholarship obliterates the "aura of scholarly exclusivity" and thereby diminishes prestige, some legal scholars may contend that publication in an OA journal could adversely affect the scholarly impact of their works and their eligibility for privilege and tenure at their institutions.[37] Another related concern stems from the reputation of authority and prestige that is inherent in traditional print law journals following decades of publication in an accepted medium. As a result, when confronted with the opportunity to publish in two or more otherwise similarly situated law journals, these legal scholars may opt for the journal that publishes under the traditional print model, rather than the OA journal.[38]

There are many ways to respond to concerns regarding prestige. First, the prestige (real or perceived) of a journal does not depend on the medium used to deliver its contents. For law journals, the same selection, editing, and publishing process (all performed by unpaid, and mostly untrained, law students) is used regardless of the manner in which content is made available. Second, while the extant literature on OA to legal scholarship has not yet addressed, in full, the question of whether publication in an OA journal would have an adverse impact on determinations of privilege and tenure (P&T), it seems unlikely that such would be the case. P&T decisions are typically driven largely by individual or article-level metrics, not journal-delivery-level metrics. Thus, a committee charged with assessing potential candidates for a teaching position (or tenure) would first look to the quality of the candidate's work and then perhaps to the quality of the journal (by reference to its affiliated institution), not the manner in which the journal was published. Third, while the literature does reflect a certain degree of apprehension toward OA on the part of legal scholars, such apprehension is by no means pervasive.[39] Many of the concerns about the medium of OA can be allayed if journals adopt an OA publication model that includes print-on-demand. And fourth, for legal scholars who are not persuaded by the many advantages of publication in OA law journals—or who are waiting for more top-tier law schools to embrace OA (a trend already underway)[40]—there remains the option of scholars publishing in their journal of choice and self-archiving preprints of their work in institutional repositories. Combining traditional publication with Green OA accommodates legal scholars' publication preferences and achieves the goal of increasing access to said scholars' works.

Law libraries, as experts in the field of information, can and should continue to serve an integral role[41] in educating our constituents about OA and thereby dispelling concerns attendant to increasing access to legal scholarship. Attitudes toward OA will improve and as more journals—particularly those affiliated with highly ranked law schools—move to OA, others will follow.

Impact on Citations

Another common concern that law faculty members may articulate in discussions of OA to legal scholarship is how OA might impact citations to their works. Here, the concern often is geared less toward the question of whether increased access via OA would increase citation counts (which seems a given, or at least would not result in a *decrease* in citation counts), and more toward whether deposits of works into OA institutional repositories will drive traffic away from an author's profile on SSRN.[42] In our experience, allaying concerns about the potential adverse impacts on personal reputational interests by OA to legal scholarship is generally easy to do, because such concerns are often rooted in a fundamental misunderstanding of how SSRN and institutional repositories operate.[43]

Moreover, there is evidence in the extant literature on OA to legal scholarship that demonstrates that the increased access to legal scholarship afforded by OA has an overall positive impact on citations and downloads. For instance, a 2011 study was reported in which researchers analyzed articles from three law journals from the University of Georgia School of Law and compared citations rates of those articles available via OA with those not available via OA. The findings indicated that that OA legal scholarship can expect to receive 58 percent more citations than non-OA articles of the same age and in the same venue.[44] A 2009 study was also reported in which a researcher compared the number of downloads of articles from an institution's Digital Commons repository with the number of downloads from SSRN. The findings indicated that some articles deposited in the Digital Commons enjoyed download rates upward of 400 percent as compared to SSRN.[45]

MOVING PRINT LAW JOURNALS TO OPEN ACCESS

Given the myriad reasons so far discussed for making legal scholarship freely available electronically via OA, it makes sense for many student-run law journals currently publishing in a traditional print-only format to want to transition to an OA format. However, student-run law journal staffs have little incentive to look beyond the current volume for which they are responsible, often making it incumbent upon less transitory stakeholders at a law school, like law librarians, to step in and ensure that the move to OA publishing is well planned and executed in a way that will be sustainable for years to come.[46]

Working in collaboration with individuals from the Dean of Students Office at the University of California, Los Angeles (UCLA) School of Law, UCLA's Graduate Students Association, UCLA Library's Scholarly Communications and Licensing Unit, and the California Digital Library, we launched a program in the spring of 2012 to assess student and faculty interest in OA to legal scholarship at the UCLA School of Law. Based on the initial results of this program, efforts are now underway to work with law faculty to further develop our institutional repository and to help our law journals shift to OA Plus publication using the free, open-source publishing platform eScholarship.[47] The latter efforts give the UCLA School of Law the distinction of being the first law school within the UC system to use eScholarship as the OA publishing platform for its law journals. The following discussion is informed by our aforementioned experiences and begins by comparing the typical print publication model for law journals and all of its inefficiencies with an OA model. Thereafter, we address several important points to discuss with law journal staff while making the decision to transition to OA.

Comparing the Traditional Print Publication Model with Open Access

The traditional print publication model for law journals involves many of the same stages as an OA publication model: article selection, article editing, formatting and publishing, and distribution. However, the print publication model often involves more steps with less efficient communication and transmission methods.

In a typical print publication model, the article selection stage begins with soliciting article submissions from authors and collecting those submissions using postal mail, fax, individual e-mails, platforms such as ExpressO (http://law.bepress.com/expresso) or Scholastica (https://scholasticahq.com/law_reviews), or a combination thereof. These submissions are then reviewed by the journal staff, who slate the issue (decide which articles to publish). Communications with authors follow as authors with accepted manuscripts need to sign copyright agreements[48] and agree to various other items such as the number of offprints they would receive once their article is published. The efficiency of this stage is improved with the use of back end software often used with electronic publication, which offers journal staff a way to organize incoming submissions and to communicate with each other. One option for this type of software is Open

Journal Systems (OJS), a free, open-source journal management and publishing system developed by the Public Knowledge Project (http://pkp.sfu.ca/?q=ojs) that is integrated into eScholarship. Upon logging into the OJS system, journal staff can easily view submitted manuscripts, share comments as they deliberate on the slating of articles, communicate with authors and each other, and track the manuscripts through each step of the publication process, including the final verification that each accepted author has signed the required agreements.[49]

The article-editing stage begins next. At this stage, the accepted manuscripts are edited and cite-checked by journal staff. The substantive editing process involves review of things like the author's organization and grammar, while the cite-checking process involves reviewing the author's footnotes to make certain that sources cited do, in fact, support the author's statements and are formatted according to the proper citation style. Cite-checking therefore involves the collection and review of the sources cited in the footnotes—a major undertaking involving numerous journal staff members. Junior staff often collect the sources and perform a first check, after which senior staff verify the work and make further corrections. In order to accommodate the use of the collected source materials by multiple staff members and to avoid the need for a source to be pulled multiple times, the electronic materials are often downloaded, the print materials photocopied or scanned, and then all are uploaded to electronic file-sharing platforms like Dropbox (https://www.dropbox.com), Box (https://www.box.com), or Google Drive (https://drive.google.com) for easy reference by all staff involved in the cite-check. These steps, and the communication among journal staff and authors required to complete them, can be accomplished more efficiently with the use of a back-end management software system like OJS.

Once the editing of articles is complete, they are formatted for print publication.[50] Formatting can involve either (a) application of a complex template macro that will format parts of the document like margins, fonts, spacing, page numbers, and running headers, ultimately turning the document into a "camera ready" PDF for submittal to the printer, or (b) application of a simpler template, keeping the document in Microsoft Word format for submittal to the printer and paying the printer to further format the document into its camera-ready state. With either option, sending the final articles and accompanying documents (front matter, table of contents) to the printer requires filling out numerous forms like a print order form, author copyright agreements, offprint order forms, the order of materials form, and any special instructions. There follows a period of back and forth communication with the printer to confirm and correct things such as contract proofs that need to be checked, and sent back with an error memo if necessary, at a cost of $5.00 per page for corrections. Second contract proofs need to be approved if changes were made to the first proofs before finally giving the printer authorization to print.

This formatting and publishing stage is where OA publishing can look much different, and much more efficient, than the traditional print model. Without the need to send files and forms to a printer involving multiple back and forth communications and additional fees, the process is mostly completed in house by journal staff. The staff apply the formatting templates, make their own corrections, organize the order of materials, and upload all finished documents to their chosen OA publishing platform, such as Digital Commons, eScholarship, or a standalone website, often created using WordPress (http://wordpress.org). Of these options, we advocated for the use of eScholarship. eScholarship is preferred over Digital Commons because use of the platform is free for schools within the UC system, and the institutional support afforded to our journals using eScholarship is invaluable.[51] eScholarship is preferred over a standard website because the platform complies with data access standards that facilitate the indexing of content by search engines, thereby making content easier to find. Moreover, the back-end program OJS is incorporated into eScholarship, providing integration that can take the journal staff from the submission stage and through the publication stage in an organized and efficient manner.

To complete distribution, the fourth and final stage in the traditional print publication model, journals often contract with the printer to deliver printed issues to subscribers on the journals' mailing lists. In this case, after the printer receives authorization to print a set number of issues, the journal provides its subscription mailing list and postage fees, and the printer then mails issues to all subscribers, mails offprints to the authors, and ships the remaining stock to the journal. The printer is also often responsible for sending the final files to commercial legal research databases like Westlaw, Lexis, and HeinOnline if the journal is included in said databases. The final steps involved with distribution include communication with subscribers. Journal staff must contact subscribers regarding invoicing issues and must reconcile any problems regarding nonreceipt with the subscribers and the printer. OA distribution with eScholarship simply involves letting interested parties[52] know that the journal's latest issue is complete, online, and ready for viewing, download, or print-on-demand (POD) (if the POD option is provided by the journal).

Benefits and Challenges of Transitioning to Open Access

Law librarians can play an important role in helping journal staff consider whether moving from a traditional print publication model to an OA publication model is appropriate for the journal. Some of the more important points to discuss with journal staff include opportunities to save money and ways to increase efficiency, and downsides such as the learning curve that comes with being responsible for the first issues published during a transition to OA.

Money Saving Opportunities

Over-ordering printed issues is expensive, yet it often happens with the print publication model because printers print in fixed amounts, with the price of printing often tied to the number of copies ordered. Journals often opt to purchase a higher number of copies than what is actually needed because the price per issue is less than purchasing a smaller number of copies. OA publishing can lower publication costs, even if the journal provides POD as a service in addition to online OA. While offering the POD option usually costs the journal some initial amount of money,[53] and the journal may still opt to purchase a limited number of copies for an archive or to offer complimentary copies of the issue (rather than offprints) to authors, the journal ultimately saves money and never has to over-order because the POD purchasers order their own copies.

Providing authors with offprints at the journal's expense is another costly practice involved with print publishing because authors often expect to be offered a number of complimentary offprints of their article. With OA publishing, including or not including a POD option, a journal can consider changing offprint culture by declining to offer authors offprints (thereby saving the cost to the journal of printing and shipping them), and instead providing authors with monthly download reports that indicate that their articles are being read.

Various fees paid to printers—like paying $5.00 per page for corrections—can add up to large expenses. Fixing errors digitally when publishing in an OA format is less expensive than fixing errors with a printer, since changes are made directly by journal staff. Additionally, postage fees paid to the printer can quickly add up. Postage must be paid for issues to be sent to subscribers, for remaining issues to be shipped to the journal's office, and for offprints to be sent to authors. Postage is saved (and as a bonus, trees are saved, and the carbon footprint from trucks and planes delivering mail is reduced) when publishing in an OA format.

Finally, collecting money owed to the journal after invoicing can be cumbersome. If invoicing and subsequent follow up is not done properly, the journal does not collect the money it is owed and thereby operates at further loss. With OA publishing, and even with POD printing, the journal is not involved in the payment process so invoicing and collecting money owed are no longer a concern.

Efficiency Increases

As discussed above, the article-selection stage and the article-editing stage processes can be unorganized when they are handled via a combination of communication media (postal mail, e-mail, and submission systems like ExpressO or Scholastica) and electronic file storage systems. Efficiency in these stages can be increased by streamlining the processes and the communications involved using back-end software like OJS.

In the formatting and publishing stage, mailing proofs and order forms back and forth with a printer takes time, especially when multiple errors are discovered and corrections are necessary. Additionally, distribution stage issues like postage, maintaining the subscription mailing list, invoicing, mailing problems such as complaints of nonreceipt, and collection of subscription fees owed are time consuming. When journal staff focus is pulled away from their article-editing tasks and toward these non-editing tasks, such redirection of focus can lead to missed editing deadlines and publication delays. Time can be saved using an OA publishing process where the journal formats and uploads the final documents, thereby avoiding all of the back and forth communication with a printer. And with OA Plus, the printing process is streamlined so editors can focus on editing and meet editing deadlines.

Finally, in a traditional print publication model, there is often no in-house institutional support to help journal staff navigate issues arising from the printing and distribution process. However, with an established OA publishing platform like eScholarship, there is help available from the developer's staff. Further, with standardized production across law journals using the same platform, cooperation is easier and editors and support staff can help each other.

Potential Downsides during the Transition Process

Since there is no money to be made from publishing in a purely OA format, one downside of transitioning from a traditional printed format is the loss of potential for making money. However, most student-run law journals are not reaching this potential anyway and are, in fact, spending more money than they take in because subscriptions are rapidly declining.[54] For journals currently losing money via traditional print publishing, the move to OA would have a positive, rather than negative, effect on their budgets because while they would no longer have money coming in from print subscriptions (or, if they opt to include POD, they would likely have less money coming in), they would no longer be overspending money on print publishing costs.

Another possible downside of the transition to OA is that there can be a steep learning curve involving extra time and effort to set up a journal on an OA platform. This is especially true if the law journal is the first to make the transition at the law school or even the first law journal within a multicampus university system to make the transition, as is the case for the UCLA School of Law's journals that are now transitioning to eScholarship. That said, typical student-run law journals have editorial board turnover every year anyway, as third-year students graduate and first and second-year students take over management of the journals. Oftentimes, so much institutional knowledge is lost during these yearly turnovers that the incoming editorial board members would face an enormous learning curve even if they were maintaining a print publishing status quo.

Finally, for law journals that had been relying on a printer to accomplish certain postediting tasks, the transition will mean that journal staff will have to take responsibility for these tasks. One such task is article formatting, sometimes called typesetting or handcrafting. This task should not be difficult after an initial setup. Templates for formatting articles into their final published state can be borrowed from another journal, purchased from the journal's former printer,[55] or created from scratch. Once the template is finalized, it can be used for subsequent articles and issues. Another task that would fall to journal staff without a print publisher would be submitting content to commercial databases. To perform this step in the workflow process, OA journals would need to notify their representatives at Westlaw, Lexis, and HeinOnline that their latest issue has been posted online so the vendors, in turn, can proceed to pull the articles and include them in their respective databases.

LOOKING AHEAD: OPEN ACCESS TO LEGAL BOOKS

Although the OA movement in the legal academy is progressing more slowly than in other disciplines, there is growing evidence of what the next frontier in OA to legal scholarship will be: open access to books. In recent years, developments in this area have included expansion of the Center for Computer-Assisted Legal Instruction's (CALI) imprint eLangdell Press (http://elangdell.cali.org), resulting in its publication of more DRM-free legal casebooks that are compatible with devices like smartphones and e-readers; the expansion of library-based publishing;[56] the creation of the Directory of Open Access to Books (http://doabooks.org), a searchable index to peer-reviewed OA books, which currently lists sixty-eight OA legal titles;[57] libraries' development of databases such as the HathiTrust (http://www.hathitrust.org) and The Online Books Page (http://onlinebooks.library.upenn.edu) that facilitate access to books that are freely available online; as well as the launch of affordable course material initiatives that aim to incentivize instructors to use low-cost or free alternatives to costly books and course materials.[58]

CONCLUSION

In the preceding pages, we have endeavored to provide an overview of the ongoing development of OA to legal scholarship and to demonstrate why OA to legal scholarship is important. Simply stated, OA to legal scholarship is important because the traditional print model of publication is untenable. It is economically untenable to adhere to a traditional print model of publication at a time when paid subscriptions to law journals are rapidly declining and the costs of production and distribution are increasing. It is academically untenable to adhere to a traditional print model of publication at a time when technology makes it possible for law schools and legal scholars to extend the reach of their work beyond the legal academy and to other disciplines and countries. And it is culturally untenable to adhere to a traditional print model of publication at a time when we are called upon to remove barriers to accessing legal information for the public good. As experts in the field of legal information, law librarians are uniquely situated to advise constituents about the benefits of OA and the challenges confronted in its implementation. Let us use this expertise to help our institutions fulfill their educational missions.

FURTHER READING

"Budapest open access initiative: Resources." *Open Society Foundations*. http://www.opensocietyfoundations.org/openaccess/resources.

"Harvard open access project." *Berkman Center for Internet & Society at Harvard University*. Last modified August 29, 2012, http://cyber.law.harvard.edu/research/hoap. The Harvard Open Access Project (HOAP) publishes a newsletter that provides information regarding HOAP projects and new developments in OA. To subscribe, click on the tab "Subscribe to the HOAP Newsletter."

Keele, Benjamin J., and Michelle Pearse. 2012. "How librarians can help improve law journal publishing." *Law Library Journal* 104(3): 383–410, http://www.aallnet.org/main-menu/Publications/llj/LLJ-Archives/vol-104/no-3/2012-28.pdf.

Suber, Peter. 2012. *Open access*. Cambridge, Mass.: MIT Press.

Swan, Alma, and Leslie Chan. "Open access scholarly information sourcebook: Practical steps for implementing open access," *Open Oasis*. Last modified September 22, 2012, http://www.openoasis.org/.

"Ten years on from the Budapest open access initiative: Setting the default to open." *Open Society Foundations*. Last modified September 12, 2012. http://www.opensocietyfoundations.org/openaccess/boai-10-recommendations.

NOTES

1. Peter Suber, *Open access* (Cambridge, Mass.: MIT Press, 2012), 7–8.

2. "Ten years on from the Budapest open access initiative: Setting the default to open," *Open Society Foundations*, last modified September 12, 2012, http://www.open

societyfoundations.org/openaccess/boai-10-recommendations. The definition of OA introduced in the original Budapest open access initiative in 2002 was reaffirmed in 2012.

3. Suber, *Open access*, 6.

4. Carol A. Parker, "Institutional repositories and the principles of open access: Changing the way we think about legal scholarship," *New Mexico Law Review* 37, 2 (2007): 431–477, http://law.bepress.com/expresso/eps/1705/. Provides an excellent overview of several repositories.

5. We use the terms "law journals" and "legal scholarship" interchangeably as shorthand to refer to both flagship law reviews and secondary law journals in this chapter.

6. "Social Science Research Network: Frequently asked questions," *SSRN*, http://www.ssrn.com/update/general/ssrn_faq.html#elec_lib_charge.

7. "HowOpenIsIt? Open access spectrum," *Public Library of Science*, http://www.plos.org/about/open-access/howopenisit/.

8. "Law journals: Submissions and rankings," *Washington & Lee University School of Law*, http://lawlib.wlu.edu/lj/index.aspx. We performed a search of this database for student-edited, online-only journals published in the United States. This query indicated that out of the 664 student-edited U.S. law journals indexed in the database, there are 74 law journals that publish solely in an online format.

9. "OA journal business models: Hybrid OA journals," *Open Access Directory*, last modified October 15, 2012, http://oad.simmons.edu/oadwiki/OA_journal_business_models. The term "hybrid OA" commonly denotes a publication model in which a journal publishes some OA articles and some non-OA articles, where the decision to afford OA resides with the author, who may be charged a fee upon electing OA. Since this arrangement has little, if any, application in the context of law journal publication, we have elected to use the term "OA Plus."

10. "Timeline of the open access movement," *Open Access Directory*, last modified September 4, 2012, http://oad.simmons.edu/oadwiki/Timeline.

11. "Open access at Duke Law," *Duke University School of Law*, October 19, 2009, http://law.duke.edu/libtech/openaccess.

12. Dan Hunter, "Walled gardens," *Washington & Lee Law Review* 62, 2 (2005): 607–640, http://law.wlu.edu/deptimages/law%20review/62-2hunter.pdf.

13. Ibid., 613.

14. Ibid., 639.

15. "Creative Commons and Science Commons announce open access law program," *Creative Commons*, June 6, 2005, http://creativecommons.org/press-releases/entry/5464.

16. "Open access law: Author pledge," *Science Commons*, http://sciencecommons.org/projects/publishing/oalaw/oalawpledge/.

17. "Open access law: Principles," *Science Commons*, http://sciencecommons.org/projects/publishing/oalaw/principles/.

18. "Open access law adopting journals," *Creative Commons*, last modified May 25, 2012, http://wiki.creativecommons.org/Open_Access_Law_Adopting_Journals.

19. "Open access law: Publication agreement," *Science Commons*, http://sciencecommons.org/projects/publishing/oalaw/oalawpublication/.

20. "Open access law: Authors," *Science Commons*, http://sciencecommons.org/projects/publishing/oalaw/oalawauthors/.

21. *Lewis & Clark Law Review* 10, 4 (2006): 733–924, http://law.lclark.edu/law_reviews/lewis_and_clark_law_review/past_issues/volume_10/.

22. "Durham statement on open access to legal scholarship," *Berkman Center for Internet & Society at Harvard University*, February 11, 2009, http://cyber.law.harvard.edu/publications/durhamstatement.

23. Ross E. Davies, "Law review circulation 2011: More change, more same," *Journal of Law* 2, 1 (2012): 179–188, http://ssrn.com/abstract=2005243. A review of U.S. Postal Service Form 3526 reported for the year following publication of Davies's article reveals that most law reviews are continuing to experience losses in paid subscriptions.

24. Ibid., 179.

25. "Free full-text online law review/journal search," *American Bar Association*, http://www.americanbar.org/groups/departments_offices/legal_technology_resources/resources/free_journal_search.html.

26. "Law school institutional repositories," *bepress*, http://digitalcommons.bepress.com/institutional-repository-law/.

27. "About DOAJ: Selection criteria," *Directory of Open Access Journals*, http://doaj.org/doaj?func=loadTempl&templ=about&uiLanguage=en#criteria.

28. "Browse by subject: law," *Directory of Open Access Journals*, http://doaj.org/doaj?func=subject&cpid=46&uiLanguage=en.

29. Davies, "Law review circulation 2011," 179.

30. "Timeline of the open access movement."

31. "Graduate student publications," *UCLA Graduate Students Association*, http://gsa.asucla.ucla.edu/services/publications.

32. Stacey Meeker, e-mail message to Vicki Steiner, October 5, 2012. The support and expertise provided by Meeker and GSA Publishing has been invaluable in helping UCLA Law's journals transition to OA.

33. Ibid. The $3,000 allocated may also have covered business cards for *Journal Y*.

34. Ibid.

35. See also Jessica Litman, "The economics of open access law publishing," *Lewis & Clark Law Review* 10, 4 (2006): 779–796, http://works.bepress.com/jessica_litman/6/.

36. Michael J. Madison, "The idea of the law review: Scholarship, prestige, and open access," *Lewis & Clark Law Review* 10, 4 (2006): 901–924, http://law.lclark.edu/live/files/9594-lcb104madisonpdf.

37. Stephanie Plotin, "Legal scholarship, electronic publishing, and open access: Transformation or steadfast stagnation?" *Law Library Journal* 101, 1 (2009) :31–57, http://ssrn.com/abstract=1350138. Provides an excellent discussion of conservatism in the legal academy in relation to the forms in which legal scholarship is embodied.

38. Richard A. Danner, Kiril Kolev, and Marguerite Most, "Print or perish? Authors' attitudes toward electronic-only publication of law journals," *Duke Law Scholarship Repository*, last modified July 2012, http://scholarship.law.duke.edu/faculty_scholarship/2393/.

39. Ibid. This survey was delivered in 2011 to 464 authors of articles published in recent volumes of flagship law reviews at institutions ranked within the top fifteen law schools by *U.S. News*. The survey had a response rate of 51.3 percent and showed that for 68 percent of the respondents, the continuation, or cessation, of the publication of

print issues would not have affected their decision to publish their article in the specific journal where it appeared.

40. Dev A. Patel, "Harvard Law Review increases online presence," *Harvard Crimson*, February 25, 2013, http://www.thecrimson.com/article/2013/2/25/hlr-doubles-online-editors/.

41. "CNI's executive roundtable: Institutional strategies and platforms for scholarly publishing," *Coalition for Networked Information*, last modified February 5, 2013, http://www.cni.org/executive-roundtable-reports/institutional-strategies-platforms-scholarly-publishing/. Provides an excellent discussion of the expanding roles of libraries in the realm of scholarly publishing.

42. "World," *Ranking Web of World Repositories*, http://repositories.webometrics.info/en/world. In July 2012, the Ranking Web of World Repositories ranked SSRN as the number one open access repository in the world.

43. For instance, access to full text may vary and versions posted in SSRN and institutional repositories may differ. It is worth noting that some agreements with commercial legal database vendors prohibit the deposit of postprints into certain third-party repositories. For instance, while HeinOnline typically will permit legal scholars to deposit copies of their articles that have been digitized by Hein into institutional repositories, it often will not permit deposit into SSRN.

44. Carol A. Watson and James M. Donovan, "Citation advantage of open access legal scholarship," *Law Library Journal* 103, 4 (2011): 553–573, http://digitalcommons.law.uga.edu/law_lib_artchop/24/.

45. Pamela Bluh, "TCO and ROI: Assessing and evaluating an institutional repository," presentation at the American Association of Law Libraries meeting, Washington, D.C., July 27, 2009, http://digitalcommons.law.umaryland.edu/fac_pubs/796/.

46. Courtney Smith, "Five steps to digitally archiving your law review," presentation at the American Association of Law Libraries meeting, Denver, Colorado, July 18, 2010, http://digitalcommons.bepress.com/repository-research/33. Provides a helpful discussion of how law libraries can help ensure the sustainability of OA legal scholarship through supportive technical services, such as assisting with the capture of metadata for journal content.

47. "Top Institutionals," *Ranking Web of World Repositories*, http://repositories.webometrics.info/en/top_Inst. The Ranking Web of World Repositories ranked eScholarship as one of the top ten institutional repositories worldwide in July 2012.

48. If a journal has been publishing using the traditional print publication model and plans to move to OA, the journal's standard copyright agreement with authors, as well as any existing agreements with commercial databases (Westlaw, Lexis, and HeinOnline), should be reviewed to determine whether changes are needed in order to publish OA.

49. Although "peer review," in the traditional sense, is not a common practice in student-edited law journals, it is worth noting that the peer-review process is built into OJS software, making it easy to manage the communications and documents involved with the review and editing process.

50. "Glossary," *Joe Christensen, Inc.*, http://jci.mightydrake.com/public/glossary.htm. Provides an excellent guide to terms related to the formatting process described in this chapter.

51. "Frequently asked questions," *eScholarship*, http://escholarship.org/help_faq.html.

52. Interested parties in the case of journals switching from a traditional print publication model to OA would most likely be those on the print subscriber mailing list.

53. "Services: Distribution," *Lulu*, http://www.lulu.com/services/distribution. Lulu is the POD service provider for OA journals affiliated with the University of California using eScholarship. The cost to participate in Lulu's "globalREACH" program is $75.00 per issue, which includes distribution via online retail sites such as Amazon.com and BarnesandNoble.com, and listings with Ingram Book Company, which enables brick and mortar retailers to purchase issues.

54. Davies, "Law review circulation 2011," 179.

55. Journals that have relied on templates created by JCI for this stage of publication can save time and money by purchasing their templates from JCI for a reasonable fee.

56. Jennifer Howard, "For new ideas in scholarly publishing, look to the library," *Chronicle of Higher Education*, February 4, 2013.

57. "About DOAB," *Directory of Open Access Books*, http://doabooks.org/doab?func=about&uiLanguage=en.

58. See, for example, "UCLA affordance course materials initiative," *UCLA Library*, January 14, 2013, http://www.library.ucla.edu/libraries/managing-collections.

5

User Services Analysis for Decision Making

Kim Clarke
BENNETT JONES LAW LIBRARY, UNIVERSITY OF CALGARY

Law librarians count everything that happens in the library. Technological advances have made it easier to gather and analyze statistics. In fact, it may be too easy to gather the data, which can lead to statistics fatigue wherein library staff get tired of tracking their activities and cease doing so. Library administrators must ensure they are not gathering statistics simply because it is possible to do so. They must clearly understand why they are gathering statistics and the data elements needed to provide them with the necessary information. They should also communicate the reasons to the library staff to get them invested in the statistics gathering.

Statistics are frequently gathered by librarians to understand the changing needs of their patrons. Longitudinal statistical studies can show trends in how patrons are utilizing the library's facility, services, and collections. Tracking requests for services or instructional programs that are not currently being offered allows libraries to determine ways of meeting those needs. Statistics and qualitative information gathered through surveys[1] and unsolicited means are frequently used to justify decisions to senior administration when requesting funds for additional personnel or technology, for example.

Law libraries are often required to submit statistics to either accreditors or professional associations to demonstrate the quality of the library's operation. In fact, a library may be required to provide statistics to multiple accreditors, which can be problematic as they likely ask for different statistics. An academic law library in Oregon, for example, may be accredited by the American Bar Association and its university could be accredited by the Western Association of Schools and Colleges, so the library would need to submit data to both organizations. Law libraries at large, research-focused universities may also submit statistics to the Association of Research Libraries.[2] Law library administrators often also use statistics to benchmark their libraries against peer institutions.[3]

Libraries gather both quantitative statistics and qualitative information about their programs and services. Quantitative statistics provide a strict numerical count. Qualitative statistics provide descriptive information regarding the library's services, collections, or programs. Essentially, quantitative statistics answer the question "What are we doing?" and qualitative information explains "How we are doing." Quantitative statistics are significantly easier to obtain than qualitative, as they simply require the tracking of activities or events. Qualitative information must be received directly from patrons either through formal or informal means.

Law libraries can gather information or statistics about all of their services, programs, and activities. Librarians must determine which statistics and data points they will track to meet internal and external information needs. In this chapter, we discuss reference, instructional, collection management, and website usage statistics, focusing on the statistics law libraries gather and the electronic means used to collect and analyze them.

REFERENCE STATISTICS

One of the key functions of all law libraries is to assist patrons with their legal research questions. Statistics-gathering for reference transactions likely has occurred from the very beginning of the service. The primary reasons for keeping reference statistics are to track patron usage of the service, to identify trends and staffing needs, and to comply with requirements of professional associations.[4] The interest in reference statistics has increased substantially in recent years as library administrators are trying to understand and explain changes in the use pattern. While quantitative statistics show that fewer reference interactions are occurring in academic libraries, reference librarians feel the questions are more complex and require more in-depth assistance, which consumes more time than previously experienced,[5] or are occurring away from the reference desk. Statistics help administration determine if these assertions are true.

Traditionally, reference statistics were kept on paper using a form comprised of the various data elements the library wanted to track, with a tick mark representing a single reference transaction. The transition to electronic reference statistics solutions has been surprisingly slow. As recent as 2002, 99 percent of the libraries who participated in an Association of Research Libraries survey still used a paper form.[6] More than half of the respondents indicated they were planning on making significant changes to the data collection practices in the future,[7] presumably including reviewing their electronic options.

Electronic reference statistics-gathering options overcome several shortcomings of the print form. Firstly, print forms are unable to capture all the data elements the reference librarians and library administration want to track.[8] Print forms generally comprise a chart with data elements along the vertical and horizontal axes. For example, the amount of time spent on the reference transaction could be along vertical axis and the various patron groups served along the horizontal axis. It is difficult to introduce more data elements without creating additional charts on the same form. The print form is also not effective for recording qualitative information, such as the questions being asked or the resources used to answer them. A comments box could be incorporated on the form; however, the information written on the form becomes invisible to the reference staff as soon as the used form is replaced.

Data analysis is difficult to perform in the print environment. It is best performed with the assistance of spreadsheet software such as Microsoft Excel or statistical products like SPSS. The data from the print forms must be manually entered into the software products, which is time consuming and subject to data entry errors. The difficulties increase with the complexity of the statistics being recorded, including the number of data elements on the form and if there were multiple service points where the statistics were recorded. Difficulty in conducting analysis on the statistics and the lack of qualitative information limit the library administration's ability to identify trends in service needs with a print form.

Librarians frequently hold in-depth research consultations or answer e-mail or chat reference questions in their office but forget to record those transactions on the statistics sheet at the reference desk. Electronic forms can be accessed from multiple locations so they can be updated immediately after the transaction occurs.

Electronic reference statistics forms allow libraries to gather information on many more data elements in a systematic and analyzable manner. Both the data elements and their variables selected are dependent on the information needs of the library. Academic law libraries likely use the most complicated form, tracking date and time of the transaction, location where the question was received, format the question was received in, type of assistance sought, who the patron was, length of time spent on the question, and the name of the reference staff member providing assistance. Each of these data elements can have multiple variables. The "type of assistance sought" data element, for instance, could be directional, referral, technical, ready reference, or consultation. The online reference statistical products allow the data to be exported into Excel, which makes analysis much easier than using the print form.

Electronic forms also allow for the inclusion of qualitative information. A comments box or question and answer boxes allow the reference staff to add information relating to each transaction. They can add additional identifying information such as course number, practice area, the identity of the patron as appropriate, or the resources used to answer the question. This explanatory information may be useful for the other reference staff members if they receive similar questions, and could also be used as real-life examples during instructional sessions. Administrators could use the qualitative information when designing employee training sessions for new employees or to fill identified gaps in the knowledge of current staff.

REFERENCE STATISTICS TOOLS

Law libraries can either develop an in-house electronic reference statistics form or use one of the available commercial products. Homegrown electronic reference statistics forms are often created in ubiquitous software products such as Microsoft Word or Excel. The forms can be loaded on the library's intranet or even on the cloud via websites like Google Drive to allow reference staff members to access the form from a variety of locations. Homegrown reference statistics forms can track multiple pieces of information in questions, such as patron type, question type, communication method, time of day the question was received, and the length of time spent on the question.

Zoho Creator is still a relatively unknown software product that can facilitate the creation of very professional-looking forms. Zoho Creator's "drag and drop" interface allows even novices to quickly create an effective reference statistics recording document simply by dragging the desired fields onto the blank form. The developer can implement a wide range of field types, including single- and multiline text fields, checkboxes or radio buttons for multiple choice variables, date and time, and drop-down lists for fields such as location. Fields can be made required to ensure that the desired information is gathered for every transaction. The form can be shared with all the reference staff members simply by entering their e-mail address, so they can access the form from any computer. The form is also accessible from some mobile devices, such as the iPhone. For a free product, Zoho Creator has extremely robust data viewing, analytics, and reporting features. One can create pivot tables or pivot charts, or present the data in almost any thinkable fashion—list, grid, calendar, summary, or chart view—and search, sort, filter, or group the data based on specified fields. An example of a professional looking reference statistics form developed using Zoho Creator is shown in figure 5.1. The image also displays the wide selection of fields that can be dragged and dropped into the form.

There are several reference statistical products on the market today. The first consideration in choosing a reference statistics software is whether the product's customization allows the library to gather information on their selected data elements and the ability to create the reports they want. Additional factors are the

Figure 5.1 A Reference Statistics Form Created in Zoho Creator

number of locations the product can be used at and whether mobile accessibility is an option. In the more complex library structures, there may also be the desire to have different data elements at separate service points. And, of course, there is always the question of cost.

Some of the commonly used reference statistical products are RefTracker, DeskTracker, LibAnalytics, and Gimlet. All of these products allow some customization of the reference statistics form in terms of the data elements and their variables. Even the free version of LibAnalytics by SpringShare (which restricts the library to one form) allows the library to select up to thirty data elements, specifically ten multiple-choice, ten free text, and ten numerical data elements. Similarly, they all have a reporting feature that allows for charts, graphs, and tables to be created, as well as data exportation into spreadsheet software such as Excel. Consideration will need to be given to the precision of the reporting of each product, and whether there are restrictions on the permutations of the data elements and variables that can be combined.

RefTracker by Altamara is an extremely popular option with academic law libraries due to its customizability, comprehensiveness, and ease-of-use. One of RefTracker's best features is its dashboard, which allows individual staff members to easily record a new reference transaction, access their ongoing in-depth research projects, review or add entries to its knowledge base or run a statistical report. From the dashboard, reference staff members can record both routine reference questions on the customizable DeskStats forms and in-depth research projects. The status of all open in-depth research projects can be reviewed at a glance through the "pool view," allowing reference administrators to review the questions received and oversee workload issues. RefTracker can be integrated with other Altamara products to facilitate and record chat and IM reference transactions.

LibAnswers, by SpringShare, is one of the newer and most advanced reference statistical products on the current market. It combines the statistical recording, tracking, and reporting capacity of LibAnalytics with social media communication options and a searchable knowledge base. The product records the questions received and answered via e-mail, chat, text, or Twitter. Its knowledge base is a database of research-related questions and answers that is available to both reference staff and members of the public. The staff member decides whether to make a particular reference question and answer public and searchable. They can also add commonly asked or introductory questions and answers to the knowledge base to meet anticipated information needs. When patrons cannot find their desired information in the knowledge base, they can complete a Web-based form with their question that is then sent to the reference staff for reply.

Mobility and self-service features are the future of online reference statistical tracking products, allowing a reference librarian to receive and respond to reference questions anywhere and at any time, which increases the patron's satisfaction with the service. Both of these features support the concept of providing assistance to patrons at the moment they need it, as does the knowledge base.

INSTRUCTIONAL STATISTICS

Legal instruction has traditionally been conducted in person, face to face. The instruction can involve a large or small group of participants or even with a single individual; it can occur in a for-credit continuing legal education or university course, a scheduled training session, on a drop-in basis at the reference desk, or spontaneously in the hallway. While in-person legal instruction continues today, it is being augmented with virtual instruction.

Librarian instructors use a variety of virtual instructional models, selecting the method that best supports the learning goals for the particular session. The librarian instructor will need to choose between synchronous or asynchronous virtual instruction models. Online synchronous instruction is quite similar to traditional in-person instruction as the instructor and students participate in the virtual course at the same time. Webinars and webcasts, conducted with Web conferencing services like GoToMeeting, are the most popular type of synchronous virtual instruction. If the webinar or webcast is later posted on the Internet or course management system for viewing, it can become an asynchronous learning tool.

Asynchronous virtual instruction allows all the individuals to participate in the course at different times, depending on their personal schedules. Asynchronous options move library instruction away from a "just in case" instructional model to a "just in time" model, allowing students to learn a research skill as they need it and will be applying it. Popular asynchronous learning tools include electronic research guides and online tutorials. The most commonly used electronic research guide product in academic law libraries is LibGuides by SpringShare. LibGuide's template is customizable, allowing librarians to determine the number of tabs and content boxes needed to impart the desired information. Links to websites, databases, podcasts, and other multimedia products can be embedded in the LibGuide. LibGuides are interoperable with the current social media networks and can be accessed on mobile devices that support the just-in-time delivery model.

Online video tutorials can be created in advance on topics librarians believe will be of interest to patrons, or can be created on-the-fly to answer a question from an off-site patron. Screencast products like Snagit and Jing allow librarians to develop instant videos with audio overlaid. If an e-mail patron needs assistance searching a database, for example, the librarian can quickly record a video on how to conduct the search and e-mail it to the patron. Camtasia is a more advanced screen recording product that allows librarians to create professional-looking videos. Camtasia has a full suite of user-friendly editing tools that allow for cutting and splicing the video segments, and importing music, photos, and other media products. Camtasia videos can be uploaded to the library's website or even to YouTube, allowing on-demand access.

What statistics or pieces of information can librarians collect regarding their virtual instruction activities? Professional library associations, such as the Association of Research Libraries, have not updated their instructional statistics questions to reflect the new virtual paradigm[9] so each library needs to consider what data are important to them.

With face-to-face instruction, librarians tracked the number of sessions they conducted, the number of attendees for each session, and how many hours spent preparing for each session.[10] They often sought qualitative information regarding the quality of their instruction and the attendees' satisfaction with the content through a short assessment tool. Librarian instructors can utilize these same techniques with synchronous virtual teaching methods; however, they are not appropriate for asynchronous online teaching.

Some data for asynchronous instructional aids are easy to collect, such as total views of online tutorials or Web-based research guides. For aids initially created for a course but then made publicly available, the librarian instructor could also compare the total views with the class size to extrapolate whether the aid's usage, and therefore its impact, went beyond the students in the course. Website analytics could also be used to distinguish between new and repeat visitors to the various aids.

In both methods, librarians can continue to record the number of hours spent in developing the virtual instructional aids. Obtaining satisfaction information and suggestions for improvement may actually be easier in the virtual environment. A widget embedded in the online tutorial or research guide could ask viewers a couple of questions on that regard.

COLLECTION MANAGEMENT

Other than salaries, the largest portion of law libraries' budgets is the collections budget. The question is whether the library is spending its collections money in an effective manner. To determine whether the money is being spent appropriately, the volume of usage of specific print and electronic materials must be reviewed against their cost.

Law librarians must periodically review how the materials in their collection are being used by patrons to ensure that they are subscribing to resources that support their patrons' research and information needs.

Print Materials

Notwithstanding the volume of research materials that are now available electronically, many legal treatises and loose-leaf sets are still only available in print, so law librarians must consider their usage. Data on circulating materials can be gathered through the library's integrated library system (ILS). The item record for each volume can provide some usage data about that specific volume, such as the number of times it was borrowed and its last checkout date, which can illustrate the book's popularity. Reports can be produced from the ILS that would provide other borrowing information and trends, such as the number of books borrowed in specific call number ranges.

A significant portion of a law library print collection is often designated "in-library use only" and "do not circulate," including loose-leaf and multivolume sets. Tracking the use of materials in the library is usually done with tick marks on paper, which provide the total number of items used but does not provide usage information for individual titles, which is critical in determining which expensive sets should be cancelled or retained. It is possible to monitor individual title usage in small collections, such as the reference or reserve collections. A spreadsheet of reference titles could be created, with the staff recording usage each time they reshelve the book. Some ILS systems, such as Innovative Interfaces Inc.'s Millennium, can also be used to record in-library use of specific titles, with a field in the item record being coded to track each time the book is scanned into the system before being reshelved.

Radio frequency identification (RFID) security systems, which are slowly being implemented by law libraries, allow for improved inventory control so misshelved books can be located and missing books identified more easily. A mobile reader, attached to a portable computer, is run across the books on a shelf and the information retrieved is uploaded into the ILS, allowing for collection reports and missing book lists to be run.[11]

E-books

Patron use of electronic books (e-books) can also be tracked by law librarians. Some law libraries can obtain usage statistics directly from their ILS. The more sophisticated systems record every time a patron clicks on the link to the e-book in catalog record, allowing the library to track the total number of e-books used as well as usage for specific titles. Other libraries can request usage statistics, including a summary of usage by subject matter and title-by-title, from their e-book vendor. The most effective trend analysis would be to compare e-book usage with the volume of e-books in the collection, as this would help determine if the apparent increase in usage was simply the result of the volume of e-books in the collection.

Many academic libraries have begun supplementing library selection of electronic monographs with a patron drive acquisitions (PDA) option. PDA models allow books to be purchased or acquired through short-term loans when they are needed by a patron. PDA models signal a change in collection development philosophy from purchases made "just in case" to purchases made "just in time" of need.[12]

In a PDA system, the law library's book vendor downloads records for new e-books into the library's catalog. When patrons conduct a search in the catalog,

PDA records are intermingled with records for purchased print and electronic books, without any indication that they are PDA records. When a patron selects a PDA record, he or she can look at it for a couple of minutes without the library being charged, allowing the patron to look at the table of contents, for example, to determine if he or she wants the book. Certain actions, however, can trigger either the purchase of a book or a short-term loan,[13] including multiple patrons selecting the e-book, a patron reading the book for more than just a few minutes, or printing or downloading any portion of the book.

All contact with PDA e-books is recorded by the book vendor and shared with the library. One of the most critical pieces of data is the call number ranges of all PDA e-books selected by patrons, as this allows librarians to determine which areas of law are of increasingly high interest to patrons, which could lead to additional focus on those areas. It could also lead to the opposite—a realization that a previously hot area of law was on the decline.

The book vendors also record patron habits with respect to e-books, including how long patrons spent reviewing the book and how many pages or chapters were printed or downloaded. This information can show the depth of interest patrons have in the book's subject, which may impact future selection decisions.

Database Usage

Database subscriptions are extremely expensive and librarians need to ensure they are investing in the correct databases. How many times has the database been accessed? How many searches were run? What specific titles or libraries in the database were used or not used?

Obtaining useful usage statistics from database vendors can be problematic, primarily due to the lack of consistency between the definitions used and the statistics kept by vendors. In 2002, a group of libraries, consortia, publishers, and database vendors came together to create a nonprofit international organization, Counting Online Usage of Networked Electronic Resources (COUNTER), whose purpose was to develop standards for database usage statistics. The first protocols "included definitions for variables such as full-text article, turnaways, searches, and sessions. It also specified the acceptable formats and delivery methods for usage statistics."[14] The Standardized Usage Statistics Harvesting Initiative (SUSHI) was developed to assist libraries efficiently gather the COUNTER usage data directly from the database vendors.[15] Commercial products, such as Scholarly Stats and 360 Counter, make it even easier for libraries to gather and analyze COUNTER-compliant usage statistics by going to the library's database vendors' websites, downloading the usage reports, and creating the usage reports for analyzing.[16]

Many vendors have become COUNTER compliant, allowing libraries to easily compare the usage between the databases. Unfortunately, none of the major legal database vendors have taken this step. In fact, it can be quite difficult to get meaningful usage data from them. Some vendors, such as HeinOnline, e-mail some usage statistics data elements to customers on a periodic basis. LexisNexis and Westlaw produce some usage statistics but they are primarily meant for billing purposes, allowing library administrators to predict or double-check invoices.[17] Law libraries can request additional usage statistics from their representatives, including the number of searches conducted in the various titles or libraries. Rarely is the usage data received in any format that allows easy exportation of the data into a spreadsheet for analysis, so manual data inputting is required.

For these reasons, many law libraries are seeking other methods to obtain database usage statistics. The actual usage data obtained varies widely between the resources used. EZProxy, a database authentication product, is layered over the databases and requires every patron to log in and be authenticated before using the database. EZProxy records who is logging in to use a database but does not track the duration of the connection. It allows smaller law libraries to track individual patrons' database usage for patterns, which would be helpful in database renewal decisions and to develop database training programs.

Electronic resource management software products, such as ResearchMonitor and Onelog, are utilized by many law firm libraries to obtain more usage data. These products allow the library to record which patrons are using the various databases and how long each connection occurred. They also allow retrieval of the search terms used and the results received for every search conducted, assisting librarians to remain aware of the legal issues and topics of interest to their patrons.

Collection Gaps

Knowing how existing materials are being used is critical, but so is identifying gaps in the collection. Tracking materials that patrons are requesting, either specific titles or legal topics, is also useful in good collection management.

Allowing patrons to suggest items for purchase is an easy way of identifying collection gaps. Suggestions can be received through an old-fashioned suggestion box or e-mail/chat. Alternatively, a "request a book not in our collection" link or widget could be added to database webpages and in the catalog to allow patrons to suggest items they were not able to find in their searches. Librarians can also gather collection gap information through their interlibrary loan (ILL) services. ILL requests, whether for books or journal articles, should be reviewed annually to determine why the request was needed. Perhaps the library owns the book or journal but it was either already borrowed or was missing, or the library might have the title in one format and but the patron needs it in another format, a request academic law libraries often receive from law journal editors during cite-checking. Of course, it is also possible that the library does not own the title. Similarly, missing book slips should also be reviewed at least once a year to determine which should be replaced.

Depending on the library's policy, the requested titles may be purchased immediately either due to its subject matter, the purpose for which it is needed, or the person who requested the book. Other titles will be added to a wishlist and tracked to see if other requests are received or if the subject becomes more popular. A spreadsheet may be needed to track expensive print sets or databases for multiple years.

Financial Reports

Law libraries require substantial financial support from their parent organizations, whether private or public entities, and must produce regular financial reports to demonstrate their fiscal management skills. Fortunately, law libraries can obtain excellent financial reports from their integrated library systems (ILS) to explain how their collections budget was spent. Annual and multiple year comparisons of spending can also be analyzed to determine trends or prepare the next year's budget request.

The specific types of financial reports that can be produced depend on the ILS system being utilized as well as how the individual library implements the system. Large libraries, such as academic law libraries, use sophisticated ILS systems, such as Innovative Interface Inc., ExLibris, or SirsiDynix, and small- to medium-sized libraries, including law firm libraries, are more likely to use products like EOS International, SydneyPLUS, and InMagic. Every ILS produces some standardized financial reports such as the amount spent for each vendor/publisher or the amount spent on each fund code. The ILS may also allow customized financial reports to be produced as needed, designed to answer a particular question.

WEBSITE STATISTICS

Websites have become an integral component of library operations. They are the primary vehicle through which patrons learn about the library's services, find contact information, obtain research assistance, and access electronic resources. Given its increasingly important role, it is critical for libraries to ensure their website design and its content allow patrons to efficiently meet their information needs.

The most basic and easily obtainable website statistic is the number of visitors who accessed a particular page, through the use of hit-counters. Website hit-counts are the electronic equivalent of library gate counts—they indicate how many people have visited the library but do not explain why the person visited the website or whether they were satisfied with the interaction. Without this qualitative information about the website's visitors, it is hard to identify trends and develop needed Web-based services and programs.

Website analytic products allow Web managers to obtain significantly more information about the website's visitors and the effectiveness of the website in assisting the visitors to obtain their desired information. The most popular website analytic software is Google Analytics, due to its "tremendous power, . . . its free availability, its ease of use, its flexibility, and its clear reporting mechanisms."[18] There are other analytic software options, such as StatCounter, Clicky, and Piwic.[19]

Google Analytics and its competitors can provide valuable information about website visitors. For example, the visitor tracking feature identifies the number of new versus repeat visitors. The software also monitors the number of webpages each visitor clicked on per visit and the amount of time the patron spent on each page. Finally, libraries can learn how visitors are arriving at their website—whether from a search engine, directed from another website, or by typing in the webpage's URL.

Analytics software can provide significant information about the effectiveness of the website's design. A primary consideration is whether visitors can successfully navigate the website to reach preidentified end pages.[20] Webmasters also consider the click density analysis, which shows the number of links that were clicked for the visitor to reach those end pages. The bounce and abandonment rates are reviewed to see the number of people who immediately leave the library's website after getting to the landing page (bounce rate) and from which pages visitors leave the website, possibly due to frustration or confusion over its navigation (abandonment rate).[21] From a mobility support perspective, the analytical products can identify which mobile devices and even mobile carriers are utilized more frequently to ensure the website design works effectively on those devices.

Librarians may also want to monitor usage of their social media endeavors, such as blogs, wikis, YouTube videos, podcasts, RSS feeds, Twitter accounts, and Facebook pages. The number of official subscribers, followers, or friends for these sources of communication, which the social media products track themselves, shows how many individuals value the information provided through these sources highly enough to take that step. However, many people read or view these materials without subscribing, so additional means of calculating total viewership is needed. The analytic products discussed above track the number of viewers for these pages and how long they stayed, which will provide insight into whether the visitors read or watched the entries. Analytics can also track viewings of specific blog entries or YouTube videos, allowing librarians to determine the popular topics which can influence future productions.

All of the above options provide quantitative, numerical data. Qualitative visitor data can provide context to the quantitative data. Widgets can be embedded on all or selected webpages and social media sites with a short pop-up survey seeking the information they desire, such as whether the visitor had located the information they needed and whether they were satisfied with their experience.

FACILITY USE

With many library services now being available on the library's website or offered through e-mail or chat, and the collection increasingly tipping toward electronic resources and away from print, patrons have less need to physically come to the library. Faculty members, judicial clerks, and practicing lawyers can access all the databases from their desktop and can e-mail library staff for assistance when needed. This suggests a reduced need for dedicated library space and an amplified need for statistics to demonstrate spatial needs. Library administrators need to know how many patrons are coming into the library and how they are using its spaces.

The gate security system records the number of visitors to the library but does not identify which areas of the library they are utilizing or how. Unfortunately, library staff must still do manual, visual sweeps of the library to determine this missing information. The staff members can also obtain explanatory information from the visiting patrons. Are lawyers using the library as an escape from the interruptions in their office? Are law students tired of carrying their laptop and using the library's computer lab instead? This information helps right-size the facility and identify the furniture and spaces that will best meet the patrons' needs.

Many law libraries provide their primary patron groups with after-hours access. In smaller libraries, access may be facilitated by distributing keys or even leaving the door unlocked. Larger libraries generally use an electronic access system, often a card-swipe system where the patrons swipe or hold their identification card against a reader to gain access.

If the identification cards have information regarding the patron embedded in them, the electronic access systems will record the identity of the after-hours visitors as well as the time they entered the library. It is possible that some patrons only come to the library when it is closed so library staff members are not aware of the extent to which they use the library. Identifying repeat library users can help library administration when library-related issues are being discussed by senior administration or the faculty, as these patrons may speak on behalf of the library.

CONCLUSION

Technology has significantly changed how law libraries are recording quantitative statistics, essentially eliminating print statistical gathering. Arguably, it has had a dramatically stronger impact on what data law libraries are now collecting. Technology assists libraries in tracking more quantitative data elements and in recording and analyzing qualitative information quickly and efficiently. Administrators can easily acquire a comprehensive depiction of the library's activities, which can be used for operational planning purposes and in discussions with senior administrators. The better the information that administrators in the law library and their parent organization have, the better the decisions they will make.

FURTHER READING

Aguilar, Paulita, Kathleen Keating, and Sue Swanback. 2010. "Click it, no more tick it: Online reference statistics." *Reference Librarian* 51: 290–299.

Bottorff, Tim, and Andrew Todd. September 2012. "Making online instruction count: Statistical reporting of Web-based library instruction activities." *College & Research Libraries*, 33–46.

Bowen Ayre, Lori. July 2012. "RFID in libraries: A step toward interoperability." *Library Technology Reports* 48 (5).

Conyers, Angela, and Jo Lambert. 2011. "The Journal Usage Statistics Portal (JUSP): Delivering usage data through collaboration." *Serials* 24 (2): 178–182.

Garrison, Judith. 2010. "Making reference service count: Collecting and using reference service statistics to make a difference." *Reference Librarian* 51: 202–211.

Grogg, Jill, and Rachel Fleming-May. August/September 2010. "The concept of electronic resource usage and libraries." *Library Technology Reports* 46 (6).

King, David Lee. August 2012. "Running the digital branch: Guidelines for operating the library website." *Library Technology Reports* 48 (6).

Marek, Kate. July 2011. "Using Web analytics in the library." *Library Technology Reports* 47 (5).

Northam, Sarah H. July/August 2012. "The reference statistic shuffle." *Infotoday*: 24–26.

Novotny, Eric. 2002. *Reference service statistics & assessment.* SPEC kit vol. 268 Washington, D.C.: Association of Research Libraries.

Turner, Steven J. 2010 . "Website statistics 2.0: Using Google Analytics to mea- ; sure library website effectiveness." *Technical Services Quarterly* 27: 261–278.

Vardell, Emily, Kimberly Loper, and Vedana Vaidhyanathan. 2012 . "Capturing every patron interaction: The move from paper statistics to an electronic system to track the whole library." *Medical Reference Services Quarterly*: 150–170.

Wiggins, Everett. "Usage statistics: What and why." *Thoughts on Librarianship* (blog). http://thoughtsonlibrarianship.blogspot.ca.

NOTES

1. Such as the Law School Survey of Student Engagement, which most accredited law schools undertake every few years. See http://lssse.iub.edu/.

2. The university library is the official member of the Association of Research Libraries but the law library usually submits its statistics to the university library to present the complete picture of the institution's library-related activities.

3. American academic law library directors primarily use the annual American Bar Association statistics for benchmarking purposes.

4. Paulita Aguilar, Kathleen Keating, and Sue Swanback, "Click it, no more tick it: Online reference statistics," *Reference Librarian* 51 (2010): 290.

5. Judith Garrison, "Making reference service count: Collecting and using reference service statistics to make a difference," *Reference Librarian* 51 (2010): 204.

6. Eric Novotny, *Reference service statistics & assessment*. SPEC kit vol. 268 (Washington, D.C.: Association of Research Libraries, 2002), 9. Some of the libraries were supplementing the print form with some technological options, such as using software to automate the data-gathering process for chat reference interactions.

7. Ibid., 11.

8. Garrison, "Making reference service count," 203.

9. Tim Bottorff and Andrew Todd, "Making online instruction court: Statistical reporting of Web-based library instruction activities," *College & Research Libraries* (September 2012): 35.

10. These are the instructional-based statistics requested by the Association of Research Libraries and became the industry standard.

11. Lori Bowen Ayre, "RFID in libraries: A step toward interoperability," *Library Technology Reports* 48, 5 (July 2012): 14.

12. This mirrors a change in the on-demand instructional philosophy discussed supra.

13. The patrons do not notice any difference between a purchase and a short-term loan, as they have the book for the same length of time and can print or download as desired. The difference is for the library as short-term loans cost less than the purchase price. An e-book can be borrowed several times before the short-term loan changes to a purchase.

14. Jill Grogg and Rachel Fleming-May, "The concept of electronic resource usage and libraries," *Library Technology Report* 46, 6 (August 2010): 11. COUNTER also developed standards for e-books in 2006.

15. Angela Conyers and Jo Lambert, "The Journal Usage Statistics Portal (JUSP): Delivering usage through collaboration," *Serials* 24, 2 (2011): 179.

16. Grogg and Fleming-May, "The concept of electronic resource usage," 15.

17. Everett Wiggins, "Usage statistics: What and why." *Thoughts on Librarianship* (blog). http://thoughtsonlibrarianship.blogspot.ca.

18. Kate Marek, "Using Web analytics in the library," *Library Technology Reports* 47, 5 (July 2011): 8.

19. Many of the commercial products also have a free version that would be satisfactory for many law libraries.

20. Steven J. Turner, "Website statistics 2.0: Using Google Analytics to measure library website effectiveness," *Technical Services Quarterly* 27 (2010): 268.

21. Ibid.

6

Law Library Management

Camille Broussard
MENDIK LIBRARY, NEW YORK LAW SCHOOL

Ralph Monaco
THE NEW YORK LAW INSTITUTE

Gitelle Seer
VISIONARY INFORMATION SOLUTIONS

Changes in the legal industry and legal education, including the consolidation and globalization of firms, increased client demand for cost-effective legal services, significant changes in legal information publishing, as well as the impact of technology, have put increased pressure on libraries, especially on firm libraries, to change their service model. We are seeing a shift in focus from library as place to library as resource. These changes affect all operational aspects of library management, space planning and design, acquisition and maintenance of the collection, budgeting, and training.

In this chapter, we examine the various aspects of law library management in both law firm and law school libraries as affected by these changes, and outline innovative ways in which law librarians can leverage their skills to provide leadership in the provision of research and collateral services to their patrons. Although the library setting may be different, when viewed through the managerial lens, many of the issues are quite similar.

STRATEGIC PLANS AND MISSION STATEMENTS

The discussion of the transformation of the library as place is not new, but the pace of change will present law library managers with many challenges over the next few years. Planning for the future of the library's space needs, collection growth, services, technology, and staffing needs is critical to ensuring the library's success within the organization.

Strategic plans helps define the library's mission and are essential to helping managers define development goals and strategies. A written strategic plan will help the library manager set the goals necessary to develop services to meet user needs, to recognize the changing nature of collection development from a predominantly print environment to a predominantly electronic one, and to offer appropriate professional development for the library staff to meet these goals.

For example, many law firm libraries no longer buy materials just to have them in case someone might ask for them. Decisions to purchase particular materials are often based on specific requests or immediate needs. Whereas law firm libraries are moving toward a "just in time" rather than a "just in case" strategy, the planning process and many decisions made by academic law libraries are driven by the accreditation of the legal education process. The Council of the Section of Legal Education and Admissions to the Bar of the American Bar Association (ABA) promulgates the "Standards and Rules of Procedure for Approval of Law Schools" with which law schools must comply in order to be ABA approved.[1]

What academic institutions buy, the services offered, the structure of those services, the design of physical space, and amount of seating all take into account the standards set forth by the ABA. Under chapter 6 of the Standards, "the law library shall provide a core collection of essential materials accessible in the library as well as provide a collection that either through ownership or reliable access meets the research needs of the law school's students, satisfies the demands of the law school curriculum, facilitates the education of its students and supports the teaching, scholarship, research and services interests of the faculty."[2]

In many respects, the ABA standards still reflect a time when quantity as measured by the size of the collection was considered the most important measure of quality. The standards are not static. New standards have been proposed to change the current language to recognize changing realities. The discussion is beginning to focus on finding better ways to measure libraries quality and to provide the services needed to train lawyers on the tools of their trade and to support the scholarship that moves the profession forward.[3]

What follows is a discussion of the challenges and consequences the digital age has placed on space planning, budgeting, collection development, the importance of statistics and metrics, vendor relations, licensing, staff management, marketing, and disaster planning, and how information professionals can adapt to the changes that will allow them to successfully redefine their relevancy and value to the organizations they serve.

SPACE PLANNING

Law librarians in all settings want to create a welcoming and inviting library environment conducive to learning and engaging in research. We are still in a transitional period regarding the collection proportions of paper, microform, and electronic resources, and libraries must strive to provide efficient access to all of these collections. The correct balance of formats is different for individual libraries and may always be a mixture of formats. As electronic resources gain predominance, efficient access to digital collections within library space includes not only easy access to the information but comfortable workspace for attorneys and students using laptops, mobile devices, and other technology. In an article on library design, Jean O'Grady, DLA Piper Library director, "speculated that in the future, law libraries may look like Apple Stores . . . while we may replace law books with desktop virtual libraries, the need for spaces for research consultation, collaboration and 'just in time learning' will continue."[4]

In large measure, libraries are about resources—electronic resources and book resources—and, as importantly, they are about the services the staff designs to connect users to the right material regardless of format, and to help them navigate those resources to find the information needed. In the academic environment, an important service goal is to provide training classes in using all information sources, in all formats, to help students make meaningful and contextual connections between classroom learning, research skills, and the practice of law. In the law firm setting, the focus changes to training attorneys to use the firm's resources. The library space needs to be a dynamic physical and virtual space to accommodate these goals.

Institutions often begin analyzing library space needs by looking at the book collection. But as less focus is placed on library book stacks and greater focus is placed on accommodating current and future technology and staff, this mind-set needs to change. Librarians should look first at the goals of the entire organization and then analyze the library operation and how it should be integrated into the institutional footprint to ensure that the space is designed to support the work of the organization.

One standard space design will not fit the needs nor meet the mission of all libraries. The trends discussed below provide important considerations for designing or redesigning library space in the age of the digital library. The annual April issue of *AALL Spectrum*, the monthly publication of the American Association of Law Libraries, is devoted to library architecture and space planning.[5]

Library Space Design Trends

The most significant library designs illustrate the change in emphasis from housing collections of books to supporting a range of library activities, from quiet study to collaborative work and training activities, while enhancing the library's value to patrons. The space should be flexible enough to change as needed and to accommodate multiple uses. Today's law library spaces:

- *Promote collaboration.* Today's libraries are designed to support and encourage collaboration among and between users and staff members. Configuring the library space to blend with other areas will underscore the importance of research and information services to the core business of the institution. Using devices such as cafés or coffee bars and casual seating areas can promote personal interaction and encourage the use of the library as a gathering place.
- *Accommodate the remaining print collection.* Libraries can save space by housing most of the collection on compact shelving or off site. For example, compact shelving may be able to house up to 60 percent more capacity in the same floor space when compared with stationary book stacks.
- *Acknowledge decreased foot traffic.* Libraries can reduce the size of the reference desk or eliminate it completely in favor of locating staff offices within eyesight of the entrance. Consider moving reference specialists to practice floors or faculty suites in an academic setting. Consolidate the reference and circulation desks into a single service desk.
- *Encourage training.* Design multipurpose space that can be used to train lawyers, students, or staff, and which can also be used as classrooms or workrooms. Equip this space with teaching equipment, including projectors, screens, and electronic whiteboards.
- *Support technologies.* Users expect libraries to provide the latest equipment on site and to support a "Bring Your Own Device" (BYOD) model. Library managers need to plan accordingly.
- *Combine technical and public service spaces.* The traditional model of separating technical services and public services no longer guides the allocation and design of space. Collaboration can be promoted by locating staff members where they can easily work with, learn from, and consult with each other.

Case Study—Squire Sanders Puts Their Library on "Main Street"

In an article on Squire Sanders's transition from a traditional library, with a large reference desk and considerable shelving, to an open small footprint "library without walls," Jean O'Grady described the law firm's innovative approach to library space planning. The library director made dramatic cuts in the print collection (from 7,000 to 200 linear feet) in favor of digital resources. "Instead of being marginalized by the space reduction, the library became the central hub of 'Main Street.'. . . [The library] includes a coffee bar that invites traffic and conversation. The staffs are near the managing partner, the business development personnel, an active international trade group and the lunch room." This successful integration into the daily life of the firm "has increased lawyer demand for research services and . . . [b]eing located within a high traffic area in this open office space has increased the visibility of our services and has remarkably increased our interactivity and workload with all of the practices."[6]

Case Study II—A Library without Borders: The Marquette University Law Library Opts for an Unconventional Design

The Eckstein Law Library, a "library without borders," is located on all four floors in one corner of the law school building. "At the center of the building is the Zilber Forum, an open four-story space that includes a student lounge, café, and a central staircase accessible from all classrooms, study rooms, and offices." The library design "complements and reinforces the open atmosphere of the building by seamlessly integrating the law library on all four floors of the building without turnstiles, doors, or, in many places, even walls. . . . Students and faculty are thus able to move freely from offices and classrooms to the library collections on each floor and use the materials throughout the building." The circulation and reference desks, as well as the reference collections and the public-access research computers, are highly visible and accessible to the student body and the faculty. The library has more than 47,000 linear feet of shelving, comprising stationary and movable shelving, and easy-to-use self-checkout terminals are conveniently located for those who wish to borrow materials.[7]

BUDGETING

Library managers are responsible for all aspects of the library budget, including both the operational and personnel budget lines. Library budget lines include collection development in both print and electronic formats, personnel (including associated professional development and travel needs), equipment and supply purchases, software and technology expenditures, marketing, and other expenditures necessary to conducting the day-to-day operations of the library.

As budgets shrink and expenses grow, library managers must consider approaches to intelligent retrenchment in expenditures while maintaining the integrity of services and necessary collections. Constant review of purchases and various budget lines allows the manager to make informed decisions and be responsive to changing dynamics. The ability to cancel items that are no longer necessary to the organization's needs and redirect monies to items that are necessary is crucial to retaining a vibrant and institutionally centered library operation.

Budgeting for personnel requires a similar approach. Library managers often have difficulty justifying staffing levels and appropriate compensation, especially for those staff members who are less visible, like technical services staff. There are several industry benchmarks that library managers can use to bolster the case for a specific salary (e.g., AALL Salary Survey), but the library manager's ability to retain staff will depend on making a compelling business case for the value of each staff member in a substantive and quantifiable fashion.

COLLECTION DEVELOPMENT POLICY

Budgetary realities impact collection development policies in law firms and academic institutions. Legal resources are migrating rapidly from print to digital format, or a hybrid of both. As the real estate devoted to library space decreases, the adoption of digital resources and the planned replacement of print volumes become even more important. Sometimes materials are more useful in electronic format; sometimes they are simply more expensive and less accessible, primarily due to license restrictions. Librarians must evaluate all options when it comes to institutional information needs, so that the library has the best resources for the best price.

A formal collection development policy should be adopted to give the librarian a structure for planning purchases to meet the information needs of the institution. A sound collection development policy prepares the library for making hard choices while ensuring that patrons have access to the resources they need.

At its most basic, a collection development policy includes a description about the constituencies supported, the types of materials needed (i.e., core, ancillary, jurisdictional, topical), and the preferred format(s) of purchase. In a law firm, these decisions are typically based on the areas in which the firm practices. However, firms with multiple offices need to allow for regional and practice area differences, not just in the content of materials but also in the format in which they are offered. Changes in practice concentration, the addition of lateral hires with a practice new to the firm, and the lateral attorneys' previous experiences and expectations regarding resources are all factors to consider when devising, and updating, a policy.

The collection development document should also address selection and approval criteria, holdings and retention, and a checklist that allows the library manager to detail (and compare) costs, upkeep, redundancy, access, and staffing implications.

Academic law libraries develop collections, through ownership and reliable access, to support the instructional and curriculum needs of the institution as well as the research, scholarship, and service activities of the law school faculty and students. Chapter 606 of the ABA Standards requires a core collection of essential materials along with other materials necessary to meet the law school's needs. The standards also state that a law library shall formulate and periodically update a written plan for development of the collection. The Academic Law Libraries Special Interest Section (ALL-SIS) of AALL has a standing committee on collection development that posts many policies, materials, and other resources on their website.[8]

The breadth and depth of a particular academic law library's collection depends on many factors, including the mission of the school; the research needs of the various disciplines supported; the number and nature of clinical, graduate, and moot court programs offered; the number and subject matter of the law journals supported; and whether the library supports the legal information needs of a broader university or public community. As in the law firm library, the many competing factors place a burden on the library director to ensure the wisest expenditure of the acquisitions budget.

Many academic libraries also participate in cooperative collection development to meet their patron's research needs. Metropolitan areas that include multiple academic institutions may have a consortia arrangement to allow members access to use the collections of different libraries. Some agreements include expedited interlibrary loan or document transmission between institutions. Other national or regional groups, such as the NELLCO (the New England Law Library Consortium)[9] allow members to maximize resources and support the missions of their institutions through various collaborative projects and consortium purchasing and licensing agreements.

E-BOOKS

E-books are an emerging format in law libraries. As with all new technologies, it will take some time before the advantages and disadvantages are fully revealed and addressed. Advantages of e-books include significant reduction in the cost of housing, maintaining, processing, and updating print resources. At this point, there is little standardization among publishers regarding platforms and pricing models. Libraries will have to create new workflows and procedures around the administration of e-book acquisitions and management.

STATISTICS AND METRICS

Quantifying an activity or function is the first step in developing a system for measuring success, or return on investment. The library must have an objective way to measure the use of services and resources. Using that information to identify gaps in services, develop creative enhancements, or justify an increased budget is the challenge faced by all managers. Library managers should develop an effective method for collecting the statistics needed to understand library activity and to plan future direction. Three important areas to measure include data on use of electronic resources, circulation statistics for print resources, and statistics on reference services provided to the library's patron base.

Law firm librarians are using online resource tracking tools like Research Monitor, OneLog, and Lookup Precision[10] to manage, monitor, control, and capture usage information for external online database subscriptions. Additional functionality may include cost recovery and/or password management modules. Academic librarians also need usage data to determine the best selection of databases to meet the research needs of their students and faculty. However, because of cost, many academic librarians rely primarily on vendors to provide these statistics. The results are not as helpful as these monitoring tools.

At their most basic, and if configured carefully, these monitoring tools can give the library manager valuable information about how often a database is being used and by whom, who is *not* using a database (which raises important questions such as "Is this resource valuable?" "Are your patrons unfamiliar with it?" "Do they not know how to use it?"), are there sufficient (or too many) licenses, and even, in some cases, what searches are being performed. Based on this information, the library manager can make objective decisions about renewing a contract for a database, where to focus training efforts, and what print resources can be cancelled based on their availability online.

Cost recovery, of particular interest in law firms, is another feature of resource tracking tools. Resources can be configured so that they can only be accessed with a valid client billing number. It is advisable to work with the finance department to develop protocols to make the internal client matter system and the resource tracking system "talk" to each other. In that scenario, charges to clients from online database use can be imported directly into client bills.

In addition to measuring the use of online databases, library managers should objectively measure the use of their print collection. The circulation modules of library management systems (e.g., SydneyPLUS, EOS, Innovative Interfaces, Aleph)[11] provide circulation statistics. Circulation statistics, whether collected electronically or manually, allow the library manager to identify the titles that have had no or very low circulation within a period of time. That information can be used to support alternative access decisions, such as using online access, borrowing the title as needed, or discontinuing automatic updating, among others.

Data on reference requests and reference staffs' client billables should also be captured. At the academic level, monitoring bodies like the ABA, state or national education agencies, and several research library consortia require the reporting of statistics of public service activity. The number of reference requests, the number of training classes taught, and the number of students trained are extremely important statistics. Many academic libraries use LibAnalytics from Springshare or Desk Tracker to capture this information. Similarly, law firms may use RefTracker, as one example, to capture the entire reference workflow, as well as statistics.[12]

Whether the library manager tracks the number of reference questions he or she handles on an index card or uses an automated reference tracking system, the manager must analyze the information and use it to justify and support programs and innovations he or she wishes to offer.

Proving the ROI (return on investment) of a law library goes beyond a numeric calculation. For example, we could measure the costs associated with a library, like the cost of books, database subscriptions, and library staff time, against the financial benefits, such as saving users' time, capturing and billing back for services, and centralized purchasing and licenses. Calculating the dollar value of these financial benefits is only one challenge. Managers must also show the qualitative value of the services provided. Involve staff in identifying ways in which tasks and processes can be done differently (i.e., more efficiently) to make the most of opportunities to save money, while demonstrating and marketing the value of the services provided.[13]

VENDOR RELATIONS

As library managers purchase an increasing amount of material in electronic format, we spend more time *negotiating* for content rather than simply *purchasing* content. The buyer–seller relationship with vendors may become contentious, as both parties have more at stake than with a print title. It is important to cultivate a relationship with vendors by understanding their content and business model and being forthright about collection needs, and the challenges that the library is facing.

LICENSES AND CONTRACTS

Before reviewing the terms and conditions of a contract, the library manager should outline some objective criteria to use to evaluate the contract that go beyond content and cost. A checklist of criteria to consider is a good idea for all purchases, whether electronic or print. Note that all of these conditions are potentially negotiable.

Criteria to consider:

- Does the contract lay out the pricing and any applicable discounts, as well as an explanation of the pricing model (enterprise, office, headcount), and who has access? If, for example, you engage in an enterprise agreement, agree on terminology that lets you add a new office at no additional cost during the life of the contract.
- Are upgrades/enhancements included for the life of the contract?
- Will the vendor provide Web parts or other widgets to enable access to their content via your portal at no additional charge?
- Are there any additional costs for equipment, training, consulting fees, traveling expenses? Typically you want to include all of the aforementioned at an initial all-in price.
- Are there discounts for multiyear contracts, prepayments? You must ensure that there is a favorable (to you) opt-out clause. Avoid all contracts with so-called evergreen clauses (i.e., automatic renewal).
- What level of support and training are available, when, by whom, and via what means (e.g., on site, phone, webinar, e-mail). Ideally, you want an experienced trainer knowledgeable about the product as well the types of research for which it would be applicable, rather than a salesperson.

Key steps in a successful negotiation are:

- Define your goals and objectives and be prepared to articulate them. Why do you want to buy this product: as a replacement for another, to fill an unmet need, or to satisfy the request of a single practitioner or small practice group? Are there viable alternatives?
- Set the stage and establish rapport. Exchange ideas and information with the vendor prior to the first negotiation meeting, especially if you do not have an existing relationship with that company. Suggest a preliminary meeting with your account manager, senior members of the sales team, upper management representatives, and a few of your senior staff members.
- Be prepared. Review all existing contracts that you have with the vendor, and be familiar with the terms of the contract under discussion. Determine if your firm already has a relationship with this vendor (you can often obtain this information from a procurement officer or the accounting department). Conduct due diligence on the vendor, through research and discreet conversations with colleagues (i.e., do not post questions on a LISTSERV!). Gain some knowledge about their pricing models and flexibility regarding customized contracts. If you already have the product in question, ask the vendor to provide usage information before the meeting. Compare that information with your own usage information, if you have it. If you work with other departments to negotiate contracts, such as procurement or fi-

nance, you should become familiar with their negotiating stance before the initial meetings.

- Separate your emotions from the process. Feelings of anger, annoyance, and frustration are not uncommon when you reach a stalemate in a negotiation. Do not let them interfere with working toward a solution. Do not attack the other side or assume a defensive posture. Remain professional and objective. Negotiation is only one aspect of your relationship with your vendors, but the tone and level of respect and compromise shown during these negotiations often colors the entire relationship.
- Be flexible but firm. Identify the terms and conditions on which both sides agree, and put those aside. Prioritize the content on offer from most important to least important, and be willing to give up some of the latter to gain the former. Consider what your alternatives are if the negotiation fails, and measure that against the best deal you can get. Be flexible and innovative, and recognize that there can be more than one option, but do not lose sight of your original goals.

STAFF MANAGEMENT

Organizational Structure

In law firms there are probably as many different library organization charts (roles and reporting relationships) as there are law firm libraries. Global expansion, reduced budgets, changes in law firms' business models, and the changing roles of the librarian all contribute to a fluid environment. The library's place in the firm's hierarchy can be fluid as well. As law firms try to adapt the corporate model to their administrative organizational structure, the library's reporting relationships may change regularly, moving from the responsibility of the CIO to the CMO, the CEO, the CKO, or any combination thereof.

As U.S. firms expand their geographic footprint, library managers are faced with the challenge of providing service to patrons who may be in different time zones, speak languages other than English, and practice within a legal system with which they are unfamiliar. In this global setting, reporting relationships are often less well-defined and more changeable than they are in the more traditional "home office" setting. For example, a librarian in a branch office may report to the office manager, a regional administrator, a partner, or a committee. Different reporting structures can lead to various challenges. In some cases, library personnel in branch offices may have difficulty negotiating loyalties and relationships with the local supervisor vs. the library supervisor. On occasion, the person tasked with library responsibilities may not have the necessary training, experience, or substantive knowledge to perform the job without considerable assistance from, and training by, the global library manager and other staff members. To the extent practicable, all staff should follow the same departmental policies and procedures.

Most large law firm libraries still retain the traditional separation of technical and reference services. However, new roles such as Electronic Services Librarians are emerging. These positions incorporate and require skill sets built from aspects of both areas. In some firms, library staff are also beginning to work within other administrative departments, such as marketing or within practice groups ("embedded" librarians). Although the embedded librarian concept may enhance the visibility and value of librarians, the model also raises questions about how the library as a *department* will function in a cohesive fashion. For example, there will be questions about who sets priorities and evaluates work in the context of best practice research, who supervises the staff member on a day-to-day basis, who conducts performance reviews, and who recommends salaries. It may also be a challenge to schedule time for the embedded librarian or staff working in other departments to participate in library department trainings, meetings, and brainstorming sessions.

In addition, the need for certain positions (e.g., mail check-in and shelving personnel) have diminished, while the need for others (e.g., researchers, analysts, database specialists, and technologists) has grown. The challenge of finding staff with the skills needed to perform the more research-intensive, analytical work has resulted in a broadening of the definition of "librarian" in law firms to encompass non-MLS personnel, and the collateral move toward using titles other than librarian (e.g., researcher, legal research analyst, electronic resources specialist, information coordinator).

There are a number of creative ways in which law firms are leveraging their staff's skills and availability, including the implementation of a "virtual reference desk" system where reference activity is shared among offices in different locations. The availability and innovative use of issue-tracking software creates an infrastructure for receiving, researching, and responding to a reference request without in-person interaction between librarian and patron. Sophisticated communication tools add many options for satisfactory contact and resolution of requests. Location becomes less relevant, although time zones, language, and subject proficiency are issues that must be addressed when implementing a virtual or shared reference desk.

Academic libraries are not as fluid in the roles and reporting relationships as their law firm counterparts. The academic library director in most settings reports directly to the dean of the law school and works with the dean and faculty to set library policy and determine staffing needs. Academic librarians are not, however, immune from the pressure to recognize changing needs and to leverage staff skills to best accomplish institutional goals. As law firms change, the traditional model of legal education may need to be redefined to meet new realities. The library director must be prepared to redefine the library's place and be ready to provide a new array of services to train students to meet the needs of a changing, more global, legal profession.

The traditional separation of technical and reference services remains in place in the academic library. However, as current staff retire or relocate, many academic libraries reorganize and restructure. As in law firms, collaboration and cross-training between the reference and technical services staffs has become more critical and more prevalent. The staff needed to process and maintain print collections is not necessarily the staff needed to work with the growing importance and dominance of digital collections and technological productivity tools. The teaching and research role of the academic librarian has become increasingly important.

Performance Evaluations

Performance evaluations are an important managerial responsibility. They provide an essential tool for ensuring that the library functions at its highest level, and they are a critical for helping library staff learn the skills necessary to the job they are asked to perform. At a minimum, managers should provide an annual performance assessment. However, a good manager gives feedback on a staff member's job performance throughout the year, both formally and informally.

Supervisors should keep a file of notes regarding the ongoing performance of their direct reports. Even the best supervisor will have a hard time writing substantive evaluations, using specific examples of outstanding work, or issues that arose during the year that affected performance, without notes. Comments from patrons and colleagues also help inform an effective review. Starting with a staff member's self-review or evaluation is another way to inform the discussion. In addition, every librarian and paraprofessional should be required to submit an annual activities report describing work-related (e.g., projects, research assignments, portal design/implementation) and professional development (e.g., publications, teaching or lecture activities, professional association committee assignments or meeting activities, skills training) accomplishments.

Use the comments section of the performance review form to highlight aspects of the employee's contributions, progress, key accomplishments and, where necessary, areas in which improvement are needed. Use these narratives as the basis for discussion. Before each annual review, ask each employee to write a goals statement for the following year. It can be as simple as "I will learn PowerPoint" or "I will co-teach a program for new associates." Discuss the process to accomplish these goals during the review.

Management Styles and Best Practices

Many characteristics define a good manager. Law library managers must lead by example to create a collaborative, enjoyable, mutually respectful, and energizing work environment. Managers should provide the resources, support, and encouragement so that each member of the staff can succeed. They should share information about the firm or school that staff members need to do their job effectively. An "open door policy" is critical and, above all, managers should always advocate for their staff.

Best practices include:

- Be present and visible. Although this can be challenging in a virtual, global environment, it is always possible to touch base with staff frequently. Meet with them in groups and one on one, using all available technology, like Skype, video conferences, and webinars.
- Take an interest in every library department and the job of every staff member. A good manager will be conversant with everyone's tasks, challenges, and successes. Encourage all staff members to be familiar with the work of others, and make sure that upper management recognizes the contributions of each staff member.
- Encourage staff cross-training and collaboration. Where practical and appropriate, provide time and structure for cross-training. You will gain flexibility in the deployment of job skills, and the work will be more interesting and challenging for the staff.
- Be organized and learn project management skills. Lack of organization is a time-waster and sets a poor example for those who are following your lead. It is very hard to keep track of issues and projects, or to implement new initiatives, without a system in place to keep you on point. Pay attention to details. Don't get buried in the weeds, and always give your staff autonomy to make their jobs interesting, but be familiar with all aspects of the operation of the department. Being a big-picture manager is important but, as they say, "the devil is in the details."
- Cultivate relationships with other department managers. Always recognize your primary responsibilities and priorities, but work with and offer your assistance to other departments. You will raise the visibility of the library and demonstrate how integral it is the operation of the institution.
- Market the library and help market the organization. Managers must ensure that the lawyers and the administration are aware of what the library does for them and, as importantly, can do for them. Library communications (e.g., database announcements, newsletters, research tips), should be relevant, brief, and branded. Law librarians should offer to speak whenever possible—at department/curricular meetings as well as orientations and all training sessions with a research component.

DISASTER PLANNING

A disaster plan provides procedures and resources for ensuring business continuity when events such as earthquakes, fires, terrorist attacks, power or telecommunications outages, flooding, and infrastructure emergencies prevent access to the organization's physical space.

In most cases, the university or law firm's IT department has a plan in place that governs how the enterprise systems are backed up and accessible in the event of an emergency, and organization-wide protocols for business continuity. Library managers should be knowledgeable about which systems and data are backed up, and how the data can be accessed in the event of power outages or hardware failure. It is imperative to include the library's critical systems (e.g., the library management system, contact databases), in the organization's business continuity plan as priority systems to restore.

If the library uses cloud-based services, it is advisable to have a backup of critical data to protect against data loss if the provider goes out of business. All backup data should be stored off site and periodically checked to ensure that it can be restored if needed. In law firms with library staffs in other offices, develop a plan to leverage that geographic diversity. Create a workflow that guides the staff and the patrons to the most effective way to contact and utilize staff in other offices. As previously mentioned, the creative ways in which law firms are leveraging their staff's skills and availability, including the implementation of a "virtual reference desk" system where reference activity is shared among offices in different locations, this same concept (tool) could be utilized for purposes of contacting staff in other offices in emergency situations.

The library staff should have paper copies of certain critical documents, including a list of all of the librarians' passwords for all databases, passwords to be used if IP authentication is not functioning (e.g., when the servers are down), passwords to get into the back-end systems of databases for access to patrons' passwords and usage information, contact information for vendors of both online and print material, and a library calling tree, including all personal contact information and instructions about who calls whom. The use of cell phone numbers allows texting, which can be a more efficient communication tool in many circumstances. Library managers need a general triage plan that lists who will be responsible for contacting vendors, performing research, working with other administrative departments, and a host of other time-critical activities.

In addition, there are basic steps that the library can take to provide their patrons with the ability to do research remotely, if necessary. Create a simple guide covering access to library and research services in the event of an emergency. This guide should include:

- Information about how to reach the library staff, and whom to call for different types of requests.
- Information on accessing the online catalog remotely, including passwords.
- A list of all databases, with URLs, shared or generic passwords, and contact information for customer support. Encourage your patrons to keep an offline record of all of their personal passwords, including those that the library does not handle, such as bar and professional association websites.
- A list of libraries available to your patrons. Bar and law school libraries, if unaffected by the event at issue, might be able to accommodate a small number of patrons, for limited periods of time.

The above discussion touches some key aspects of planning, but a truly robust emergency preparedness plan, including responding to localized situations such as mold, mildew, leaks, and pests, must include prevention and protection strategies, a checklist for pre- and post-disaster priorities and steps, and instructions for recovery.[14]

INSTITUTING CHANGE

Law libraries are not immune to changes in the legal profession and in legal education. The significant shifts in the traditional models of legal publishing combined with new economic challenges are creating an increasingly complex environment for law library managers to navigate. Times of transition are often tumultuous—and they are often times of great opportunity. Successful library managers will be leaders ready to accept the challenge of change, while working within the culture of the organization they serve. They will be proactive, creative, and innovative in transforming the library—physical and virtual—into critical research and information centers that contribute to the success of their organizations. To keep library collections and services valuable and viable, managers must recognize that everything is connected. All library activities—buying, processing, organizing, describing, maintaining, moving from print to electronic where appropriate, providing reference service, integrating technology, teaching, and programming—must interconnect and flow smoothly.

FURTHER READING

AALL Private Law Libraries (AALL-PLL). *Resource Guides.* http://pllresourceguides.wordpress.com/guides.

AALL Spectrum. http://www.aallnet.org/main-menu/Publications/spectrum.

"Risk, value, strategy, libraries, knowledge and the legal profession." *Dewey B Strategic*, http://deweybstrategic.blogspot.com.

"Practice Innovations: Managing in a Changing Legal Environment." http://info.legalsolutions.thomsonreuters.com/signup/newsletters/practice-innovations.

Humphries, LeJean. March 13, 2013. "Cheaper online? Our firm library's gradual move to all electronic." *AALL Spectrum* 17 (5). http://www.aallnet.org/main-menu/Publications/spectrum/vol-17/No-5/mar13.pdf.

Stilwell-Tong, Jocelyn. 2012. "Generational divides: Paper and electronic resources." http://www.aallnet.org/main-menu/Publications/aall-ilta-white-paper/paper-electronic.pdf. (This article was first published in AALL/ILTA's October 2012 white paper titled "The New Librarian" and is reprinted with permission.)

NOTES

1. "ABA Standards and Rules of Procedure for Approval of Law Schools," http://www.americanbar.org/groups/legal_education/resources/standards.html.

2. "ABA Standards Chapter 6 Library and Information Resources," http://www.americanbar.org/content/dam/aba/publications/misc/legal_education/Standards/chapter_6_2012_2013_aba_standards_and_rules.authcheckdam.pdf.

3. "Proposed Revisions to Chapters 6 and 7," meeting of the Standards Review Committee, American Bar Association Section on Legal Education and Admissions to the Bar, November 30, 2012, http://www.americanbar.org/groups/legal_education/about_us/leadership/council_meetings.html.

4. Jean P. O'Grady, "Law libraries transformed: Crowded collaboration and social solitude (the Apple Store vs. the Commons)," *Dewey B Strategic*, July 18, 2012, http://deweybstrategic.blogspot.com/2012/07/law-libraries-transformedcrowded.html.

5. *AALL Spectrum*, April issue. Annual. http://www.aallnet.org/main-menu/Publications/spectrum.

6. Jean P. O'Grady, "Squire Sanders puts their library on 'Main Street.'" *Dewey B Strategic*, November 20, 2012, http://deweybstrategic.blogspot.com/2012/11/squire-sanders-puts-their-library-on.html.

7. Julia Jae and Patricia A. Cervenka, "A library without borders: The Marquette Law Library opts for an unconventional design," *AALL Spectrum*, May 2011, http://www.aallnet.org/main-menu/Publications/spectrum/Archives/Vol-15/No-7/pub-sp1105-Marquette.pdf.

8. AALL Academic Law Libraries Special Interest Section (ALL-SIS), http://www.aallnet.org/sis/allsis (login required).

9. NELLCO, http://www.nellco.org.

10. Research Monitor, http://www.priorysolutions.com; OneLog, http://www.onelog.com; Lookup Precision, http://www.lookup-precision.com.

11. SydneyPLUS, http://www.sydneyplus.com; EOS, http://www.eosintl.com; Innovative Interfaces, http://www.iii.com; Aleph, http://www.exlibris-usa.com.

12. LibAnalytics, http://www.springshare.com; Desk Tracker, http://www.desktracker.com; RefTracker, http://www.altarama.com.

13. Erica Anderson, "Demonstrating value," *In Formation*, August 16, 2012, .http://information-transformation.tumblr.com/post/29548742462/demonstrating-value.

14. Katherine Coolidge, "Program outlines elements of an effective disaster plan: Communication, role, assignments," *AALL Spectrum*, September–October 2003, http://www.aallnet.org/main-menu/Publications/spectrum/Archives/Vol-8/pub_sp0309/pub-sp0309-disaster.pdf.

Part II

TECHNOLOGIES

7

Digitization

Michelle M. Wu
GEORGETOWN UNIVERSITY LAW CENTER

In exploring the future of law libraries, the issue of digitization invariably arises. Growing user disinclination to use nonelectronic sources and the rapid development of e-book technology makes digitization a requirement for libraries seeking to maximize use of all of their collections. In this chapter, we explore digitization, its role in the future of law libraries, examples of current forward-thinking projects, and some challenges libraries face in the digitization of materials.

WHAT IS DIGITIZATION?

We start by looking at digitization itself. What is it and why is it important to libraries and their users? For the purposes of this chapter, digitization is the conversion of an object in analog form to an image-based form that can be read by and transmitted through electronic means, readable by computers and other devices. While non-image-based transcription has played a role in database content for decades (e.g., Westlaw), rekeying is not the same as facsimile reproduction.

Digitization produces a portable file that takes up no physical space, relatively little storage space, and allows for greater manipulation of the information contained within. For example, a printed book is typically accessed through the table of contents or an index, and is only available to the person physically in the same space as that book. A digitized version of the same book, if run through an optical character recognition (OCR) program, retains both print access features, and in addition, becomes fully searchable. Using other tools, it is also linkable to other texts or information stores, may be cut/pasted, may be "added" to multiple collections even when residing primarily on a single server, and may be accessed from any location in the world (with an Internet connection, that is). At its heart, digitization enables greater access and use of a text.

The advent of affordable technologies for this type of work has created opportunities for libraries to share information broadly at lower costs; have expanded the library's role from information provider to information developer and distributor; and have provided additional options to ensure the redundancy of information, though in a manner that preserves only the content of a text and not its physical form. Let us explore each of these opportunities in greater detail, honing in on some costs and benefits of these new library roles.

EXPANDED ROLES FOR LIBRARIES

Libraries have long stood for equal access and collaboration, encouraging users to take advantage of their resources and building networks for the sharing of resources (e.g., interlibrary loan). For these reasons, widening the scope of sharing through digitization is probably the easiest leap for libraries to make. The goals are the same as with print materials; the format just makes the sharing less time and space dependent. Unlike interlibrary loans or traditional collaborative collection development, there are neither shipping costs to bear nor any delays between request and retrieval. However, there are equipment and staff costs,[1] especially in a coordinated plan where libraries contribute preservation-level digital copies of materials to be used by a collaborative.

Although preservation requirements differ from institution to institution,[2] all of them share one feature in common: high quality, achieved through strict technical standards and careful quality control. In other words, scanning the materials through your generic office copier/scanner, while certainly a form of digitization, is unlikely to meet these standards. The elevated technical standards typically require more expensive scanners with cameras able to achieve a high video dot density (dpi)[3] and with special features (e.g. book cradles, high-speed capacity, automated page turning) to process materials expeditiously without breakage.[4] Skilled staff are also necessary for indexing, production of metadata, and digital document preparation and cleanup, and servers are needed for storage and redundancy. In collaborative projects, there are usually additional costs to consider. For example, HathiTrust (http://www.hathitrust.org/) has over sixty partners, each bearing the costs of digitizing their works according to deposit standards (http://www.hathitrust.org/deposit_guidelines) as well as a portion of the cost to host the joint collection (http://www.hathitrust.org/cost). Those costs include storage, maintenance, backup, physical servers, staff, and space for servers/staff.

The benefits of a collaborative approach generally outweigh the costs, though. First, digitization is spread across all partner libraries, reducing duplication of effort and enabling greater coverage through this more efficient use of resources. Second, local staff, equipment, and space need not be acquired for storage and maintenance. Third, all partner libraries (and society!) can enjoy the fruits of the collective labor, facilitating resource savings even for those institutions outside the collaborative. If they have funds set aside for such projects, they can expend them on digitizing unique materials instead of scanning already-converted materials. Although there are some materials in projects like HathiTrust that are available only to the owning libraries, there are many more available for the world to use and enjoy.

This open access highlights the second new role of libraries, as developer and distributor. Instead of waiting for publishers to reissue a new edition of an out-of-print title, libraries can scan and make widely available works in the public domain. Where copyright holders agree or where use is carefully circumscribed, as in Internet Archive's OpenLibrary, libraries can expand this access to in-copyright works as well. They also have the option of feeding the resulting content into federated systems or search engines to enable easy discovery far beyond their own search engines or catalogs. Further, since technology is not limited to any particular industry, libraries may partner not only with other libraries but with any entity able to contribute labor, equipment, or programming to the cause.[5]

As it makes information broadly available, a library can also enhance its collection or expedite its identification in ways that would have been too costly if the documents were in print alone. Archival materials have long wallowed in the depths of libraries for lack of staff to research their background, provide finding aids, and otherwise provide context traditionally necessary to make such collections usable. However, digitization has made it possible to tap into a much larger resource for this task. Libraries have posted pictures from their archives and used crowd-sourcing to identify the people, events, and locations portrayed. They need not rely only on their staff's research or expertise but can call on the collective knowledge of the world, and can even invite participants to contribute any additional pictures or documents related to the existing collection.

With such user/library interaction, materials that might otherwise have remained buried for years are revealed and made useful to the public, and the public has the opportunity to engage with the library in information creation. The National Archives' NARAtions (http://blogs.archives.gov/online-public-access/?p=5379) and Dartmouth College's Metadata Games (http://www.tiltfactor.org/metadata-games) are two examples of archives or libraries inviting users to contribute to record enhancement. NARAtions allows users to tag any record with descriptions that are meaningful to the user, giving each record additional access points. Metadata Games allows teams to work together to develop more meaningful tags for any given picture in a library's collection.

Digitization has also elevated libraries' profiles as authenticators and philanthropists, both fitting adjuncts to their existing mission of information preservation. Although libraries have long existed for the benefit of society, they are more traditionally solicitors of donations as opposed to donors themselves. However, digitization allows a library to "donate" its collection without losing access to it, nor losing the original copies owned by it. The digital donation is easily replicable, requires little space, and if properly distributed, is insulated from loss through natural disaster. Using this method, libraries can assist in efforts to establish objective, stable, accurate, and complete records of nations' cultural or political development. This can best be illustrated through example, such as the earthquake damage that Haiti suffered in 2010. Every aspect of the nation was devastated, and the loss included their written history. Since copies of their government works were housed only in physical form, destruction of them left the nation without any record of their own legal history. Led by LLMC (http://www.llmc.com/HaitiOverview.asp), libraries banded together, taking advantage of the inexhaustibility of digital products. They pooled together their scattered Haitian resources into a single, complete digital library for Haiti, while retaining ownership of the originals. In doing so, they also created the single most complete record of Haitian government documents for the world.

These roles of philanthropist and authenticator will be particularly important for the world in future years, both to ensure that that the digital world of information is not skewed only toward those countries with the greatest resources and to provide certainty for users that the documents they are using are authentic and not sophisticated fabrications.

ESTABLISHED EFFORTS

There are some tremendously successful digitization projects currently in operation, and while the space for this chapter cannot accommodate them all, I wanted to highlight a few particularly ambitious ones. It should be noted that the most successful projects are as much about access as about digitizing content itself. All of the projects below are collaborative, as the advantages of digitization are most clearly realized through the combination of resources.

No mention of book digitization can pass without mentioning Google Books,[6] which was the world's wakeup call on mass digitization.[7] Prior to the announcement of the Google Print Library Project in 2004, libraries had experimented with digitization (e.g., Library of Congress's American Memory) but very tentatively and languidly, leaving the greatest progress to commercial vendors willing to take greater risks with developing systems. Libraries focused on the perfect, the best image, while Google focused on as much content as fast as possible. Google changed the world's mind-set, with both positive and negative consequences. For users, the scales fell much more heavily on the positive side than negative; they have access to more edited and reviewed materials than ever before, and even where quality control is questionable, the information that they have is more reliable than what they had previously.

For institutions within the United States, despite the controversy over legality, Google Print (now Google Books) spurred the realization that if nonprofit organizations could not mobilize on this issue on a much larger scale, the entire world of information would be co-opted by private, commercial entities whose whims would then determine future access. For nations beyond the United States, there were looming fears that soon the world's history would be the world's history only from the American standpoint. After all, though foreign titles were to be included in the initial project's scope, the vast majority of the partner libraries were in the United States with the majority of materials in English.

Other nations responded in force and have since enthusiastically and rapidly embraced the promise of digitization, developing robust systems with search features supporting the needs of their primary population. Europe has been particularly aggressive in its digitization efforts with strong state, public, and private support.[8] The nations have together created the European Library (http://www.theeuropeanlibrary.org/tel4/), containing more than 20 million digital items from over fifty leading research libraries, and Europeana (http://www.europeana.eu/portal/) with over 23 million objects from over 2,200 institutions, including not only libraries but also museums, universities, and governments, with texts, manuscripts, sound recordings, video, pictures, and more. They support not only established field searching but also map searching and language searching, as their objects are so varied in both of these areas.

Individual state contributors have made even more lofty goals known, with the National Library of the Netherlands committing to digitize all "all Dutch printed books, newspapers and magazines since 1470, a total of 730 million pages" in 20 years.[9] Initially a completely state endeavor, they have since partnered with Google and Proquest to enable them to meet their desired deadline. Although they have identified some characteristics to be used for prioritization, they see the process as fluid and accommodate digitize-on-demand requests.

In the United States, as referenced earlier, digitization efforts started long before Google Books. Due in part to timing, U.S. efforts were until recently much more labored and less coordinated than efforts abroad. Still, it served as a leader for the world prior to Google Print, with the Library of Congress undertaking digitization in 1990 with a pilot project now known as American Memory. The pilot ended and the official endeavor launched in 1995 as the National Digital Library Program, an initiative intended to digitize and preserve primary source materials for the nation. Since then, the Library of Congress has morphed into a much more integrated program (http://www.digitalpreservation.gov/), not only digitizing books but also preserving born-digital content, developing customizable user tools, and providing educational materials for individuals seeking to digitize their own collections.

However, the Library of Congress's greatest contribution to digitization at present is not in its digitized content but in their coordination. They have formed the National Digital Stewardship Alliance (NDSA) and the National Digital Information Infrastructure and Preservation Program (NDIIP), both committed to a sustainable preservation policy for digital materials and the "open exchange of ideas, services, and software." NDIIP has built a distributed digital library, with books, maps, architectural plans, photographs, sound recordings, video files, and websites from almost 300 partner institutions.

The Internet Archive,[10] Lyrasis,[11] and HathiTrust provide other examples of robust and diverse collaborative models within the United States. Internet Archive's OpenLibrary (http://openlibrary.org/) partners with other libraries for content while providing the equipment and staffing for the production itself. Further, it not only supports libraries in digitizing their collections through scanning services and permanent content hosting, it also expands the public's access to the materials, by making them (including in-copyright materials) generally available. An e-book lending library program, facilitated through 1,000 participating libraries and Overdrive, ensures careful control of in-copyright titles.

HathiTrust is equally ambitious but takes a slightly different approach, aiming to build a comprehensive library of digitized materials where ownership and maintenance is the shared responsibility of all member libraries and access is more circumscribed. With over 60 partners, over 3 million volumes in the public domain, and over 7 million additional volumes scanned, it is one of the largest collaborative collections available in the United States. Its model limits access to some materials to contributing and/or owning libraries, thereby respecting copyright as well as the financial contributions of its members. It is more structured than Internet Archives' effort and submissions must comply with strict requirements, but both efforts are geared to long-term preservation and facilitating easy access.

Lyrasis's Mass Digitization Collaborative takes yet another approach, on a regional level, focused on reducing the burden of digitization on libraries by directing workflow, from selection through digitization to maintenance of produced digital files. Participating libraries need not acquire equipment or have any expertise beyond what they need for their usual operations, relying instead on Lyrasis to fill any deficits.

Regardless of approach, the most successful models share a common feature: perpetual evolution. They constantly adapt, reviewing missions, content, and services, and collaborating in ways that maximize contributions.

EMERGING TRENDS AND THE FUTURE OF COLLECTION DIGITIZATION

The future of libraries and digitization will bring a closer melding of needs and expertise of the individual and the powers of the collective. The future of libraries resides not only in the digitized documents themselves but even more in the design of mechanisms to access, use, and manipulate them.

If the Digital Public Library of America (http://dp.la/) achieves its goals, digitization will be possible for every library, regardless of resources or size. "Scanabegos,"[12] mobile scanning units, will enable even the smallest libraries or even individuals to scan their materials. As technology continues to advance at its astonishing rate while dropping in prices, it is conceivable that eventually scanning materials at preservation levels will be possible for anyone, making the preservation of our history everyone's responsibility and not just the purview of libraries or archives. Newer inventions or open source tools such as Gado (http://projectgado.org/about-the-project/mission/) and Google's linear scanner (http://code.google.com/p/linear-book-scanner/) attest that this ambition is not all that distant, and the willingness of volunteers to contribute to such endeavors has long been established.[13]

As digital collections grow through the contributions of many, so will technologies to mine them. Information repositories will be networked into enormous data pools in which the individual pieces may carry little information but where the whole could identify patterns or provide empirical support for long-questioned hypotheses. For example, if all of the world's maps were uploaded, one could imagine that it would be possible to create comprehensive time lapses of population movements or create visual representations of changes (e.g., redistricting). Sophisticated interfaces could mine legal texts to map the changing in meaning and importance of specific legal terms at an even more detailed level than Google's *n*-gram. While largely just a vision at this time, the Social Networks and Archival Context Project (SNAC) (http://socialarchive.iath.virginia.edu/) is one current project aimed at facilitating mining among dispersed collections.

Users will also be empowered to customize digitized collections as they develop. They will create their own personalized digital collections from the online libraries that they frequent, or even just by "adding" individual titles to their collections as they browse sites, showing no preference for any library or information source. The results will be deduplicated and merged automatically. To future users, location of the source and the providing library will become increasingly less important. In accessing these personal libraries, users may not always be able to read a given title (i.e., if all available copies are being used by others), but the interface will identify all possible sources for them. They will not need to check each individual owning institution to determine what is available and what is not. They will be able to tag materials both with notes just for themselves as well as those that can be shared with the public. Interfaces will become intelligent, improving on predictive searching and better anticipating how a given user searches, allowing the interface to point out items of interest. These technologies already exist in emerging forms, both within consumer-based services like Amazon or Netflix and through scholarship tools like Mendeley and Zotero.

COSTS OF DIGITIZATION

We have explored the benefits of digitization, but this path is not without costs as well. This chapter does not discuss financial costs in detail, as they are changing too rapidly to capture, but I have provided links to organizations likely to update cost estimates along with their current projections in the endnotes and suggested readings.[14] Instead, this subsection seeks to understand to explore nonfinancial costs of digitization.

One of the most apparent losses is the loss of sensory, historical, and social context. A physical item has a particular texture, look, smell, and how it is constructed tells us much about the era in which the item was produced: what materials were available, what chemicals and inks were used, what binding techniques were common, and what media choices an artist made for the illustrations. Digital copies cannot convey features like illuminated texts, chromolithograph plates, or scrolls even through the best technology available. For this reason, print repositories for "last copies," such as PalmPrint and the Western Regional Storage Trust (WEST),[15] will continue to be necessary, as well as individual libraries' commitment to special collections.

LLMC (http://www.llmc.com), a nonprofit cooperative of libraries, is an example of how digitization projects can be integrated with print preservation. LLMC concentrates its efforts on government and legal information, and takes a two-pronged approach in project design. It seeks to digitize and make accessible online key materials while at the same time preserving original materials. Through partnerships, it has digitized 5,744 titles, totaling over 40 million pages. Upon completion of scanning, print originals are preserved in an ideal dark-storage archive with Underground Vaults and Storage Inc. Using this approach, it both ensures easy access to content while also preserving all that is unique in the printed originals.

There is also a fear that digital copies will make libraries and individuals more comfortable in discarding their nonunique print titles, copies that take up space and are not nearly as portable as their electronic counterparts, and that the destruction of materials will happen at such a rate and with such indiscriminate abandon that we will lose unique materials and be unable to recover them if the digitized version of it turns out to be faulty (e.g., missing pages). Information unique to particular copies of texts, like marginalia and notes tipped in by their readers, will disappear.

One could hypothesize, though, that exactly the opposite will happen, that instead of discarding materials, users will donate them to libraries earlier than they might have when print was the only method of access. If this turns out to be the case, then the potential for preserving unique information remains, so long as libraries continue to have space to store and staff to process them and the inclination to preserve them.

Whichever direction our future takes, the costs of digitization are real and cannot be ignored. A thoughtful and thorough plan for digitization will contain provisions both for the preservation of unique original physical sources as well as produced electronic documents.

CHALLENGES

A chapter on this topic cannot leave challenges unaddressed. Digitization, especially mass digitization, faces many hurdles, including copyright, financial trends in libraries, library differentiation, migration, effective data mining, authentication/security, privacy, and standards.

Copyright is the obstacle that is likely most familiar to readers, as recent challenges,[16] victories,[17] and defeats[18] in the courts have played out in the popular press in the past decade. While history has already demonstrated that vendor concerns for loss of profit are generally overstated, copyright owners have a long history of forcefully challenging new technologies that have any potential—real, attenuated, or illusory—to affect the individual's ability to sell the item himself or herself. In most of these instances, those feared technologies have actually opened new income streams or made it easier for authors to reach an audience,[19] but such empirical historical information has not been effective in stemming the defensive outpouring that arises with each new development.

Libraries need to be both creative and determined in their approach to copyright, respecting its spirit but resisting the overstepping of copyright owners, particularly corporate copyright owners, in asserting copyright in a manner not suited for this technological age and not in line with societal interests or constitutional principles.[20] Copyright owners unquestionably deserve to reap the fruits of their labor, but the heart of copyright was not about author advantage over all else. It was to encourage creation by making it possible for authors to make a living, while still permitting society to access the works for education, entertainment, and enjoyment. It is not a particularly good use of society resources for libraries to pay multiple times for the same text, simply because the format changes multiple times in its life span. They should have the ability to covert these materials to different formats, while constraining circulation in a manner consistent with the initial purchase, and recent judicial developments[21] support such a direction.

Two other challenges often slowing digitization are tied into self-perception, ranking, and reputation. Law libraries have traditionally sought to develop unique offerings through which to differentiate themselves from their peers. If we digitize collections and make them available via interlibrary loan for other libraries, our collections will no longer serve as a distinguishing factor, with exceptions for specialized collections that defy digitization. All libraries will have access to the same content,[22] even if only one library owns it. Two related concerns then arise. The first is that we will lose our identities as our collections become collective instead of individual. The second is that our administrations will stop funding our libraries once the majority of content is online; with collections being "invisible," it becomes harder for those outside the library to understand that funds are still necessary to digitize new materials, make them accessible, and enhance them for discovery.

The uneasiness about differentiation is a very personal one and must be conquered by recognizing that distinguishing ourselves cannot come at the cost of the best interests of our users. Our users' research and scholarship become stronger the more they have scholars with equal access to relevant information with whom to debate; making legal materials widely available to all researchers will only improve the education of our students and the quality of intellectual engagement in our schools. We have other ways (e.g., service), or will find other ways, to distinguish ourselves.

The related anxiety over funding must be counteracted with serious education of our administrations by library leaders. Nonlibrary administrators and our users, the latter of which can serve as our strongest advocates, need to understand where cost savings are possible, where costs will increase, and what the funding enables them to do. Since so much online is seen as being costless, nothing less than clear, convincing data otherwise will be enough to counteract this automatic reaction, and the data have to be packaged in easily absorbed messages (e.g., executive summaries not detailed reports) and provided to stakeholders repeatedly in different forms.

The last series of challenges for digitization are general technology challenges, not restricted to libraries or their content: effective data mining, privacy, authenticity/security, data migration, and standards. With such rapid proliferation of information—and not just traditional scholarly publications but also video, social media, judicial or legislative materials, just to name a few—digitization adds to information chaos even as it makes content accessible to the public.

This is why digitization cannot merely be the transformation of one format to another but must be accompanied by tools for easy discovery, customization, and manipulation. Developing effective tools for a diverse user population is a never-ending process. Google, Westlaw, Lexis, and other search-engine designers are constantly adjusting their algorithms to more effectively surface the most relevant results, and if libraries seek enter this field, they will need to consider whether they are equipped to engage in this process themselves or if there is a greater benefit in partnering with more experienced industries.

As we build better tools for discovery, we must also be wary of unintended consequences, the greatest of which is user privacy. In digitization, particularly of documents containing personal information (e.g., administrative forms), libraries must be careful to protect user privacy as much as possible. Authenticating materials in electronic format presents unique concerns, as control of them is more complex than for print documents. If a document can be accessed from multiple locations, it can also be hacked from multiple places, so repositories of such information, particularly controversial information, must always be vigilant of security and providers of content must have assured means of authentication. Authentication methods are varied, encompassing digital stamps/watermarks, cryptosystems, offline digital masters, or print copy comparisons. Authentication will also require a tracking of an electronic document's trail through its life, as technology is so protean that information stores will be migrated to new technologies or rescanned for better quality periodically.

The final general challenge may be the most difficult to overcome, as it speaks to society's need and libraries' desire to always provide the best authenticated resource possible. Libraries have been gatekeepers to information, both allowing users access as well as ensuring that the information is protected from damage or manipulation. Librarians are detail oriented and meticulous about the content placed in their care, following the gold standards in quality, indexing, and preservation. In taking the same approach to digitization, a field facing constant refinement, libraries may risk obsolescence.[23]

There is too much information to digitize and currently not nearly enough resources within libraries to scan even a significant portion of this information at the preservation level.[24] By insisting on only the highest standards, we necessarily limit ourselves in participants and content. The question libraries will face is whether lowering our standards and preserving more information at a lesser quality might better serve society. Authenticating or getting best-copies could be end goals, with immediate goals being content.

As a cautionary example, I'll point to Google's search engine, which has essentially displaced every catalog or index known as the predominant way to search.[25] Libraries spent (and continue to spend) a tremendous amount of effort on controlled terms, and those terms undoubtedly provide great value. They were much more relevant in the predigital age, though. Google has been more effective in meeting most user needs than library catalogs, their structures, and the standards used to support them. Even if our higher standards can convey a great benefit, if the tools created by those standards are never used, what value are we obtaining for our investment?

CONCLUSION

It has undoubtedly not escaped your notice that while this book is about law libraries, very little of this chapter focused on law-library-specific efforts. This is because digitization strips specialization and evens all playing fields among content providers. Law libraries have contributed to many of the projects discussed, though, usually as part of a larger effort, such as university-wide efforts.

Going forward, law libraries should make an effort to play a much more prominent role in this arena. We have specialized expertise, especially as so many in our institutions have law degrees as well as information degrees. We can lead the discussion on copyright as it applies to digitization, as well as the lobbying to reform it.

FURTHER READING

Bulow, Anna, Jess Ahmon, and Ross Spencer. 2011. *Preparing collections for digitization* London: Facet.

Dale, Robin L. April 2012. "Leveraging expertise: Outsourcing digitization activities." *Lyrasis.* http://www.connectingtocollections.org/wp-content/uploads/2012/04/Leveraging_Experience-Outsourcing_Digitization.pdf. Succinct comparison of the advantages and disadvantages of outsourcing versus in-house scanning, and steps for soliciting and evaluating digitization proposals.

Ng, Kwong Bor, and Jason Kucsma. 2010. "Digitization in the real world: Lessons learned from small and medium-sized digitization projects." New York: Metropolitan New York Library Council.

Primary Research Group. 2011. *Survey of library and museum digitization projects.* New York: Primary Research Group. Annual or biannual publication that provides detailed information on costs, staffing, and methods of digitization on projects at over 100 libraries.

NOTES

1. Although somewhat dated, a sample flowchart of the process and the types of equipment and staff needed for a digitization process can be found in the *U.S. EPA*

library collections digitization process report (September 24, 2007), also available at http://www.epa.gov/nscep/DigitizationReport.pdf.

2. For one example, see *Technical guidelines for digitizing cultural heritage materials: Creation of raster image master files*, Federal Agencies Digitization Initiative Still Image Working Group, August 2010, http://www.digitizationguidelines.gov/guidelines/FADGI_Still_Image-Tech_Guidelines_2010-08-24.pdf.

3. A list of digitization terms and definitions is available through the University of Michigan's Digital Conversion Unit at http://www.lib.umich.edu/files/short_DCU_Terms_Definitions.pdf.

4. The field of equipment vendors is changing rapidly, but some established current vendors are Kodak, Fujitsu, Minolta, *Kirtas*, *Bookeye*, Zeutschel, and Atiz.

5. Partnering with nonlaw libraries can create some difficulties, as illustrated by the Google Book Project. While library and archive-only projects have special copyright protections, as detailed in 17 U.S.C. §108, this protection may not always extend to uses by third parties.

6. A brief history of Google Book's development can be found at http://books.google.com/intl/en/googlebooks/about/history.html. Litigation documents on the Google Books Project can be found at http://thepublicindex.org/.

7. Robert J. Lackey, "From Google Print to Google Book search: The controversial initiative and its impact on other remarkable digitization project," *Reference Librarian* 49, 1 (2008): 35–53.

8. *The new renaissance: Will it happen? Innovating Europe out of the crisis.* Third and Final Report of the European Research Area Board 2012, http://ec.europa.eu/research/erab/pdf/3rd-erab-final-report_en.pdf. Additional information on Europe's digital agenda can be found at http://ec.europa.eu/information_society/activities/digital_libraries/index_en.htm.

9. Olaf D. Janssen, "Digitizing the National Library of the Netherlands: 730 Million pages in 20 years," http://b12.cgpublisher.com/proposals/20/index_html.

10. http://archive.org/index.php. More information about the Internet Archive can be found in Michele Kimpton and Jeff Ubois, "Year-by-year: From an archive of the Internet to an archive on the Internet," Web *archiving*, ed. Julien Masanès, 201–212 (New York: Springer, 2005).

11. http://www.lyrasis.org/. More information about Lyrasis can be found in Kathy Anderson and Laurie Gemmill's "Step easily into the digital future," *American Libraries* (July/August 2011): 36–39.

12. For the origins of the "Scanabego" term, see http://www.archivalmediaconsulting.com/blog/thoughts-on-the-digital-public-library-of-america-plenary-meeting.

13. See the National Archives' use of volunteers in digitization at http://www.archives.gov/press/press-releases/2012/nr12-113.html.

14. See "Summary of digitization & metadata creation prices" at https://wiki.library.oregonstate.edu/confluence/pages/viewpage.action?pageId=19327; Nick Poole's "The cost of digitising Europe's cultural heritage" (November 2010) at http://ec.europa.eu/information_society/activities/digital_libraries/doc/refgroup/annexes/digiti_report.pdf; or the Council on Library and Information Resources, *Comparative costs for book treatments*, http://www.clir.org/pubs/reports/pub103/appendix6.html.

15. PALMprint is a joint effort of the Legal Information Preservation Alliance (LIPA) and NELLCO (http://www.nellco.org/?page=palmprint). Information about WEST can be found at http://www.cdlib.org/west/.

16. *Authors Guild v. Google*, 770 F.Supp.2d 666 (2011) and *Wiley v. Kirtsaeng*, 654 F.3d 210 (2011) (currently on appeal to the Supreme Court).

17. *Cambridge University Press v. Becker*, 863 F. Supp. 1190 (2012); *Authors Guild v. HaithiTrust*, 2012 WL 4808939 (2012).

18. *Eldred v. Ashcroft*, 537 U.S. 186 (2003); *Golan v Holder*, 609 F.3d 1076 (2010).

19. In re Aimster, 334 F.3d 643 at 650 (2003).

20. U.S. Const. Sec. 8, Cl. 8; *Sony v. Universal Studios*, 464 U.S. 417 at 429 (1983).

21. *Authors Guild v. HaithiTrust*, 2012 WL 4808939 (2012).

22. Collections will not be the same, as licensed content and born-digital content will also play significant roles, but for the purposes of this chapter, we focus on the content derived from existing physical resources.

23. Jill Hurst-Wahl, "Digitization: How many best practices, guidelines, and standards do we need?" *Information Standards Quarterly* 21, 4 (2009): 22–25.

24. For just a glimpse of the task that stands before libraries in terms of the amount and scope of information that needs to be preserved, see http://www.publicintegrity.org/2011/08/02/5456/millions-federal-court-records-are-being-destroyed-save-money (describing the destruction of court records) or http://www.tahrirdocuments.org/ (detailing the types of materials that are being generated as part of Egypt's revolution).

25. Karl V. Fast, and D. Grant Campbell, "'I still like Google': University student perceptions of searching OPACs and the Web," *Proceedings of the American Society for Information Science and Technology* 41, 1 (2004): 138–146; Michelle M. Wu and Leslie A. Lee, "An empirical study on the research and critical evaluation skills of law students," *Legal Reference Services Quarterly* 31, 3–4 (2012): 205–238.

8

E-books in Law Libraries

Ellyssa Kroski
THE NEW YORK LAW INSTITUTE

The popularity of the e-book has skyrocketed over the past several years with the introduction and mainstream adoption of e-book readers and tablet devices such as the Amazon Kindle and Apple's iPad. Market forecasts project a steady increase in American spending on electronic books, and library patrons are requesting these resources more than any other form of downloadable media, including music and videos. Today, 95 percent of academic libraries are offering e-books to their patrons, and private law libraries are not far behind.[1] According to the International Legal Technology Association's 2012 Technology Survey of nearly 500 firms and 90,000 attorneys, iPad devices were in use at 88 percent of respondent's law firms and Android tablets at 26 percent.[2] This new e-book demand is one that libraries of all types and sizes are meeting in order to accommodate the evolving needs of their users.

E-books have been around since 1971 when Project Gutenberg founder Michael Hart at the University of Illinois started the first collection of free electronic books, which now numbers over 40,000. But they've come a long way from those plain-text roots to today's robust digital offerings capable of syncing between multiple devices via e-reader applications. Academic libraries have been offering students online access to e-books since the late 1990s though subscription packages like Books 24x7 (founded in 1994) and netLibrary (founded in 1998). These e-books were available in html format only and could merely be read online, not downloaded or transferred to other devices. But in 2002 the Overdrive e-book distribution service introduced its digital download platform for public libraries and made it possible for library patrons to download e-books to their mobile devices. And they remain one of the major providers of e-books to libraries today.

Although consumers are increasingly embracing e-books, not all libraries (especially law libraries) are convinced that they are a good idea. Here's a look at some of the benefits and challenges that e-books pose in law libraries.

BENEFITS OF E-BOOKS IN LAW LIBRARIES

No Shelf Space Required

Shelf space in law libraries (especially law firm libraries) is at a premium, and e-books offer a viable digital alternative to storing large collections of print materials.

Instant Access

Much of the work that patrons do is time sensitive, and research is no exception. When attorneys are conducting research for a case, and law students for a paper, they want access to relevant resources as soon as possible. E-books provide immediate access to relevant resources and can be obtained at the user's convenience.

Increased Portability

Library patrons can carry an unlimited amount of e-books with them at any given time, as opposed to their oftentimes-heavy print counterparts.

Full-Text Search

Having access to a full-text search of materials is an invaluable tool when conducting research. Nearly all e-books come with the ability to be searched, and many e-book publishers also offer a full-text search across their entire collection of e-books as well. This can really come in handy when researching whether a particular brand name or topic is mentioned in any e-book in the collection.

Added Features

Most e-books come with the ability to add digital bookmarks, make annotations, highlight portions of text, access built-in dictionaries, change the font size of the text according to the preference of the reader, copy and paste portions of the books, and print a certain number of e-book pages.

Enhanced Publisher Content

Publishers that are making their e-books available directly through their own platforms are also offering enhanced functionality and content with their e-books. For example, Thomson Reuters allows readers to transfer their notes and highlighting to new editions of their e-books. Additionally, publishers that also offer legal database subscriptions are linking directly to these resources from within their e-books, such as Thomson Reuters linking to WestlawNext from within e-books, and LexisNexis linking to Lexis Advance.

CHALLENGES OF E-BOOKS IN LAW LIBRARIES

Long Waiting Lists

Many e-book packages limit e-books to a single simultaneous user at a time, making for potentially long queues of people waiting for the same e-book. All it takes is one or two visits to your website where patrons cannot get an e-book they're looking for in a timely fashion that they stop thinking of the library as a viable source for e-books.

Different Vendors/Interfaces

Since there are so many e-book vendors and publishers available, each with a different interface and different rules of use and access, it can quickly become very confusing for library patrons and staff alike to learn how to access the library's e-books.

E-book Device Support

Right now there is a wide variety of e-book reader devices on the market, including different versions of the Kindle, Sony e-Readers, Nooks, iPads, Android tablets, and more. If the library is offering e-books that are compatible with all of these different devices that their patrons may be using, they also need to be prepared to support them and know how to assist their users with loading e-books onto each of them.

E-book Formats

Right now there is no unified e-book standard format. Although the market is leaning toward the ePUB open standard, it has not yet gained complete adoption. Therefore, it becomes essential to know which e-book format each vendor is using so that you can guarantee compatibility with the widest number of devices.

Access and Pricing Models

Probably the biggest, most intimidating factor of implementing an e-books program in a library is all of the different pricing models that you need to navigate.

It really makes evaluating e-book programs like comparing apples to oranges. And oftentimes vendors will offer multiple pricing models and packages, making it even more confusing. But here's a look at the main ways that e-books are being offered to libraries today:

- *Single user vs. multiuser vs. unlimited users.* E-books may be restricted to the number of readers who can access them at one time. Even though the title is digital, publishers want to be sure not to lose out on potential sales of copies, so they will often charge more for e-books that are accessible to multiple users at the same time. It is important that you are clear about how many users your library is allowed per e-book so that you can pass this information on to patrons.
- *Individual e-book purchases vs. subscriptions.* Although many e-book publishers and vendors are offering e-book sales, there are a few who are offering subscriptions to their e-book libraries for a flat rate. This is potentially one way to offer a large number of e-books to your patrons in one fell swoop, although an analysis of the subject area coverage and titles in the collection is an absolute necessity.
- *Perpetual ownership vs. ongoing access fees.* It's important to be aware of the fact that not all e-book vendors are offering perpetual ownership of e-books that are purchased through their packages. You may be stuck paying ongoing fees to be able to access the e-books you have purchased if you switch vendors, so this is something to be sure and clarify.
- *Short-term loans (STLs).* This is a concept that is unique to e-books and is also sometimes referred to "pay per use" (also known as "pay per view"). It is oftentimes offered by e-book aggregators and is essentially the rental of an e-book rather than a purchase. The idea behind it is that a library patron can borrow titles from an aggregator's catalog that the library doesn't own for a short-term period, ranging anywhere from one to twenty-eight days. The library is charged a percentage that can range from 5 percent to 30 percent of the title price of the book for a rental of the e-book, so it costs the library less than purchasing it.
- *Patron-driven acquisitions (PDA) or demand-driven acquisitions (DDA).* With a demand-driven acquisitions model, libraries make their e-book purchases based on exactly what patrons want to borrow. Books are purchased when library patrons issue a loan for a title and not before, eliminating wasteful or unwanted purchases. A huge benefit to this model is that libraries can offer their patrons access to a very large collection of e-books that they haven't actually bought yet and make their purchase decisions based on just what their patrons want and need. It works by proving access to a range of hundreds or thousands of titles, which they decide upon based on

all sorts of criteria such as price, publisher, and so forth. The library can set up either mediated or unmediated access to books so that if a patron wants access to a book, a mediated request can be sent to the library and librarians have an option to purchase that book for the patron, or patrons can be allowed unmediated loans of books that are purchased upon loan request. Some e-book packages offer libraries the option of STLs in conjunction with this model, which enables them to rent the book to patrons a few times before making a purchase, in this way ensuring its demand beyond just one instance. EBL refers to this as "just-in-time" collection development.

- *Nonlinear lending*. EBL has a unique offering called nonlinear lending. While many e-book vendors provide only single simultaneous reader access to their e-books, EBL e-books are available to unlimited users simultaneously. Instead of limiting usage by number of readers, usage is limited to a total number of loan days per year, which is 325 days. So any number of people could be reading the same e-book at the same time and all of their loan days would be added to the total of allowed days per year. For example, if twenty-five people checked out the same e-book for a 1-day loan, that e-book would be available for 300 more loan days within that year. After that number of loan days has been exceeded, the library must either purchase another copy of the book or just make it available via short-term loan (rental).

Publisher–Library Tension

There have been a lot of hiccups with this new format and its distribution by libraries. Publishers, afraid of losing potential e-book profits, have made some very unpopular decisions regarding e-book availability in libraries. And librarians have been quite vocal about their dismay over these choices. In February 2011, Harper Collins announced the decision to change their licensing agreement with the Overdrive service, allowing only twenty-six checkouts per e-book title by libraries. Later that year, in November 2011, the Penguin Group pulled all of their e-book titles from Overdrive, citing security concerns. Although they did reverse that decision following a serious backlash from the community, they announced that no new titles from Penguin would be made available to libraries. Hachette, Simon & Schuster, and Macmillan do not make e-books available to libraries at all. Random House is now the only "Big Six" publisher to allow unrestricted access to all of its e-books in libraries.

Software Installation

E-books often necessitate some sort of e-reader software application to be installed on a user's computer, such as Adobe Digital Editions. This can pose a problem in law firms and other libraries that must go through the IT department to have new applications installed on all desktops in the organization.

Firewall Issues

Although many libraries have access to remote resources such as library databases and e-books via their local area network, there are oftentimes issues when they try to access e-books via their tablet devices because of restrictions on their wireless network. This is an additional hurdle that must be dealt with through IT.

Mediated Loans in Law Firms

Most law firms rely on the librarian to check out or recommend books to attorneys. Most e-book packages do not have the functionality built in to deal with mediated loans, and so workarounds must be put into place, especially within libraries with only single-user access to e-books.

WAYS TO PURCHASE E-BOOKS

When it comes to purchasing e-books, libraries have a myriad of vendor choices at their fingertips, each with different and sometimes multiple packages and lending models available, making the landscape difficult to navigate. These are the different ways that libraries can go about acquiring e-books.

Aggregators

One of the easiest ways to purchase a large number of e-books in one fell swoop is to sign up with an aggregator. They provide access to e-books from hundreds of different publishers in many different subject areas. And the best part is that they negotiate the licensing agreements with the individual publishers, leaving you to just sign one agreement. The most well-known aggregator out there right now is Overdrive, which has over 22,000 library and educational subscribers. Here are some of the major e-book aggregators.

Overdrive

This digital distributor (http://www.overdrive.com) provides libraries of all types with e-books, audiobooks, music, and video files. Resources are compatible with thousands of mobile devices, including iOS and Android devices, Kindles, Sony Readers, and more. They offer over 1 million digital titles from 1,300 publishers, including over 5,000 law-related titles. Their service is used by 22,000 library subscribers, schools, and colleges worldwide. They have recently partnered with LexisNexis to offer the LexisNexis Digital Library of 1,800 primary law, deskbooks, code books, treatises, and other titles.

EBL—Ebook Library

This innovative Australian aggregator (http://www.eblib.com) offers nearly 300,000 titles from over 500 publishers in a wide range of academic and professional subjects, including 8,600+ law related titles. They also offer some very unique lending models, all of which are discussed later in this chapter, including demand-driven acquisitions (aka patron driven acquisitions), short-term loans, and the nonlinear lending model. At the time of this writing, EBL is being acquired by Proquest, so these details will likely be changing (http://www.proquest.com/en-US/aboutus/pressroom/13/20130122.shtml).

*E-books on EBSCO*host

EBSCO Publishing (http://www.ebscohost.com/e-books, formerly NetLibrary) now offers e-books and audiobooks on EBSCO*host*. They have made subscription and ownership options, including a patron-driven acquisitions model, available to libraries that wish to purchase e-books. They offer over 300,000 e-books, more than 8,000 of which are law-related titles.

MyiLibrary

Offering nearly 250,000 titles in all major disciplines including 8,000 law-related titles, from nearly 1,000 publishers, MyiLibrary (http://www.myilibrary.com) is a major provider of e-books to libraries. They offer multiple pricing models to all types of libraries.

Proquest Ebrary

Proquest (http://www.proquest.com/en-US/products/brands/pl_ebrary.shtml) offers a subscription to its 70,000-title e-book collection through ebrary, which includes over 6,000 law-related titles. Other lending models such as patron-driven acquisitions are also available and may be combined with short-term loans to fit the library's needs.

Publishers

Libraries can go directly to the source and find out what e-book packages individual publishers are offering. They may also buy individual e-books that might not be available in the packages that aggregators offer by going this route. However, you then must juggle multiple agreements with different vendors and have multiple interfaces and usage restrictions for your staff and patrons to learn how to use. Additionally, some publishers will not sell to libraries at all.

Thomson Reuters Proview

Proview (http://thomsonreuters.com/proview/en/) is Thomson Reuters' e-book platform, which can be accessed via free iPad and Android tablet apps. They currently offer over 400 e-books, all of which are law-related. In addition to typical e-book features, the notes, comments, and highlights in these e-books are automatically transferred to new editions. Links within Thomson Reuters e-books connect seamlessly with WestlawNext.

LexisNexis Digital Library

LexisNexis (http://www.lexisnexis.com/e-books/lending) has partnered with the Overdrive digital distribution service to provide their 1,800 primary law, deskbooks, code books, treatises, and other e-book titles. The platform provides detailed usage reports to help track e-book usage patterns.

PLI Discover Plus

Discover Plus (http://www.pli.edu/Content/_/N-/Term-e-booklibrary?Ntk=All&Ntt=e-booklibrary&Nty=1) is PLI's new e-book platform, offering access to their catalog of over 1,600 treatises, course handbooks, and answer books. This subscription service also provides access to transcripts of PLI seminars and downloadable legal forms.

Wolters Kluwer International E-book Store

Wolters Kluwer (http://kluwer-law-e-boek.mijnboekhandelaar.com) has recently launched an e-book store through which they are offering over 100 e-book titles for purchase in the areas of law and business.

Fastcase E-book Advance Sheets

Fastcase (http://www.fastcase.com/e-books) is making Advance Sheets, including both published and unpublished opinions, available as e-books for iPad, Kindle, Kindle Fire, or Android tablets for free. They have over fifty sheets ready for download, with plans for hundreds more to be published under the same Creative Commons license.

Cambridge Books Online

Cambridge Books Online (http://e-books.cambridge.org) offers over 14,000 e-books for purchase from this authoritative publisher, 900 of which are law-related.

Wholesalers

Wholesalers such as Baker & Taylor offer content from both publishers and aggregators. By going this route, you can pay one vendor for all of the e-books you want from the different sources. However, they most often do not offer any sort of unified contract arrangement—you still need to agree to the terms of each individual vendor's contract and package restrictions. It's also important to note that most wholesalers don't actually host anything or have their own platform, so you'd have the same issue of different interfaces for everyone to learn, combined with different access restrictions for different e-books because of the multiple vendors. This could potentially be very confusing for library patrons who may be able to access one set of books easily while having to wait in a queue to access others.

Yankee Book Peddler

YBP Library Services (http://www.ybp.com/acad/core1000cover.htm) is the academic division of Baker & Taylor. They offer over 600,000 e-books from Proquest ebrary, EBL, e-books on EBSCO*host*, Gale, and a host of publisher-direct titles from university presses and primarily academic publishers.

Baker and Taylor

This wholesaler (http://www.btol.com/axis360.cfm) covers mainly the school and public library market. They have their own e-book platform called Axis360 and they offer e-books from Proquest ebrary, e-books on EBSCO*host*, and Gale, as well as many individual publishers.

Consortia

Many library consortia offer discounts for e-book packages. Purchasing an e-book package via this route guarantees that the vendor has been vetted and the packages they're offering to their members provide a good value for the cost. Libraries should check with their local library consortia to see if they offer such deals with any e-book vendors.

HOW TO EVALUATE E-BOOK PACKAGES

There are many questions libraries will want to ask of e-book vendors when they are comparing different packages for purchase. The library staff at Wellesley College Library has put together a stellar spreadsheet encompassing many of these criteria, available at http://tinyurl.com/6om7plj.

Organizational Issues

When starting to evaluate e-book vendors, think about the requirements of the initiative and begin to develop a summary of what exactly your library wants to offer its patrons. Some useful questions to ask at this stage include:

- What devices do the majority of our users have?
- Are we out to replace what we have in print or complement it?
- What's our budget for this and will it become part of our print acquisitions budget or our e-resources/database budget?
- Will we lend out e-readers/tablets or support what patrons have?
- Will this initiative complement our collection development policy or go in a different direction (e.g., patron-driven acquisitions model vs. librarian selected resources)?
- How will the necessity of additional software programs be affected by firewalls and our organization's internal policies?

Technical Specifications

It is crucial to acquire a good grasp of the technology involved in each e-books package so that you can determine compatibility and potential conflicts. Some good questions to ask about the technical aspects of each e-book package include:

- Which browsers can the e-books be read on? Internet Explorer, Firefox, Opera, others?
- Are there plugins or additional software programs required?
- What specific types of mobile devices can readers use to access the e-books and can they also be read on desktop computers?
- Are the e-books available in the EPUB format, which is becoming the industry standard? If not, what format are they available in?
- How are updates—additions, deletions, updated MARC records—dealt with?

Functionality

You'll want to get to the heart of things by asking specific questions about the features and functionality that your patrons and library staff will want with regard to e-books.

- Can library patrons copy and paste portions of e-books into Microsoft Word or other programs? If so, how much?
- Can readers print out sections of e-books? If so, is there a limit to how many pages?
- Can e-books be downloaded for offline reading?
- Can readers interact with the text by highlighting sections, making notes, and adding comments—a practice valued by attorneys?
- Is there a full-text search of the e-books available? Can users also search the full text of the entire e-book collection?
- Are there analytics and reports available with the e-book platform so that we can view usage patterns, etc.?
- Are there any cost-controlling features or settings of our package that we will have access to?

Content Specifications

Libraries will want to be clear about exactly what types of content e-book vendors are offering. Most have publisher lists and searchable title lists available, as well as lists of categories or subjects covered. Here are some helpful questions to determine content:

- What is the scope of the e-book collection? What is the subject coverage and is the focus academic, general, or a combination?
- How many e-book titles are available? How frequently are new titles added?
- How many publishers (legal and other) do you offer titles from? How frequently are new publishers added?
- Will MARC records be provided free of charge?
- Is there an embargo period or a pattern of delaying availability of e-book editions once the print book has been published?

Sales Model

Many vendors offer different packages and ways to purchase e-books, so it's important to nail down specifics of exactly how pricing will work with each potential solution so that you can make the best decision based on your budget.

- Is there an implementation or set-up fee to get started?
- Are there ongoing annual fees to maintain this relationship? Can these be waived after purchasing a certain number of e-book titles?
- Are e-books available for single, multiple, or unlimited simultaneous users?
- What is the e-book title cost relative to the print book cost? Is there an additional percentage added for digital versions?

- Do e-book purchases include perpetual access, even if we no longer subscribe to the service, and/or is there a fee involved to access the platform?
- Does the package include individual e-book purchases or access via a subscription?
- Is there a patron free-viewing period of a certain number of minutes or pages before e-book loans are initiated?
- Likewise, is there an additional free viewing period for staff/librarians to view and assess titles for acquisition?
- Are short term loans (STLs) available?
- Is there a demand-driven acquisitions (or patron-driven acquisitions) model available?
- Do you offer nonlinear lending? (Although this pricing model is currently only offered by EBL, this may be a sales model that other vendors adopt in the future.)
- Is there a minimum annual commitment?
- Do you offer any pricing discounts for members of consortia?

WHAT ARE LAW LIBRARIES DOING?

Law libraries are offering e-books to their patrons in a variety of ways, some combining multiple strategies to make these digital resources available to their users. Here are some of the approaches that law libraries have decided to take.

Purchasing Direct from Publishers

Many law libraries, especially those in firms, have decided to go directly to the source and purchase individual titles from publishers that are making digital versions of their books available. The libraries at Cooley LLP and the Cleveland Law Library Association have opted to purchase e-books from LexisNexis, while Patterson Belknap Webb & Tyler LLP as well as the Con Edison Law Department Library are making PLI Discover Plus e-books available to their patrons.

Highlighting Existing Collections

The New England Law Library and many others are simply highlighting e-books that are already available to them through their existing database subscriptions. For example, HeinOnline Legal Classics Library contains several hundred historical legal texts; LLMC, the Law Library Microform Consortium, provides access to thousands of full-text historic legal publications; and Oxford Scholarship Online is a group of scholarly treatises on international and foreign law. None of these e-books can be downloaded to devices, but all can be read within a Web browser. The University of Oklahoma Law Library is highlighting the Gale Making of Modern Law Collection, Hein Online's Legal Classics, and LLMC's digital books, as well as pointing to the main university's subscription to Proquest's ebrary.

Purchasing E-book Collections through Aggregators

The trend seems to be that academic libraries, rather than law firm libraries, are choosing to go with aggregators. The Florida State Law Research Center (FSU) has partnered with the Lawton Chiles Legal Information Center at the University of Florida to offer a joint e-books program powered by MyiLibrary. Both the Hofstra Law Library and the Ava Maria School of Law are offering e-books on EBSCO*host*, and the CUNY Law School Library has subscribed to Proquest ebrary.

A DEMAND-DRIVEN E-BOOKS CASE STUDY

At the New York Law Institute, we decided to go with EBL–Ebook Library. This was largely due to the fact that their patron-driven acquisitions model allowed us to offer a substantial collection of e-books immediately to our patrons without any upfront costs to us for e-book purchases. We also really appreciated that their nonlinear lending model meant that we could offer our e-books to an unlimited number of readers at the same time, so it would mean no waiting lines for our members.

After discussing our budget range with EBL, we determined that we could conservatively offer 54,000 e-books to our members right away, and still have access to EBL's entire collection of nearly 300,000 titles should we decide to add more to our initiative. We worked with them to determine a customized list of e-books that fit our criteria by choosing EBL categories we wanted to include, such as law, engineering, pharmacy, and so forth, as well as LC classes such as HG–Finance and HD–Industry/Labor, and so on. We were even able to include e-books that matched specific keywords that we specified such as "genome," and exclude others such as "for dummies." Our goal was to provide not only law-related e-books to our patrons but also e-books in peripheral subject areas that would meet the ancillary needs of multidisciplinary research.

Once this profile was developed, we were provided with our own branded EBL portal website that lets our patrons access all 54,000 e-books that matched our specifications. We were also provided with free MARC records that we uploaded to our EOS OPAC so that our members could easily find our new digital titles 24/7. We were given access to LibCentral, the staff-side administrative portal, which allows us to manage and customize our e-book collection. This is where we chose to offer one-, seven-, and fourteen-day loan periods to our members. We decided to let our patrons have unmediated access to our e-book collection and prepared our settings to initiate six short-term loans of each e-book before automatically purchasing it on behalf of the library. We ran a one-month beta test with six law firms before opening it up to all of our members.

We have a wide range of up-to-the-minute reports at our fingertips, which has helped us gauge usage patterns and forecast costs. The one challenging aspect of this type of pricing model is that costs are open-ended depending on how much or little patrons use the e-books. And since there is no flat rate involved, it's difficult to budget and determine what the end cost will be. But access to our administrative portal gives a wide array of cost-controlling options if we find that we are nearing our budget limit. For example, we could set price limits on short-term loans, set shorter STL periods for smaller costs, reset the trigger point for our autopurchases to be higher than six STLs, or set it up so that no purchases ever occur and finish out the year with rentals only. We could place limits on the number of STLs per patron per day, reduce the amount of e-books that are visible to patrons from 54,000 to 40,000 or even 20,000. Or alternatively switch to a completely mediated access model.

Accessing E-books through Membership Libraries

Law libraries that are members of The New York Law Institute (NYLI) and/or other membership libraries offering e-book packages are able to provide access to e-books as a service to their patrons with just the cost of their annual membership. For example, the library at Debevoise & Plimpton issued an announcement publicizing the availability of 54,000 e-books to their attorneys firm-wide through a partnership with The New York Law Institute. Additionally, we recently partnered with Debevoise and EOS International to develop a Seamless Catalog module that adds functionality that in essence transforms a members' OPAC into a one-way union catalog in which a library can view all of NYLI's holdings while keeping their own holdings private. Members who use EOS can instantly add all of our e-books right into their own catalogs. The Cleveland Law Library Association is also offering e-books to their member libraries by preloading LexisNexis titles onto library laptops, which they're making available for loan.

Combining E-book Strategies

Law libraries such as the Squire Law Library at the University of Cambridge are combining multiple strategies to offer their patrons the best access. They have signed on with multiple vendors, including e-books on EBSCO*host* and MyiLibrary, and are also offering e-books from Oxford Scholarship Online, Cambridge Companions Online, and Westlaw Commentary. Similarly, the Peter Rodino Law Library at Seton Hall is highlighting collections from Gale's Virtual Reference Library while subscribing to e-books on EBSCO*host* and MyiLibrary. The University of Chicago D'Angelo Law Library is acquiring e-books through the Yankee Book Peddler wholesaler, including e-books on EBSCO*host*, Proquest ebrary, and Cambridge Books Online. The Lincoln Memorial University Duncan School of Law is using most of the major aggregators including EBL–Ebook Library, e-books on EBSCO*host*, Proquest ebrary, and MyiLibrary, as well as key publishers LexisNexis and Cambridge Books Online. Others who are using this combined approach to offer substantial e-book collections to their patrons include the Vermont Law School and the Lillian Goldman Law Library at Yale University.

Creating Their Own

The librarians at the Massachusetts Trial Court Lowell Law Library led by head law librarian Cathy Alves decided to create their own. They have made current Massachusetts Court Rules and Documents into free e-books that they are making openly available. They can be found at http://www.lawlib.state.ma.us/source/mass/rules/e-books.html.

FURTHER READING

For even more on the topic of e-books in law libraries, you can refer to these valuable resources.

Center for Research Libraries. *JISC compare e-book platforms tool.* http://tinyurl.com/7ankb7e.

Freeman, Daniel A. April 11, 2012. "Continuing the conversation: Integrating ebooks and ereaders into your library." *ALA TechSource.* http://www.alatechsource.org/blog/2012/04/continuing-the-conversation-integrating-ebooks-and-ereaders-into-your-library.html.

Glassmeyer, Sarah. January 2012. "Making ebooks for fun and no profit." *SarahGlassmeyer.com.*

Lenares, Deborah. April 7, 2011. "Patron driven acquisition vendor criteria w responses." http://tinyurl.com/6om7plj.

Polanka, Sue. 2012. *No shelf required 2: Use and management of electronic books.* Chicago: American Library Association.

Polanka, Sue. 2011. *No shelf required: E-books in libraries.* Chicago: American Library Association.

Polanka, Sue. 2011. *The no shelf required guide to e-book purchasing.* Chicago, Ill.: ALA TechSource. http://tinyurl.com/6o6qweg.

Reynolds, Bess. October 2012. *The challenges of e-books in law firm libraries.* AALL/ILTA the New Librarian. http://read.uberflip.com/i/87421/.

NOTES

1. Sue Polanka, *The no shelf required guide to e-book purchasing* (Chicago, Ill.: ALA TechSource, 2011). http://tinyurl.com/6o6qweg.

2. ILTA. *ILTA'S 2012 TechnoLogy Survey* (Austin, Tex.: International Legal Technology Association, 2012). http://www.iltanet.org/techsurvey.

9

Tablets and Mobile Device Management

William R. Mills
NEW YORK LAW SCHOOL MENDIK LIBRARY

"Ubiquitous mobility" is more than just a catch phrase. It describes both a trend in American society and an expectation among those engaged in the modern practice of law. Law librarians who wish to retain their credible status as information professionals in service to legal institutions must master the skills necessary to deliver ubiquitous mobility to their patrons. Central to the modern concept of ubiquitous mobility are the tablet and the smartphone—devices that modern legal professionals use—as well as the app, the software application that delivers the information.

THE MARKET FOR MOBILE DEVICES AND APPS

Librarians must be cognizant of the burgeoning market for mobile devices, which function as information appliances in a vast and expanding array of settings. In the third quarter of 2012, the tablet market grew by 49.5 percent worldwide over the same period in 2011, with 27.8 million shipments. During that the same period, 179.7 million smartphones shipped, showing growth of 45.3 percent and beating 'analysts' expectations. Mobile device use in the legal profession has mirrored this trend. According to an ABA survey, 33 percent of lawyers reported using tablets for law-related tasks in 2012, up from 15 percent in 2011.

Apple's iPads dominate the tablet market, while Samsung's Android devices and Apple's iPhones are dominant in the smartphone segment. Microsoft, Google, Motorola, Asus, Lenovo, Sony, and BlackBerry also compete in as major players in one or both categories. Amazon's Kindle and Barnes & Noble's Nook go head to head as tablets designed for use primarily as e-readers. The market for mobile devices is recognized as volatile, with new product introductions and system upgrades offering enhanced features released several times each year.

Meanwhile, an industry research group predicted that smartphone and tablet users worldwide would download more than 45 billion apps in 2012, nearly twice the number that was downloaded the year before. This same research group has estimated that by 2016 more than 300 billion apps will be downloaded annually. In major law firms, a survey of technology chiefs indicates that the number of lawyers using iPhones rose to 44 percent in 2012, up from 31 percent the year before. And, of course, with the rise in the number of lawyers using mobile devices, the number of apps developed specifically for lawyers has increased at a steady pace, with new ones being rolled out all the time.

MOBILE APPS FOR LEGAL PROFESSIONALS

These gross statistics, however, provide little insight into the variety of purposes to which apps are put, particularly as they relate to the use of mobile devices as information appliances. To get a sense of this, you need only visit the various Internet markets where apps are made available for download. The app store found on Apple's venerable iTunes now competes with Google Play, Microsoft's Windows Store, BlackBerry World, and app stores for Kindle Fire and Nook. Never to be outdone, Facebook recently launched its own App Center. The app stores organize their offerings into various useful categories such as finance, productivity, and lifestyle.

Law librarians, serving as information professionals, must become familiar with the purposes for which mobile devices are relied on by their patrons in the workplace. Apps can be seen along a spectrum of importance to law library patrons. At one end of the spectrum are apps that serve a purely diversionary purpose, such as games. As a general rule, librarians are not expected to provide training and support for lawyers who wish to become proficient in playing Angry Birds, for example. At the opposite end of the spectrum are apps that fit squarely into a law library's traditional model. The HeinOnline app, for instance, provides mobile access to large collections of digitized law books. Most law librarians should become sufficiently familiar with this app so as to be able to promote and support its use among patrons who will find it valuable. It's worth noting that a great many librarians, as well as a great many of the lawyers they serve, are likely to have both Angry Birds and HeinOnline among the apps installed on their personal mobile devices, possibly making frequent use of both. This illustrates both the diversity of purposes to which mobile devices are put, and the extent to which they have pervaded modern life.

Along the spectrum of categories of apps important to law library patrons there is a vast middle ground. Found within this middle ground are apps that do not necessarily fall into the traditional model of law library service but that are nonetheless useful, or even vital, to law library patrons. A prime example is the category of social media apps for professionals, such as LinkedIn. LinkedIn is acknowledged as an invaluable resource for legal professionals interested in following people, news, and discussions relevant to their area of practice. The purposes for which legal professionals use the LinkedIn app are not found among purposes traditionally associated with a law library. But who within any legal institution is better positioned than the librarian to promote, explain, and support apps such as LinkedIn? At its heart, LinkedIn is just another information resource. In nearly all legal institutions, librarians are the only professionals trained to recognize, evaluate, promote, explain, and support information resources to their patrons.

This middle ground of apps important to law library patrons should be seen as both a challenge and an opportunity for modern law librarians. The middle ground includes not only social media apps but also a vast array of others that provide useful information to legal professionals and support their productivity. These are the apps that define the mobile device for them as a vital information appliance. Apps in this middle ground serve a multiplicity of purposes, and as a general rule were not developed with lawyers in mind. The popular Dropbox app stores your files in the cloud and lets you access them on your mobile device. The Evernote app allows you to write yourself notes, clip articles, and capture photos and graphic files from the Web, storing and organizing everything into scrapbooks. Readdle's PDF Expert App helps you work with PDF documents, allowing you to edit, annotate, copy, paste, fill out forms, sign and stamp PDFs, saving changes in formats compatible with Preview and Adobe Acrobat. The Documents app, also from Readdle, works as an all-in-one document viewer, media player, and file manager. Take a picture of any business card, and LinkedIn's CardMunch app will automatically convert it into a contact on your mobile device.

The apps described above were not chosen at random; all of them were cited on legal websites and blogs as being among the top-ranking useful apps for practicing lawyers. Yet none of them support purposes traditionally associated with law libraries or supported by law librarians. As a law librarian called on to support such apps for your patrons, you'll probably have to move a bit outside your comfort zone. Certainly, you must look far beyond the traditional librarian's roles of just organizing collections of books and other resources, and supporting research. And in a world where more and more library patrons expect ubiquitous mobility, you're likely to discover that support of mobile devices and apps is rapidly becoming an important professional responsibility.

Beyond the middle ground of general-purpose productivity apps, you'll find a large and growing number of apps developed specifically for legal professionals. Many law librarians will already be familiar with legal research apps from Fastcase, Hein, Lexis, and Westlaw. Several other apps do nothing more than provide convenient mobile access to the full text of various federal and state codes, regulations and court rules. One of these apps, called rulebook™, is the exclusive mobile source for *The Bluebook* legal citation manual. Trial lawyers can select from a variety of apps that automate preparation of trial exhibits, *voir dire,* and graphical trial presentations. Several developers have produced apps that help lawyers review and annotate electronic transcripts. Smart Dockets is a date calculator app that takes into account legal holidays, and "knows" the court rules for calculating dates in a number of state and federal jurisdictions. Many more examples could be given, but it makes little sense to compile a thorough list of apps for lawyers in this chapter. By the time you read this book, many of the apps you'd find on today's list will have been rendered obsolete by better ones, and many newer, more innovative apps will no doubt be available.

MOBILE DEVICES IN THE WORKPLACE

Another set of challenges facing the law librarian engaged in supporting mobile devices in his or her workplace is ensuring that these devices function reliably both inside and outside his or her institution's network. Once you've succeeded in introducing your patrons to mobile devices, they are certain to expect ubiquitous access to resources, both in the office and at home. This is a central aspect of these devices' lightweight portability. Any patron who has found relevant material on a mobile device at home will want to save this material for access at work, or vice versa, and will also want to share findings with colleagues in the workplace. Simple as this may sound, in practice it is often complicated by technical issues that involve network IP authentication and security restrictions associated with a library's subscription resources.

In the era of the desktop PC, enterprise data security was tightly controlled through uniform computer hardware and software running on a firm's local area network. Microsoft's Windows operating system, its Office suite of productivity applications, and its Exchange e-mail servers were all developed with an eye toward data security. Microsoft's offerings thus found wide acceptance by enterprises. As users began to demand access to corporate information on PCs and laptops at home and from other remote locations, enterprise turned to virtual private networks. When cell phones caught on, BlackBerry's devices were embraced as the ones most friendly to an enterprise's data security concerns. With the dawn of the ubiquitous mobility era, enterprises such as law firms face the challenge of offering their users secure access to corporate files over a wide variety of mobile devices. Enterprise has now turned to mobile device management (MDM) as the way to meet user demands. Modern MDM proceeds from the recognition that uniformity of hardware and software is no longer a reasonable expectation. It adopts a BYOD approach—"bring your own device." Law firms are now rolling out MDM software that allows IT administrators to select the apps that are visible in app stores and can be downloaded to mobile devices. MDM software lets them manage all devices in one interface, easily wiping data when a user exits the firm. MDM also allows IT managers to set and deploy enterprise-wide policies, offering a level of security to protect client information on mobile devices.

The many technical issues that surround network security, IP authentication, and MDM are ordinarily handled by an institution's IT department, rather than its librarians. But IT personnel are more comfortable troubleshooting issues that involve personal computers instead of mobile devices. And only the most knowledgeable of IT staff have even the slightest appreciation of how information resources are used by an institution's legal professionals. This is the librarian's realm of expertise. Any librarian who remains ignorant of technical issues involved in network support does so at his or her own peril. It is the peril of becoming irrelevant to clientele.

To stay on top of developments in the ever-growing-and-changing mobile device industry, the law librarian should turn to websites and blogs that support current awareness about this industry. Table 9.1 lists ten excellent sites you can use today, though keep in mind that these are likely to change over time. Favorites include LLRX.com, which is an excellent site dedicated to providing up-to-date information on a wide range of Internet research and technology-related issues, applications, resources, and tools. Founded and edited by a law librarian, it offers a unique perspective. Updated frequently and hosted by a practicing lawyer, the iPhone J.D. blog is chock full of news and detailed information, with a strong slant toward Apple's iOS devices, the iPhone and the iPad. Product reviews and posts are lavishly illustrated with screen shots and product photos. The TechnoLawyer Community blog follows the mobile device industry from a practice management point of view. This is the site where you're likely to find the most thorough coverage of tablets and apps that run on the Android and Microsoft platforms. In her Dewey B. Strategic blog, law firm librarian Jean O'Grady consistently demonstrates a knack for nailing the most significant practice innovations. And for a well-rounded understanding of mobile devices for lawyers, you should pay regular visits to Jason Wilson's rethinck postings. Wilson's blog is a great place to discover fresh, independent, and creative thinking about all aspects of legal publishing. Wilson may not blog regularly about mobile devices and apps, but when he does you can count on posts that are fresh, original, and thought provoking.

Table 9.1. Ten Current Awareness Sources for Law Librarians

All Things D	http://allthingsd.com
Court Technology and Trial Presentation Blawg	http://trial-technology.blogspot.com
Dewey B. Strategic	http://deweybstrategic.blogspot.com
iPhone J.D.	http://www.iphonejd.com/iphone_jd
Law Technology News	http://www.law.com/jsp/lawtechnologynews/index.jsp
LLRX.com	http://www.llrx.com
Pogue's Posts	http://pogue.blogs.nytimes.com
rethinck	http://www.jasnwilsn.com
TechnoLawyer Community	http://www.technolawyer.com
WalkingOffice: Mobile Solutions for Legal Professionals	http://www.walkingoffice.com

MOBILE DEVICES VS. PERSONAL COMPUTERS: ADVANTAGES

No doubt the best way for librarians to begin learning how to support mobile devices in their work setting is by simply making use of the devices themselves. Most law librarians' experience of mobile devices proceeds from their own experience of the personal computers they already use, and this is also true of the legal professionals that librarians serve. Indeed, one of the principal purposes for which mobile devices are now being used or considered for use in legal practice, according to postings in websites and blogs, is as a desktop or laptop replacement. This is particularly true of tablets. Many librarians may be fortunate enough to work for institutions that provide them with a tablet for use at work. If you are not this fortunate, but still affluent enough to own your own tablet, you are well advised to make the most of it in learning about your patrons' use of tablets. Understanding the utility of a tablet to a legal professional begins with a comparison of the tablet experience versus the personal computer experience. It can be useful to divide this comparison into advantages and disadvantages.

Certainly the most obvious advantage that mobile devices hold over personal computers is their lightweight portability, but the advantage doesn't stop there. In the evolution of modern mobile devices, tablets grew from smartphones, much as smartphones grew from cell phones. Thus voice input along with audio output are essential ingredients in the genetic makeup of today's mobile devices. Similarly, as cameras came to be common features in early cell phones, so photography is deeply ingrained in mobile devices' DNA. This combination of lightweight portability, along with voice input and photography, is an important factor in making mobile devices attractive to legal professionals. A lawyer can easily carry a tablet into a pretrial conference, snapping photos of opponents' evidence exhibits, recording voice notations about what she sees, and reviewing and annotating deposition documents, all the while checking e-mail, texting her dinner date, and taking occasional Angry Birds breaks. Plus, if the tablet the lawyer brings to the conference is one of the newest models, with a high-resolution screen and all the latest bells and whistles, she's likely to exude coolness to everyone in the room, maybe even the judge. In this and similar situations, no laptop or netbook computer can hope to compete with a mobile device when it comes to utility and versatility, not to mention confidence building.

MOBILE DEVICES AS E-READERS

Another area where mobile devices stand out over personal computers is in their function as e-readers. When it comes to e-reading, all mobile devices are definitely not created equal. One important aspect of any e-reader's utility is screen size. Mobile device screens range in size from 3.5 inches on an Apple iPhone to 14 inches on a Windows Home tablet from Acer. The Apple iPad's 9.7-inch screen may be seen as a standard for tablets, while the recently introduced iPad Mini's screen comes in at 7.9 inches. Tech pundits have spilled oceans of ink and launched torrents of pixels mulling over which size is best overall. But there's general agreement that devices with larger screens—that is, tablets—deliver a better e-reading experience than do smartphones. Even the largest smartphone currently on the market, the Samsung Galaxy Note III, with its 6.3-inch screen, may be too small to deliver an optimal reading experience. Some smartphone users are also put off by other aspects of the largest devices. Tech guru Walt Mossberg has likened using the Galaxy Note as a phone to "talking into a piece of toast."

Shoppers in the e-reader market are faced with yet another daunting choice, namely the choice between LCD tablets and E Ink tablets. Tablets that evolved from smartphones with LCD displays, such as the iPad, developed more or less simultaneously with E Ink tablets from Amazon, Barnes & Noble, and Sony, which were initially designed as devices just for reading e-books. At first the E Ink devices were praised as superior e-readers because they were easier on people's eyes than LCD screens, used less battery power, and could be read in bright sunlight. But when Apple introduced the iBooks app in conjunction with its launch of the iPad, many readers found that their reading experience on the tablet was more than adequate, bright sunlight notwithstanding. And as time has passed, the distinction between the two types of tablets has blurred, with LCD tablet makers rolling out e-reader apps and e-bookstores, while the E Ink tablet makers have embraced apps, such as Angry Birds, that have nothing to do with e-reading. LCD tablets now compete head-to-head in the market with E Ink tablets, and both Amazon and Barnes & Noble have introduced versions of their Kindles and Nooks with LCD screens.

In observing the market of e-reader devices and assessing various alternatives, a law librarian must take into consideration the particular reading habits and propensities of her clientele. In any legal institution, the librarian should be well positioned to make assessments of this sort. E-readers, whether found in LCD or E Ink devices, smartphones, or tablets, are typically oriented toward a broad-based model of reading that targets the monograph, such as a novel or nonfiction work. But this is not the type of book that law library patrons are likely to be using in most professional reading situations.

Astute law librarians recognize that the e-reading needs of their patrons do not conform to a broad-based model. Legal professionals will most often be consulting materials such as cases, statutes, court rules, regulations, treatises, and current awareness sources. Sources of this sort require sophisticated e-reading tools. In contrast to general-purpose users, legal professionals are far less likely to be reading their sources from cover to cover. They are better served by e-readers that offer sophisticated tools for navigating to relevant sections within legal sources and then reading, printing, copying, pasting, and hyperlinking from those sections to relevant sections in other legal sources. The best e-readers for legal professionals support bookmarking, highlighting, and annotating of sources. Patent lawyers need an e-reader that displays graphic diagrams. Trademark lawyers need color. Lawyers of all stripes want an e-reader that can clearly display charts and tables for various purposes. Websites and Internet browsers have been engineered to handle all these retrieval and display functions very well, and modern lawyers are habituated to using an Internet browser for their professional information-seeking tasks. The same cannot be said of the e-book readers found in mobile devices. Of course, just about all mobile devices also support some form of Internet connectivity, and so a browser can usually substitute in situations where an e-reader is not up to the task of retrieving or displaying the information a user needs on a mobile device. But librarians must keep all these factors in mind when recommending and supporting the use of an e-reader to their patrons.

Once a lawyer has discovered relevant or persuasive material on her tablet's e-reader, her next task often is to cite that material, be it a statutory provision, quote from a court opinion, treatise section, or law review article's analysis. This is another task in which the law librarian is likely to become involved. By almost universal custom, citations in legal writing must conform to the rules set out in the *Bluebook,* currently in its nineteenth edition. Rule 18 of this edition, which is devoted to the Internet, electronic media, and other nonprint resources, offers no guidance for citing material found in an e-book. Indeed, this edition of the *Bluebook* makes no reference to e-books whatsoever. For e-book content that duplicates a print source, lawyers can follow the time-honored tradition of citing material they read in electronic format as though they'd read it in print. But in situations where the e-book is the exclusive source for the cited material, the legal writer must analogize her citations. Law librarians, many of whom are accustomed to assisting their patrons in confronting the often intractable problems involved in conforming citations to *Bluebook* form, must prepare themselves to meet the challenges associated with citation of sources found in an e-book on a mobile device.

MOBILE DEVICES VS. PERSONAL COMPUTERS: DISADVANTAGES

The advantages that mobile devices hold over personal computers in realms of lightweight portability and e-reader facility must be weighed against a variety of perceived disadvantages. These disadvantages can be summed up with the single aphorism that mobile devices in general are oriented toward the consumption of media, rather the creation of media. To the extent that legal professionals are engaged in the creation of media, in the form of briefs, memoranda, opinions, contracts, client letters, and an endless variety of other documents, this aphorism would lead to the conclusion that mobile devices do not represent tools of choice.

One obvious disadvantage resides in mobile devices' input mechanisms. While these devices are typically well suited for receiving voice input, they are relatively poorly suited for text input from a keyboard, the type of input mechanism with which legal professionals are most comfortable. Apple's introduction of the iPhone in 2007 was met with widespread skepticism about whether consumers would accept a method of text entry that involved touching glass. Cell phones produced by rivals such as BlackBerry had physical keyboards with keys that, while small, gave a comfortable and familiar resistance to the touch. Although glass-touch keyboards are now accepted as a standard in all sorts of mobile devices, it is widely acknowledged that most people do not find this type of keyboard optimal for the creation and editing of lengthy and complex text documents. Many mobile device users purchase an add-on physical keyboard that connects to their tablet by cable or wirelessly using Bluetooth. Keyboards such as these can serve as workarounds to make mobile devices more acceptable as appliances for media creation.

Another drawback that inhibits the acceptance of mobile devices by legal professionals is the fact that smartphones and tablets were not originally designed for creating, editing, storing, organizing, and printing documents. Witness the fact that three years after Apple's launch of the iPad, Microsoft has yet to release an app version of its popular Office suite, which dominates the legal market. The latest mobile devices can be fitted with apps that accommodate storage and sharing of files, as well as printing capabilities. But most of these devices do not come equipped, out of the box, to perform these sorts of functions, which are basic and essential to a legal professional's role in media creation. Even the simple task of sending a print job from a mobile device may involve purchase, installation, and configuration of a print app. You have to make sure to choose the right app for the printer you want to use. And older printers—that is, printers more than two or three years old—may not work at all. Most legal professionals will benefit from advice and instruction in fitting and configuring their mobile devices with apps for use in media creation, and in many workplaces the law librarian is well positioned to play the role of advisor and instructor.

Luckily, the support of mobile device apps is not a terribly daunting responsibility, thanks to the very nature of the app. We all know the catch phrase, "There's an app for that." Apps are wonderful little things, ingenious in their conception, each one serving its own little purpose. Thanks to this, few of them fall prey to feature glut, of the type found in personal computer programs such as Microsoft Word. With most apps you can quickly learn all you need to know by perusing the developers' descriptions and screenshots you find in the app store. App stores also compile user reviews, which can be helpful in deciding which ones are best to use and to recommend. The quality of an app is a direct function of how much it gets used. The best ones will be used all the time; the worst ones quickly forgotten. When you recommend an app to a lawyer in your workplace, he'll probably either "get it" almost right away, or he never will. If he gets it, you'll have done him a solid favor. If he doesn't get it, then neither of you will have lost much, since most apps are free or sold at a nominal price.

CUSTOM APPS FOR LAW LIBRARIES

A recent article posted to ALM's *Law Technology News* website described the development of custom apps for law firms as "the next big wave." This article listed several purposes for which law firms might consider building apps, including marketing, billing, and budgeting. The article went on to assert that many large firms already have the development talent in house to make their own apps. To the extent that this article is accurate, it reflects a growing acceptance of the app in law firms. The idea of a custom law firm library app was not mentioned in the article, but most firm librarians can probably begin to envision what their app might look like. If there is in fact a big wave of custom law firm apps now building, then the savvy librarian should start waxing down her surfboard.

If there is a trend toward development of custom law library apps in law firms, that trend is in its pre-infancy. One searches in vain the professional literature and the shelves of the app stores for one such example. There are, however, segments of the library world for which custom mobile apps have been successfully developed. These segments are the academic and the public library. The New York Public Library, for example, has a smartphone app that supports catalog searches, allows users to manage their circulation accounts, and also features book reviews, and locations and hours of branches. UCLA's library offers a smartphone app that searches the catalog and also provides information about library hours, contacts, and laptop availability at campus locations. Similar sets of features can be found in free apps offered by several other public and college libraries. Such apps start with features found in the library's online public access catalog (OPAC), and deploy them in mobilized form. The apps have been developed primarily for use on a smartphone, and while they can be used on a tablet, the design of their features does not take advantage of a tablet's larger screen. In contrast to Web-based OPACs running on personal computers, apps of the current generation do not support linking from the mobile OPAC to the library's electronic subscription resources, which would give a user seamless access to articles, e-books, and databases on the mobile device. One can expect that features offering better integration of electronic resources will arrive in the next generation of library apps, to satisfy information seekers in a world where ubiquitous mobility has become an expectation.

CASE STUDY: MENDIK MOBILE

In the summer of 2011, a team of librarians on the staff of New York Law School's Mendik Library in Manhattan began planning a custom library app for our institution. One of the first questions we considered was whether to try to develop our app in house or outsource it to a contractor who specialized in building apps for libraries. No one working in our library had any experience in app programming, nor did anyone on the staff of New York Law School's Office of Information Technology. We concluded that in-house development of our app was not a feasible option. We also considered AirPAC, the out-of-the-box library catalog app marketed by our integrated library system (ILS) vendor, Innovative Interfaces, Inc. (III). But we found the features of AirPAC to be too limited for the app we had in mind, and the III option was also quite expensive.

So we turned our attention to the services offered by two contractors, BiblioCommons and Boopsie, who are responsible for the development of many of the custom apps currently offered by public and academic libraries. After reviewing these contractors' proposals, we decided to go with Boopsie. This contractor had extensive experience in developing the type of app we wanted. Their proposal assured us that Boopsie's development team would work closely with our staff to incorporate features that we specified into our app. Boopsie produces versions of its library apps that work on all major mobile platforms, including Apple's iOS, Google's Android, and BlackBerry. This was an important consideration in light of the variety of mobile devices used by our library's patrons. An even more important consideration was the fact that Boopsie had successfully developed several libraries' apps that were powered by the OPAC of our ILS vendor, III. As public and academic library apps are all freely available at no cost, we were able to demo a variety of the apps that Boopsie had produced. In general, we liked what we saw. And, most important of all, the cost of Boopsie's proposal fit well within our budget.

Development of a Boopsie app starts with a set of standard packages of features that the developer offers to all its library clients. Boopsie refers to these feature packages as "channels," and they display on the home screen of the app. Many of Boopsie's standard channels were obvious choices for our app. These included a channel for searching our OPAC, a channel that provides information about library hours and links for contacting library staff, and a channel that enables users to access and manage their circulation account. Other standard Boopsie channels were irrelevant to us, and these were dropped out of our development set. For example, a channel for locating and getting directions to library branches is useless to a library such as ours, which is housed in a single facility. Beyond the standard set of channels, Boopsie will work to create custom channels and offer features specified by its clients. One of the custom channels we requested was a directory made up of tables that indicated on which floors of our library particular collections and call number ranges could be found. We provided the directory information to Boopsie, and their staff built out the channel, which displays well on a smartphone screen.

A benefit of working with a contractor such as Boopsie to develop your app is their ability to build out complex custom features to meet your particular requirements. For our app, we had a specific set of course reserve features in mind that were not found within Boopsie's standard channels, nor had we seen these features in any of the apps we reviewed that Boopsie had developed for other clients. In our library, students make heavy use of books that faculty place on reserve for their courses. Reserve books circulate for short loan periods from behind our reserve desk, and students are constantly reviewing circulation records to determine when borrowed reserve books are due to return. Students also make frequent use of holds and renewals for reserve books. Our III OPAC has a Course Reserves module that handles all these functions. The OPAC also enables students to search for course reserve books by course and professor name. In planning our app, we felt it imperative that all of these course reserve features be made available to our mobile users. So, in response to our request, Boopsie custom built the features into a Course Reserves channel for our app. Our students' ability to access and manage course reserves remotely on their mobile apps is a major convenience. Indeed, we like to think of Course Reserves as the "killer channel" that proves the core value of our app, much as early computer programmers termed programs that were central, necessary, and desirable to the technology they were developing as "killer apps."

Once all the features were selected and the channels built out, we worked with Boopsie to brand our app and make it available to our users. The app launched in early October 2011, slightly more than four months after we began planning it. We named it Mendik Mobile, in honor of the major donor for whom our library is named. Boopsie helped us get Mendik Mobile listed in the iTunes app store, as well as in the app stores of other mobile platforms. Boopsie also hosts a mobile website that will automatically detect a device's platform and download the appropriate version of the app from Boopsie's own servers. We have invested significant effort in publicizing Mendik Mobile to our library users, and this effort was rewarded when our library won the 2012 award for Excellence in Marketing: Use of Technology from the American Association of Law Libraries. Our app has grown in popularity since its launch; it is currently logging over one hundred unique users most months when classes are in session.

Librarians considering a custom law library app for their workplace may view the story of Mendik Mobile a model for the successful development of this kind of app. But if you are contemplating such a project, you should take heed of certain caveats. Our library staff recognizes that our beloved app is not a portal to all our library's resources and services, and we do not promote it as such. Nor is Mendik Mobile a comprehensive research tool. It is not even an especially sophisticated search engine for our OPAC. It runs just fine on a smartphone, but having been developed for smartphone devices, it does not take advantage of the larger screen when running on a tablet. And as is true of all other current generation apps for academic libraries, it does not provide on a mobile device the level of electronic resource integration that can be achieved on an ILS running on a personal computer. Our librarians are well aware of Mendik Mobile's limitations, and you should be too, since you're likely to encounter a similar set of limitations if you begin development of yours today.

CONCLUSION

The story of mobile apps in law libraries is just beginning. The best apps, with the "killer features," are yet to be developed. Law librarians have exciting developments to look forward to in the next decade, and onward. Successful librarians will get out ahead of the trends in the mobile device and app markets for legal professionals. They will do so by staying on top of the current awareness sources that report these market trends. They will pay close attention to the devices and apps their patrons are using, as well as to the reading and information-seeking habits of these patrons, just as they've always done. And, perhaps most important of all, successful law librarians will take care that their employers remain aware of their special value to the work of legal professionals, as we move on into the world of ubiquitous mobility.

FURTHER READING

The Market for Mobile Devices and Apps

Black, Nicole L. January 5, 2013. "2012: The year of the mobile lawyer?" *LLRX.com*. http://www.llrx.com/features/2012theyearofmobilelawyer.htm.

Fiegerman, Seth. September 11, 2012. "Mobile app market to almost double this year to 45 billion." *Mashable*. http://mashable.com/2012/09/11/45-billion-apps-downloaded-201/.

International Legal Technology Association. 2012. "2012 ILTA/InsideLegal Technology purchasing survey." http://insidelegal.typepad.com/files/ILTAInsideLegalTechnologyPurchasingSurvey2012.pdf.

Musil, Steven. April 24, 2012. "Tablets expected to become preferred computing device." *CNET News*. http://news.cnet.com/8301-13579_3-57420541-37/tablets-expected-to-become-preferred-computing-device/.

Mobile Apps for Legal Professionals

Barrett, Josh. March 8, 2012. "Just two." *TabletLegal*. http://tabletlegal.com/just-two/.

———. May 2, 2012. "Better, cranking and quitting," *TabletLegal*. http://tabletlegal.com/cranking-quitting/.

Brooks, Ted. August 6, 2012. "Best LinkedIn groups for attorneys and legal professionals (part 1)." *Court Technology and Trial Presentation Blawg*. http://trial-technology.blogspot.com/2012/08/best-linkedin-groups-for-attorneys-and.html.

———. August 13, 2012. "Best LinkedIn groups for attorneys and legal professionals (part 2)." *Court Technology and Trial Presentation Blawg.* http://trial-technology.blogspot.com/2012/08/best-linkedin-groups-for-attorneys-and_13.html.

Dean, Rob. November 12, 2012. "Top 6 free time-saving apps for lawyers." *WalkingOffice Blog.* http://www.walkingoffice.com/applications/top-6-free-mobile-apps-for-lawyers/.

Lawson, Emily. September 26, 2012. "Good apps aren't hard to find: Resources for finding legal apps." *AALL Spectrum.* http://www.aallnet.org/main-menu/Publications/spectrum/Spectrum-Online/legal-apps.html.

McNichol, Tom. May 2012. "In tablet form." *California Lawyer.* http://www.callawyer.com/clstory.cfm?eid=922126.

Squillante, Neil J. October 30, 2012. "A special report on tablets in the legal industry: iPad, iPad mini, and Surface with Windows RT." *TechnoLawyer Blog.* http://blog.technolawyer.com/2012/10/tablets-legal-industry.html.

Taylor, Jeffrey. November 9, 2012. "Tablets and the legal industry: The Android angle and a rebuttal." *TechnoLawyer Blog.* http://blog.technolawyer.com/2012/11/android-tablets-legal.html.

Mobile Devices in the Workplace

Brooks, Ted. August 22, 2012. "Print from iPad, iPhone, Android phone & tablets." *Court Technology and Trial Presentation Blawg.* http://trial-technology.blogspot.com/2012/08/print-from-ipad-iphone-android-phone.html.

Davis, Michael. November 28, 2012. "Why mobile device management isn't enough," *InformationWeek.* http://www.informationweek.com/global-cio/interviews/byod-why-mobile-device-management-isnt-e/240142450.

Weiss, Todd. May 30, 2012. "Mobile device management: Getting started." *Computerworld.* http://www.computerworld.com/s/article/9227346/Mobile_device_management_Getting_started.

Mobile Devices vs. Personal Computers: Advantages

McKenna, Frank. June 2012. "The mobile lawyer, knowledge gathering in a connected world." http://www.lexisnexis.com.au/lexisnexisred/downloads/the%20mobile%20lawyer.pdf.

Mobile Devices as E-readers

Mossberg, Walt. February 15, 2012. "Mobile device that's better for a jotter than a talker." *All Things D.* http://allthingsd.com/20120215/mobile-device-thats-better-for-a-jotter-than-a-talker/.

Stern, Joanna. November 12, 2012. "iPad Mini review: The perfect size comes with some sacrifice." *ABC News, Tech This Out*. http://abcnews.go.com/Technology/ipad-mini-review-perfect-size-sacrifice/story?id=17699824.

Mobile Devices vs. Personal Computers: Disadvantages

Madden, Brian. October 15, 2012. "8 reasons why Windows 8 on a tablet won't fix our legacy Windows desktop apps problem." *Brian Madden's Blog*. http://www.brianmadden.com/blogs/brianmadden/archive/2012/10/15/8-reasons-why-windows-8-on-a-tablet-won-t-fix-our-legacy-windows-desktop-apps-problem.aspx.

Warren, Tom. November 7, 2012. "Exclusive: Microsoft Office for iPhone, iPad, and Android revealed." *The Verge*. http://www.theverge.com/2012/11/7/3612422/microsoft-office-mobile-ipad-iphone-android-screenshots.

Custom Apps for Law Libraries

Martin, Sean. November 9, 2012. "'There's an app' for law firms." *Law Technology News*. http://www.law.com/jsp/lawtechnologynews/PubArticleLTN.jsp?id=1202577816049&Theres_an_App_for_Law_Firms&slreturn=20130111114125.

Case Study: Mendik Mobile

Boopsie, Inc. "Case study, the Mendik Library New York Law School." http://www.boopsie.com/library/the-mendik-library-new-york-law-school/.

Innovative Interfaces, Inc. "AirPAC." http://www.iii.com/products/airpac.shtml.

RECOMMENDED READING

International Legal Technology Association. "2012 ILTA/InsideLegal Technology Purchasing Survey." http://insidelegal.typepad.com/files/ILTAInsideLegalTechnologyPurchasingSurvey2012.pdf.

LexisNexis. "LexisNexis eBooks, frequently asked questions." http://www.lexisnexis.com/ebooks/questions/.

Matamoros, Alex Berrio, and Mary Ann Neary. September 26, 2012. "Librarians, legal research, and classroom iPads—A winning combination." *AALL Spectrum* 17 (1). http://www.aallnet.org/main-menu/Publications/spectrum/vol-17/No-1/ipads.pdf.

McKenna, Frank. June 2012. "The mobile lawyer, knowledge gathering in a connected world." http://www.lexisnexis.com.au/lexisnexisred/downloads/the%20mobile%20lawyer.pdf.

Weiss, Todd. May 30, 2012. "Mobile device management: Getting started." *Computerworld.* http://www.computerworld.com/s/article/9227346/Mobile_device_management_Getting_started.

10

The Law Library Website

C. Andrew Plumb-Larrick

THE JUDGE BEN C. GREEN LAW LIBRARY, CASE WESTERN RESERVE UNIVERSITY

Law library websites are a core part of delivering library services in a digital age. For law school libraries, a platform to deliver content and services primarily to student and faculty patrons of the library is best thought of as a "website," open to the public, rather than an intranet. Even though the primary target audience is well defined and fairly close, that audience is still diffuse enough to be served best through publicly viewable websites with only select services or resources behind authentication barriers. And, besides, revealing the full site to a wider public (and even devoting some level of effort to optimizing for web search engines like Google) is likely the best advertisement for the law school of its library offerings. Indeed, the overwhelming majority of academic libraries *do* strive to maintain extensive, public-facing websites. Law firm libraries, however, do not have full websites, are unlikely to have more than nominal presence on the firm's website, and the tools they provide for services to their patrons will all be visible only internally. The actual target audience for the firm's web presence is law firm clients, for whom library services are behind-the-scenes infrastructure. Public, county, and subscription law libraries do tend to have public-facing websites. While most observations pertaining to the law school library website will apply, such libraries tend to be even leaner on technical staff and resources and may have fewer electronic content databases and online-oriented services to integrate through a website.

Law librarians in a digital age need to be equipped to deliver library content and services online, through the Web. An increasing proportion of the content to which libraries subscribe is digital content delivered through the Web. While these resources are presented through their own online platforms, management of access to and integrated discovery of these resources (especially with the broader patron audience of the academic libraries) is easiest through a well-designed library website. Many law library patrons are also accustomed to interacting with service providers online and carry this expectation to their library interactions. Tools ranging from contact-form submissions, to chat reference, to easy e-mail contact, to online research guides and tutorials can provide the form of remote library contact that the digital native or digitally inclined patron expects. A law library website serves as the "connective tissue" that knits together access via both traditional catalog-based and alternative finding tools to subscribed or owned *content*; any online *services* delivered by the library or content self-published by the library; and service or policy information *about* the library.

This chapter details models of both administrative and technical support for the website that are frequently encountered in law libraries. Most of the attention is devoted to academic libraries as they are numerous, visible, and generally the best-equipped law libraries for deploying an online, Web-based, presence. Administratively, few law libraries have their own staff positions dedicated solely to website design or development, though librarians tasked with "electronic services" in general may have significant Web-oriented responsibilities as part of their duties. Even law schools as a whole are likely to have relatively few staff, and few formal roles, involved in website production—at least in proportion to the relative complexity of the websites. An early subsection of this chapter briefly introduces some of the stakeholders in and typical contributors (outside of the library) to law school websites, and discusses ways that successful library website efforts have interacted with these nonlibrary contributors.

First, we look at some of the typical functions and purposes of law library websites. The uses we encounter in practice can broadly be divided into three categories: provision of textual (sometimes visual) information *about* the library and its collections or services; online delivery of tools for resource discovery (catalogs, discovery layers, database finders, "A to Z" lists); and contact points for the direct delivery of library services in either an immediate sense (form submissions, appointment requests, chat reference) or at a slight remove (research guides, online tutorials, major-treatise lists). We begin with a closer look at how some library websites provide functionality within each of these three categories.

INFORMATION ABOUT THE LIBRARY

Library information is not always permanent. The library website is a prime location for the display of news and notices related to special events, calendar changes, or other announcements related to the library or its users. Such announcements can be displayed as text information. Some libraries make good use of visually appealing banners to present changing, up-to-date, information.

In addition to accessibly presenting the overall calendar or schedule for the library, the library should have some means of prominently announcing holidays, closures, and other interruptions to regular service. Delay or friction in posting and changing the library hours are a source of frustration mentioned by many librarians whose staffs lack the ability to quickly make changes to website content without the involvement of webmasters outside of the library. One advantage of the trend to dynamic content management systems (CMSs) for institutional websites is the ability to devolve the deployment of such frequently altered editorial content to the staff people responsible for generating the information.

The website may include information about specific library personnel, especially those (such as reference librarians) in key visible roles and perhaps also those in less inherently visible roles that should be highlighted and acknowledged.

Although all law libraries should prominently include basic contact information for key services of the library (e.g. "reference help" or "circulation desk" phone numbers or e-mail addresses), most also do include at least basic staff contact lists. Additionally, full bio pages for the professional librarians and possibly for other key personnel can help patrons understand the library and, if photographs are included, match faces to the names they might see on e-mail or other library communications. Where individual librarians are given some control over their individual profile pages, they may choose to use them for purposes that include pushing information out to students in particular courses they teach, or for including calendar or office-hours information. For example, some of Yale's reference librarians embed Google Calendar information into their profiles.[1]

Many law libraries use at least one social media platform. More than half of all academic law libraries have a presence on at least one social media platform, and use of Facebook and of Twitter are each reported by nearly half of the libraries.[2] Integration of a library's social media presence with its website moves in (at least) two directions: the social media profile and feed should "highlight" and commonly refer back to the conventional website, and the website should itself promote the library's social media mission. Social media outreach can be represented on the website through techniques ranging from a simple badge or image on the front page to insertion of social media posts or statuses into the website, usually by means of an RSS feed or other technique for syndication. Conversely, it may be useful to syndicate frequently updated website content (such as blog posts) onto social media platforms. Accessible, commonly used tools to cross-syndicate include Facebook's own "social plug-ins" to feed Facebook content onto a blog or other website. Networked Blogs.com is a service that allows blog-like content to be easily placed into a Facebook feed.

Library websites may also be organized to present certain clusters of information to certain defined subgroups of users. Specific pages, or submenu structures, for students or for faculty members are common.

ONLINE TOOLS FOR RESOURCE DISCOVERY

The business of libraries is the business of resource discovery. The largest collection is of little use without a means to locate the resources of most usefulness and interest to the research query. Most of the professional work in a typical library is focused on supporting information discovery, as is the "traditional" technology deployed in libraries. Discovery systems from cards and files to early online public access catalogs (OPACs) to the most recent "discovery layers" have long been the major technical systems that have in many respects defined and structured the work of librarians.

Modern "integrated library systems"—the automation tools driven by the process alluded to above—all include an online-accessible catalog interface delivered through Web protocols: the WebPAC. Clear access—at least through some kind of embedded search box interface—to the WebPAC and any discovery layer software in use is a key requirement for the library website. But while WebPACs can be powerful tools, they probably seem most powerful to those of us in this profession that has traditionally cohered around the infrastructure that makes them work. It is outside of the scope of this chapter to ask how completely Web search engines and related tools have disrupted the market for the bibliographically rigorous library catalog and the processes that support it, but it is clear that other tools are required.[3]

DATABASE FINDING/LINKING TOOLS

Law library websites commonly offer tools to browse or search for databases or other online resources made available by the library. Here, we are not referring to a search across material within various subscribed databases, but rather to website tools to sort, describe, and link to some set of online resources provided through the library at the "platform" level of granularity. This is also distinct from title-level cataloging of resources that exist within subscribed databases. It is instead a more specific tool to help library patrons locate the various online database platforms and interfaces provided through or selected by the library.

These vary from simple lists of key database platforms provided for the library to more sophisticated tools that allow information about databases to be sorted or displayed according to certain criteria, or allow searching across the titles, and perhaps short librarian-authored profiles, of the included databases.

The Georgetown Law Library lists frequently used databases on its main "Research" page (one hierarchical level below the front page) while providing links to access the full list organized by either "title" or "subject." When either item is selected, the user leaves the main, CMS-delivered, website. The full list is delivered on Innovative Interfaces' Encore platform with data from Innovative's ERM (electronic resource management) module.

Case Western's Judge Ben C. Green Law Library's databases page, linked as a feature item directly from the front page, allows short descriptions of subscription databases to be located alphabetically or by "type," as well as providing scoped search across short, librarian-written, descriptions. Here, the content, navigation, and indexing is all developed within the main website content-management system, in this case Wordpress. Other law libraries offer examples of sortable, navigable, database lists built on other CMS platforms, such as Drupal.[4]

TREATISE LISTS, RESEARCH GUIDES, AND TUTORIALS

"Major treatise" or edited topical "book" lists (if the overall web design succeeds in drawing patron attention to their presence) offer a technologically simple way to address one of the major weaknesses of the library catalog for certain very frequent user interactions. Try searching a subject heading like "Contracts" or even "Contracts—United States" in the WebPAC of any significant law library. The result set will be excessively large and will offer no way to distinguish works offering the broad, structuring guidance of the traditional treatise from other materials that might range from study guides to monographic treatments of a specific argument in the field. Theoretically, discovery layer algorithms with the sophistication to recognize and handle broad searches differently than narrow searches might offer one way around the problem, though the major discovery layer algorithms that actually exist do not, in fact, help with this particular problem. But another way is to draw patrons with such broad questions away from the catalog query altogether, using web design to lead them to librarian-authored content that is designed to help users find other resources efficiently.

Ideally, such tools provide direct user access to precisely the types of major treatments of relatively broad topics that are hard to locate successfully using a library catalog, while leveraging the selectiveness and guidance of reference librarians into an online anytime format. A number of academic libraries successfully deploy such lists, with examples including:

- Georgetown (http://www.law.georgetown.edu/library/research/treatise-finders/index.cfm)
- Virginia (http://libguides.law.virginia.edu/treatises)
- Michigan (http://libguides.law.umich.edu/book-lists)
- Case Western (http://lawlibrary.case.edu/treatises-by-subject)[5]

The resources included in such lists may be strictly limited to just single- and multivolume treatises (and their online, computer-assisted, legal research [CALR] platform, and equivalents) or may also list additional material types such as specialist databases or "looseleafs" and other print or online current awareness services on the subject. Each of the library websites mentioned above reflects slightly different assumptions about how target users are likely to search for and find the treatise lists, and in the other or secondary uses intended for these documents. For example, Michigan's print-style PDFs are easily distributed off-line (and were used that way for many years before the advent of the website), but are perhaps slightly less amenable to search-engine discovery than the semantically marked up HTML content used at Georgetown or Case Western. Such "treatise" lists can also be made to work well in both print and online, either by including a downloadable "print version," of the document alongside the HTML version, or (even better) by using a print cascading style sheet (CSS) that would produce consistent, high-quality, printed output from such website pages. They are also a type of content that may merit a special style sheet for quality representation on smaller mobile devices, like smartphones.

Similarly, most academic library websites include at least a few topical research guides, though there is a lot of variety in what passes for a "research guide" on a law library website. The granularity of the topics addressed can range from quite broad ('International Law') to very specific ('International Intellectual Property Law'). The nature of the treatment can also range from a simple bulleted list of resources (not so different than what another library might characterize as a "treatise list") to narrative research articles on how to conduct research in the subject area under consideration. In selecting both a technical platform and a design principal for "research guides," you must know what levels or styles of treatment you anticipate, or select an approach that allows for wide variety. Many libraries use external products, like Springshare's LibGuides, to host their research guides.[6] This may be motivated by the need to work within administrative models that foreclose easy or quick direct authoring of website content by individual librarians, by a desire to easily impose a standard style and a user interface (UI), by a requirement for an author learning curve even gentler than those of the better general-purpose CMS platforms, or to gain some advantages in maintaining links and other updates that such platforms offer over general-purpose Web CMSs. In other cases, such products are specifically avoided in favor of the authorial flexibility that can be provided for pages of content within the law school's or library's own content management system or because they lack the flexibility for effective integration into the navigational architecture and user interaction design of the broader website.[7]

In addition to the text-based research guide, interactive tutorials may be valuable Web content. Harvard Law Library and Duke's Goodson Law Library are among a number of examples of libraries that integrate short video tutorials into their websites, consisting of narration over video capture of interactions with research tools such as database interfaces or photographs of printed material.[8] Georgetown Law Library hosts video tutorials that also integrate a review "quiz" at the end.[9]

These lists, guides, and tutorials are on the whole "low tech" but "high touch" applications. Each list or guide is a static document, manually updated on some modest frequency, and embodying the human guidance of a professional librarian without the intervention of any computational or algorithmic ranking or selection from a database. This is their strength. By design, the list must be closely tailored and highly selective. The very objective is to sidestep search—with its potential for high recall of results, but accompanying challenges in sorting those results optimally for the human researcher directing the query—in favor of a preselected set of results precisely tailored to a topic that the authoring librarian has also preselected and predefined. Such documents are real embodiments of librarian expertise. We write them largely with the beginner—either to legal research or to the specific topic at hand—most in mind, and the very point is to substitute our mediating expertise for either the assumptions baked into a novice researcher's query or into the weightings of a computer algorithm (or, into the interaction of the two). They also provide a tool for librarians, both within and between institutions, to communicate expertise among themselves.

When effective, guides, lists, and related publications are good outreach for the library. Well-structured HTML research guides, in particular, can feature very prominently in search engine results: The combination of quality guidance and advice, links to high-quality external resources, the favorable weighting of .edu domains, and (over time) inbound links from others who have relied upon their guidance can quickly drive an academic law library's well-written guide to Google's front page for queries on the topic.[10] These lists, guides, and tutorials are fully accessible to the outside world but also tailored to the particular portfolio of resources offered by the specific library. When of high quality and visibility, they are perhaps the best advertisement the library can make, on behalf of itself and its law school, to the broader Web-content-consuming world.[11]

DIRECT ONLINE SERVICE DELIVERY

In addition to materials like treatise lists, research guides, and tutorials that embed librarian expertise into the website for asynchronous (anytime) access by online patrons, law library websites can also include tools to facilitate more direct, and often real-time, interactions with the library and librarians. For example, Case Western uses a variety of contact forms to structure routine interactions with the library—such as, for a student to request a research appointment with a librarian or for a faculty member to place a request for delivery of any known (citable) item. These are direct service points in a way that research guides or tutorials are not, but they are still asynchronous. Another synchronous, or real-time, direct service that can be integrated into law library websites would be a box or window through which to initiate a "chat" with a librarian on duty. While Google's own website "badge" for initiating chat through a website was discontinued, there are a number of providers of "live chat" tools that can be embedded into websites and several examples of their use by law libraries.[12]

SITE SEARCH

The website should offer some capacity for site-specific search. No level of quality or consistency in navigational logic and information architecture of your website will substitute for the ability to search across the site. Many CMSs and other platforms on which your site may be built include built-in search tools. These tend to perform very badly. The major alternatives used by law library websites are enhanced-search "plug-ins" that essentially add their own search indexes to the CMS database and search functionality to the UI; standalone enterprise search engines (typically only relevant if you are also responsible for a large document library); or Google Custom Search (which uses Google's index of your website but presents a search box and result screens that can be tailored for site integration).

A NOTE ON THE DESIGN PROCESS

The library's Web presence needs to be tailored to specific, discernible, needs of a defined audience. Librarians tasked with website design projects need to be ruthless self-editors when it comes to what information site users actually need or want to see. The famous xkcd cartoon about university websites should be taken thoroughly to heart by law librarians.[13] Long descriptive passages about the library's history and collections are unlikely to merit prominent website placement. Ideally, user surveys and even actual observation of user interaction with test versions of the website should help to determine what information library users actually seek, and how and where on the site they expect to find it. Although such studies are often not practical, librarians do need to think about what we can do to ensure that we rigorously consider both what information about the library users will seek, and what organization and hierarchies in information presentation will make actual sense to our users.

But even once a clear user definition and use cases for the library's website are articulated, there can be a number of major trade-offs in implementation. Having a librarian available with some sense of web design and some general familiarity with interface design can be helpful here. Outsiders with expertise in information architecture, user-interface design, and related disciplines are also helpful—but the key is to find someone who can bridge the gap between the librarian's knowledge of the business purpose and role of the website and the Web professionals' understanding of both technical design principles and end-user interaction with Web interfaces.

Librarians may also have much to say about Web design for the entire law school, as there is an alignment of core professional competencies based around the organization of information. Information architects and user experience professionals may even have attended some of the same graduate information schools as librarians. It may be helpful for the library director to advocate that a librarian be included on any advisory committee that is created to work with an outside firm, and librarians may be particularly good at communicating with such professionals. In addition to providing valuable insights into the broader law school website, the librarian can raise issues specific to the library site and other "subsites" that deviate, at least in terms of audience, from the mission of the central law school website.

Assuming some level of autonomy in either a process of internal design or the provision of specifications to an outside designer, the library should focus on clarifying its intended Web users and their anticipated information needs. The library front page, in particular, should be designed to push forward the information that the library wishes to highlight. Depending on the particular design professionals used, the appropriate communication from the library of its Web design needs could vary from narrative descriptions of use cases to be supported, to collaboration on detailed charts and drawings or "wireframes" of the anticipated site design and flow or navigation from one page to another, to complete in-house development of the website.

ADMINISTRATIVE SETTING

Library websites typically must not only meet the needs of several different categories of library users but also harmonize in some fashion with the broader institutional Web presence of which they are but a part. The goal is to develop websites that can allow the librarians to easily provide their content and services online, while adhering to the brand images and design specifications of larger institutional websites. Although the specific discussion that follows is focused on academic law library websites, the broader challenge of creating a website that functions coherently in and of itself and also as part of a broader institutional Web presence is a challenge faced by other law librarians as well.

In a law school, the responsibility for the overall law school website is frequently divided between IT staff and marketing and communications staff. The latter may work as an administrative department or directly out of the dean's office. It is fairly common for IT departments to report ultimately to the administrator who also directs the law library, though even in these cases it remains likely that the law library and IT personnel work in somewhat separate silos below the level of the director or associate dean tasked with strategic leadership of both departments. To my knowledge, no marketing or communications staffs report to the same senior administrator (below the dean) who is responsible for the library. The result is that while the law school website is perhaps one of the IT-intensive functions of a law school that most closely aligns with "library" competencies and interests (it is, after all, about end-user discovery of information), it is one of the IT-intensive functions that is less likely to rest under the primary leadership of the chief law librarian.

In most cases, the law school website is now rightly recognized as a core part of institutional visibility and marketing that is critical to both student recruitment and ongoing alumni engagement, though the actual resources devoted to these websites and level of professionalism surrounding their administration may still lag behind this recognition of their importance. It is challenging enough to retain clarity of design and interface while entertaining the wishes of various administrative departments to reach out online. But the library may want not only a free hand to edit and arrange information about itself but also the flexibility to use the website to conduct library transactions with the existing students and faculty of the law school—that is, not the main audience to whom the overall site is even primarily targeted.

Many of the law library directors and other librarians with whom I communicated during the preparation of this chapter mentioned their difficulty making those responsible for the overall law school site understand the different mission and audience of the library site. It is a daunting design challenge to create a law library website that shares not only design branding but that may also be navigationally and logically integrated within a law school website that has a largely different audience than the library. How can a library's Web presence make sense as part of a law school website while still making sense as its own coherent whole website targeted to the existing patrons of the law school library?

Perhaps the first decision for law librarians—along with the deans, marketing professionals, and IT staff involved in the law school website—is how fully separate the law library website should be from the central website of the law school. There are at least three possibilities: complete integration of the site's user interface and navigation; complete separation (so that the main law school site merly "links out" to a law library website that is a fully autonomous design); or sharing of design elements with some limited use of site-wide navigation that still anticipates "subsites" with significant variety in internal organization. I've provided successful examples of each in figures 10.1 through 10.6. The first option works well enough for quite simple, informational, library Web pages. But it can either prevent the library from developing a full-featured website or result in convoluted, conflicting, or overly complex navigation within the library portion of the site. The second option is used by a number of successful academic law library websites—typically with menus and navigational architecture untethered from the law school website but with visual integration of colors, type, and other design elements specified by the law school or by central university administrations. This approach requires either a law library with its own resources to deploy and maintain a website, or a law school with central staff who have the capacity to support both a "core" law school website and a functionally distinct library website. The third option can be an effective approach where some integrating elements—perhaps a shared header containing high-level navigational tools or menus—can be adopted by not only the library but by other departments or projects that need their own websites (clinics, student journals, and other law school units or programs that may want to use the Web to reach different audiences and different use cases than the main central law school website). It is easier to integrate a single shared design and navigational element into a flexible subsite design than for a coherent central organizing principle of a core website to extend all the way down from a law school home page to, say, each research tip published by a librarian.

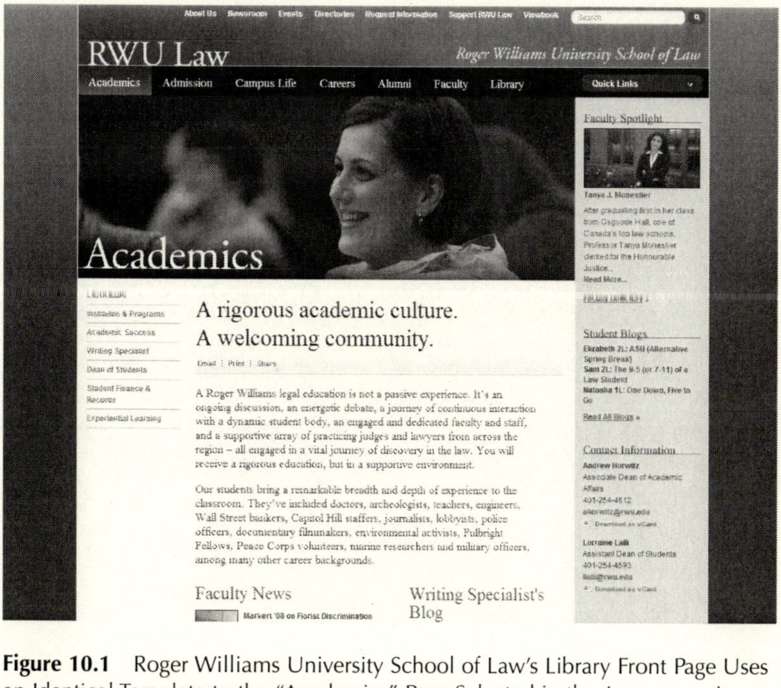

Figure 10.1 Roger Williams University School of Law's Library Front Page Uses an Identical Template to the "Academics" Page Selected in the Accompanying Screenshot. Note Particularly the Shared Top Menu and Structurally Identical Left Sidebar Submenu and Right Sidebar Features.

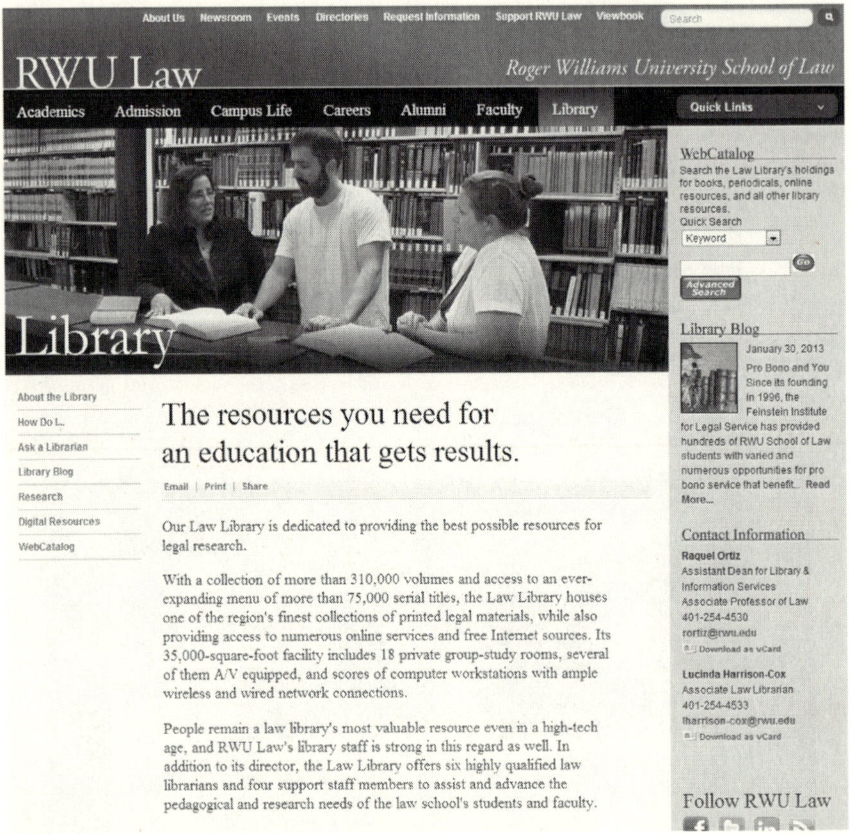

Figure 10.2 Pervasive integration Website Design Screenshot—Roger Williams University School of Law Library.

Figure 10.3 Yale's Law School and Law Library Websites Share Colors and Some Design Elements, but Are Entirely Separate.

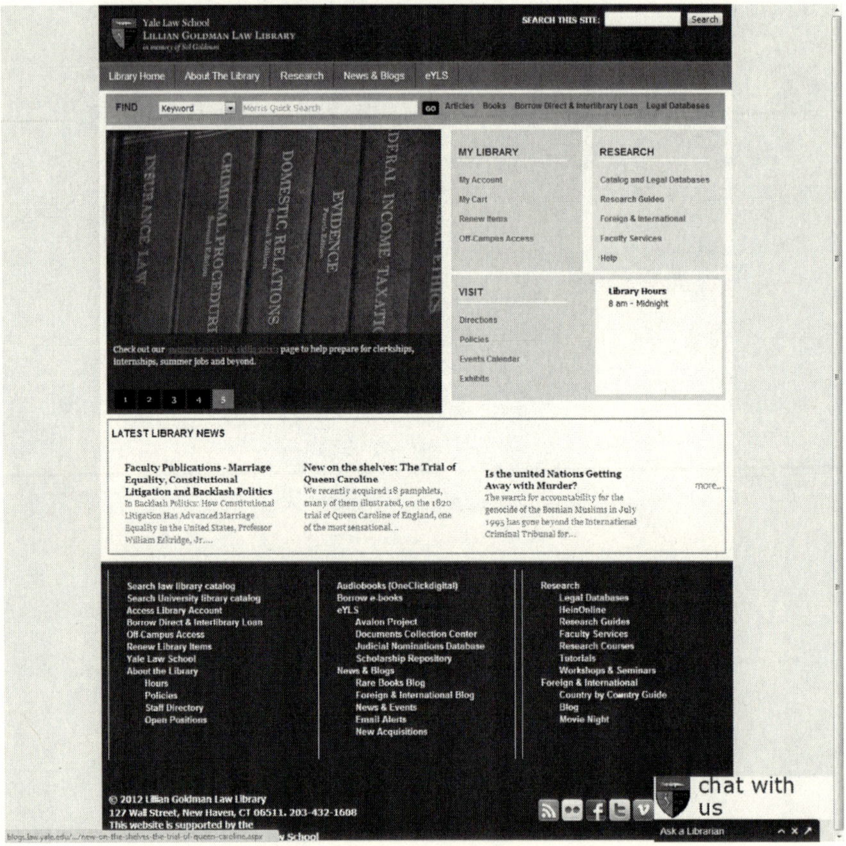

Figure 10.4 Completely Separate Website Design Screenshots—Yale Law School Library.

Figure 10.5 The Law Library at Case Western Reserve University shares a Navigational Header with Other CWRU Law School Web Pages, but Is Otherwise Functionally and Navigationally Distinct.

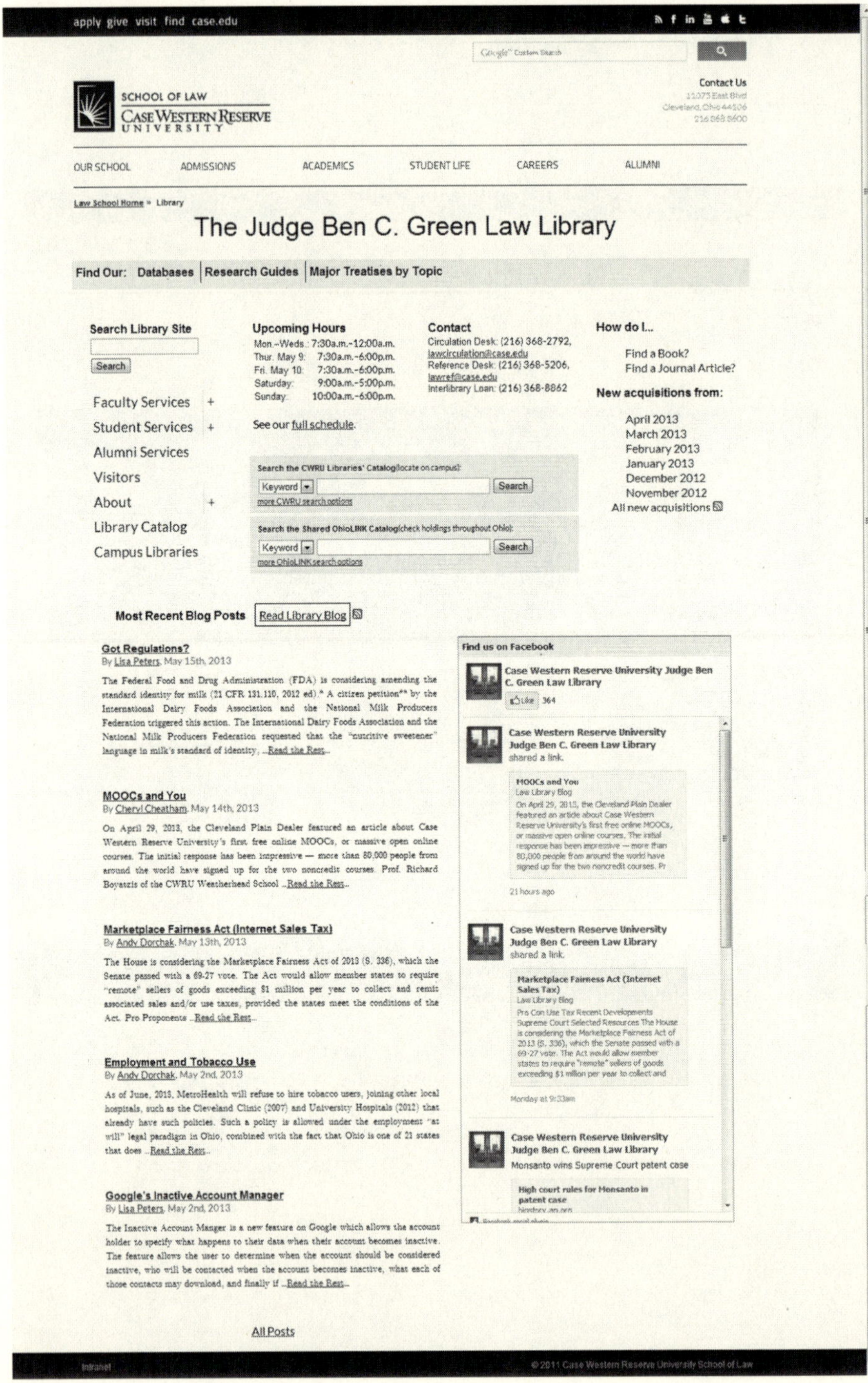

Figure 10.6 Subsite Website Design—Case Western Reserve University Law School Library.

Choosing complete navigation and user interface integration with the broader law school website might seem to *imply* also sharing a technical platform (or may be motivated by the need to share a platform). But the choices regarding a technical platform are ultimately separate from those related to user experience. Nothing prevents a library site that interacts as a seamless part of a broader law school website from being separately hosted and built on a completely independent content-management system or other platform. At the same time, a library might share a CMS with its law school or even the whole university but use it to build operationally independent websites, each with its own user interface and navigation, and each targeted to its own use cases.[14] Conversely, a library might opt to use a separate technical platform for reasons unrelated to user experience or site navigation. For example, at Case we opted for a platform (Wordpress) that provided our staff with a built-in administrative and authoring dashboard that we preferred over those integrated with the other tools used on campus at that time.

Additionally, just as the functional integration of the library website is a separate question from technical integration on a single "platform" (or website-producing application), selecting the technical infrastructure to support that platform is also a separate set of decisions. Models range from "self-hosting" (taking responsibility for one's own Web server, database, and application stack) on the library's or law schools' own hardware, to fully managing the software infrastructure on the hardware of a remote hosting provider, to consuming a content-management solution in a "software as a service" model in which all software infrastructure "below" the CMS application itself is effectively outsourced.

TECHNICAL SETTING

As a law librarian working in a digital age, you need to be equipped to think effectively and innovatively about how your library can deliver content and services through the Web. You may never have responsibilities on the technical side of website design or deployment; depending on the organizational structure of your institution, the library itself may not have technical responsibility for its website. But a basic knowledge of modern technical frameworks for websites will help you ground your thinking. Some familiarity with technical standards and terminology will equip you to ask informed and useful questions and issue better-informed, more-useful, business specifications to the personnel (in or out of the library) who are responsible for the actual web development in your organization.

Librarians who are informed of basic technology standards (HTML, CSS, Javascript, and the core technologies of server-side automation) can also serve as advocates for modern and robust Web development practices within their organizations. Although authored content from database-driven applications, like the content-management systems used for most library websites, can be somewhat flexibly deployed (and rearranged) in most any website template (including wholly table-based layouts), correct use of CSS-driven layout will make it much more possible to develop websites that accommodate fairly frequent small redesigns of the type likely to be important to librarians. Because CSS adoption in a truly uniform and standards-compliant manner by all major Web browsers was not widespread until quite a while after the Web was popularized, early authors and designers had no choice but to use HTML to control layout and appearance. A number of what are now bad habits, ranging from use of invisible spacer images, to image file replacement of text to control fonts, to the use of complicated HTML tables to control site layout, were common. As recently as 2009, a study of law school homepages observed a substantial minority still using HTML tables for layout.[15] To the extent possible, librarians should resist this. Standards-compliant websites using HTML for structure and CSS for presentation and layout are more accessible, easier to maintain, easier to redesign in the future, generally better optimized for search, and the content on them is more easily adapted to special mobile versions and other alternative views.

WEB CONTENT MANAGEMENT SYSTEMS

Most modern websites of institutions like law schools or their libraries are built using a content-management system, or CMS. In essence, a Web CMS is simply a tool to allow a wide variety of content—informational text pages, blog posts, and audiovisual contributions, for example—to be stored and displayed to viewers of your website using a mere handful of "templates" that control what actual markup and code (typically HTML, CSS, and Javascript) is delivered to the browser software of visitors to your site. CMS-generated sites are among most common types of "dynamic" websites. This means that the full webpage for each possible configuration of your content is generated "dynamically" by software from a combination of a model page or template that contains most of the HTML and style markup for the page, and separately stored page content (along with only the markup that is unique to the core content of that specific page). Depending on the system used, these "dynamic" pages may actually be generated in real time as a user navigates your website, or they may be software-generated ahead of time and stored, or cached, as traditional "static" HTML pages. In no case, however, does the author of a page need to manually create the HTML, CSS, and other code for each and every individual page of the website.

A Web CMS enables author and editor roles to be somewhat separate from the design and programming of a site. While some hard-coded "static" webpages are still used in law schools, librarians are most likely to contribute at least informational content about their libraries to a CMS-driven, "dynamic," website, whether as part of a broader platform shared with the law school or university or one dedicated to the library itself.

While there are alternative technologies, a Web CMS generally consists of a software program that generates one or many templates for the display of your content and determines which content and templates to use in response to specific incoming Web queries, and a database in which to store the specific units of content. Two of the most common CMSs used in law schools and libraries, Drupal and Wordpress, are examples of open-source software written in an open-source scripting language called PHP and using a relational database, typically MySQL, for the storage of units of content. DotNetNuke is an example of an open-source CMS that is built on Microsoft's proprietary .NET technologies.

This ability to present large, multiple-page sites without overwhelming duplication of markup is one very compelling argument for using CMSs and similar Web applications. The CMS framework also allows organizational roles to be differentiated. Because all of the content is stored in a database and rendered into a markup document for delivery by the Web server only after processing by the CMS application, the same content can be presented and used in many different ways. This can include anything from slightly different interfaces for "logged in" users, to "previews" or front-page teasers to timely content hosted elsewhere on the site. When this inherent flexibly to reconfigure content from the database is combined with standards-compliant Web design that thoughtfully separates structural markup from CSS used for layout and design, CMS-driven websites can also be built to be fairly amenable to relatively pain-free redesigns. This can allow for sites to be adjusted quickly to user feedback or new needs.

A CMS may be hosted locally (on hardware actually on site with the library or law school); hosted remotely (i.e., in the "cloud") but on a technical software stack (Web server, database, etc.) directly managed by the law library or law school; or fully administered as a service by an outside provider. Many CMSs, especially those commonly used at the enterprise level, are also capable of running many separate websites. Many law schools currently self-host open-source CMSs, especially Drupal and Wordpress, and these are well supported in the law school community through CALI and the Teknoids e-mail list.[16] Administering an independent law library website within a CMS already centrally administered (or provided as "software as a service") might alternatively offer the best of all worlds for some law schools and libraries that require substantial in-house control (not only of content but also of the design and programming of site-specific functionality) while leaving issues such as integration with user-authentication tools or other components of the university-level software infrastructure in the hands of central enterprise IT operations.

USING A CMS IN THE LIBRARY

An advantage to libraries of using a CMS is that more of the staff can usefully be involved in contributing roles for the website. Deploying a page in a "static" website requires that an author be able to write or copy the required HTML (and at least invoke any CSS or Javascript to be applied to the page), and have system permissions to actually get that page on to a production Web server. In CMSs, roles are divided so that contributors have no need to perform all these tasks. It becomes possible to separate content authorship, website design, the engineering of website functionality, and the operation of website infrastructure into roles that can be separated as required. Developers, designers, editors, and authors can theoretically all work in isolation with little access to one another's roles (or permissions) with respect to the operating website.

But while authors and editors can be somewhat removed from technical issues, someone still needs to be responsible for the more technical aspects of the website. The sheer open-ended capability of the major CMS platforms means that selection of a CMS and deployment of a particular website or Web application upon that platform will usually require specialized expertise. Some, especially those libraries relying on a CMS-based website technically administered outside of the library, may find that (depending on organizational competencies and on the access of your librarians to technical staff responsible for the CMS-driven website) there are better alternative solutions for presenting some of your library content than the general-purpose CMS.

Authors and editors may also still need a basic knowledge of *some* HTML. In my own institution, in which almost all library content exists within the Wordpress content-management system, we've discovered that while individual librarians can produce blog posts and basic news items without knowing or using any HTML, more structurally complex documents (such as research guides that include headings, lists, or navigational enhancements) require that the authoring librarian have basic acquaintance with HTML—either to mark up the document manually or to troubleshoot the HTML that the CMS is capable of generating automatically. But conversely we need to make certain they are using their own HTML only for limited and appropriate purposes. Nothing technically bars our authors from using their own HTML to attempt to select fonts, alter text sizes or colors, and do a number of other things within their posts that, if allowed, would break the CSS-specified display of our website. Our approach has been to circulate, and insist on compliance with, a house style guide dealing with these issues. Where authors have wanted to use "special" formatting for their content pages, we have worked to incorporate tools for that formatting within our CSS and our PHP page templates. For example, we use CSS "classes" that enable our research guide authors, with very minimal and specific HTML training, to invoke automated tables of contents for longer guides, content "sidebars," and Bluebook-compliant smallcaps typography.

CMS deployments do typically offer content authors both a "WYSIWYG," or visual ("what you see is what you get"), editing mode and an HTML authoring mode. In either case, the individual authors are providing their text content with some amount of basic HTML markup—the difference is whether they manually provide it or whether a text editor incorporated into the CMS software provides it. Libraries determined to completely avoid the need for authoring staff to manually employ any HTML at all might be better served by producing library content in alternative, purpose-built platforms (such as the popular LibGuides).

Besides the CMS, the other technical platform that is near-universally part of the law library Web presence is the public online catalog, generally the most visible component of the ILS, or integrated library system, that most libraries also use for cataloging and other internal process automation. The Web catalog may belong uniquely to the law library or may be shared with other libraries within a university. The ILS-driven, catalog-oriented Web interface and CMS-driven websites may each be used to provide a fair share of a library's Web presence. In other cases, particular externally hosted and administered Web-based services and tools are aimed at the library market to provide flexibility and functionality that many libraries, for administrative or technical reasons, may lack within either of those core platforms. Lastly, a few library websites may also need to incorporate their own custom Web applications developed outside of a CMS, Web catalog, or other standard platform such as catalog-system add-ins. Some libraries use custom-developed database-driven applications for things like interactive lists of electronic resources or for searchable databases of faculty scholarship originating with their law school. In other cases, such an application might be possible within the CMS platform used for other content but might (depending on the particular configuration of local resources and talents) be easier through a custom or specialized application.

ANALYTICS

One advantage of online, Web-delivered services can be their measurability. Libraries struggle to define meaningful output measures of their performance. The website is one area where at least some meaningful performance measures can be gathered; it is possible to use Web metrics to gain meaningful insight into how users interact with content like research guides and tutorials, and with online services ranging from chat to e-mail to form-based transactions.

Google Analytics, in particular, has rendered it pretty inexcusable for a library (or other institution) not to measure and set goals for its online performance. Google Analytics requires only the insertion of a snippet of Javascript into the site in order to collect data about user interaction.[17] Beyond basic visitor counts, analytics can help to shed insight into the actual patterns of navigation within a website. One of the simpler impressions a manager of a law school website can gain is an idea of relatively how much traffic is arriving via search, through links from elsewhere within the Web environment of a law school or university, or from what other linking sources on the Web. Library personnel given access to the Google Analytics information for the site can not only get a sense of what pages are viewed often but also some insights into how they are used (e.g., are particular research guides generally accessed directly or through a Web search, or are patrons clicking through a library's own list of available research guides).

Google's Webmaster Tools[18] are another type of analysis tool useful for library websites and other complex, multipage websites. In particular, the site "health" data can spot problems with a site that could originate in technical configuration but can also represent problems such as hyperlinks incorrectly formatted by an individual author within a CMS system. The Webmaster Tools also provide information on inbound search queries that can provide good insight into how search engine users are reaching the site.

ACCOMMODATING MOBILE USERS

There are two ways to deliver your Web content to users of mobile devices like iOS (Apple) or Android phones. The first is through the use of a website or website variant that is amenable to display and interaction in the browsers of mobile phones (with the main constraint being the small screen size). The second, out of the scope of this chapter, is the development of native phone "apps" to interact with library content outside of the browser. Such native apps can leverage built-in features of the phone (or tablet) interface but are probably an unnecessary undertaking for all or most libraries.[19] Considering mobile in your website development is essential, however.

Some "regular" websites are fairly usable on small devices like phones. But using a special variant of the website that is tailored specifically for smartphone and other small mobile displays can add value for mobile users while allowing the "main" site developed for computer screens and larger tablet devices to avoid making "lowest common denominator" design choices to accommodate small displays. (Note that "small" can mean low pixel count or small actual physical size. The two restrictions are obviously correlated but do not map perfectly from one device to another.)

This is not quite as simple as specifying a different CSS for smartphone browsers. It is very unlikely that a conventional "phone" version of the site should display all the same features, and very much of the same user interface, as the version of the website created for PCs and larger tablet devices like the iPad.[20] Many CMS deployments include functionality, either as part of the core application or in popular plug-ins, to easily create mobile-specific templates. Other fairly simple mobile site-builder apps can create a mobile site from content drawn from your main website using RSS feeds or other syndication protocols. An essential design consideration, however, is to make sure that it is always made very clear to users of devices that default to the mobile version of a site how they can "toggle" to the full website version and back. This is especially vital if there is content or functionality (as there usually should be) that is omitted from the mobile version.[21] Mobile users will invariably need to get to the full version from time to time, even if the interface isn't one that was designed to fit especially elegantly into the confines of their small touch-screen devices. Increasing adoption of HTML5 will also help with standardized deployment of mobile websites by further documenting/standardizing the behavior of all browsers, on all devices, around the same core standards.

Several examples of law library deployments of mobile-specific variant websites include those of the University of Indiana–Indianapolis, of the University of North Carolina, and of the University of Minnesota. Because the popular LibGuides also includes a mobile "site builder" plug-in, a library that provides research guides or other services on that platform can deploy a mobile Web variant. Typical functionality to include in a dedicated mobile version of a library website includes quick catalog searching; display of basic contact information, and frequently of an interface for "chat" or other immediate contact with library staff; and access to some types of information content, including treatise lists or research guides. Mobile websites might also experiment with services specifically targeted to users actually located within, but moving around, the library: barcode interfaces, sophisticated use of maps, and text or audio content responsive to library locations can all be integrated into such Web applications.[22]

A more flexible alternative to separating "regular" and "mobile" versions of a website is to embrace mobile "responsive" website design.[23] Although this will not be the appropriate technique for all applications, a compellingly attractive model in the library space can be seen at Princeton University Library's beta site;[24] here, a library front page rescales dynamically for a "narrower" browser window as necessary (either because of the native resolution of the display, or because the user is contracting the window) *without* discarding any information (nothing is omitted, but some content recedes pretty far "down" the smartphone screen). While site designs that are somewhat responsive to screen width are not at all uncommon, this model goes a step farther in explicitly accommodating the anticipated needs of mobile users once the resolution is sufficiently contracted. Note that while all information is presented (avoiding the most important drawback of segregated mobile templates), smartphone or other small-device users also are given no way of opting for a "shrunken down" version of what the site looks like on a full-resolution device.

Note, however, that this example is a beta interface that currently exists only for a rendition of the library front page. Successful integrating all of the other layout, typographical, and navigational issues for internal content within a framing template as radically responsive as this prototype could be challenging for the most talented designers, though the experiment is very intriguing.

Ultimately, law library websites are the launching pad for the full range of digital or online library services and initiatives. Promoting a core law library website on an equivalent basis to a law library's "brick and mortar" home creates the baseline for online, digitally mediated, services—knitting together finding tools to subscribed or owned *content*, online *services* delivered by the library or content self-published by the library, and service or policy information *about* the library. The organizational configuration of particular libraries and (for academic libraries) their parent law schools, and the technical platforms and services available to the institution, must be brought to bear in determining the particular design and implementation services appropriate to each library.

FURTHER READING

A List Apart. http://alistapart.com/. Site provides information for general Web design guidance.

Lawson, Bruce, and Remy Sharp. 2012. *Introducing HTML5*, 2nd ed. Berkeley, Calif.: New Riders.

Morville, Peter, and Louis Rosenfeld. 2008. *Information architecture for the World Wide Web*, 3rd ed. Sebastopol, Calif.: O'Reilly.

NOTES

1. See, for example, http://library.law.yale.edu/john-b-nann.

2. Ashley Ahlbrand, "The social side of law libraries," *AALL Spectrum* 17, 5 (2013): 12–14.

3. For a fairly early take on these issues, see Georgia Briscoe, Karen Selden, and Cheryl Rae Nyberg, "The catalog versus the home page? Best practices in connecting to

online resources," *Law Library Journal* 95 (2003): 151–174. For a take more specifically oriented to "disruptive innovation," see Ken Chad, "Disrupting Libraries: The potential for new services," *Proceedings of the Charleston Library Conference*, 2009, http://dx.doi .org/10.5703/1288284314783. See, generally, Clayton M. Christenson, *The innovator's dilemma: When new technologies cause great firms to fail* (Cambridge, Mass.: Harvard Business Review Press, 1997).

4. Yale Law School, Lillian Goldman Law Library, "Legal databases," http://library .law.yale.edu/research/legal-databases.

5. Georgetown Law Library, "Treatise finders," http://www.law.georgetown.edu/ library/research/treatise-finders/index.cfm; University of Virginia Arthur J. Morris Law Library, "Treatises and services by subject," http://libguides.law.virginia.edu/treatises; The University of Michigan Law School Library, "Book lists," http://libguides.law .umich.edu/book-lists; Case Western Reserve University, The Judge Ben C. Green Law Library, "Major treatises by topic," http://lawlibrary.case.edu/treatises-by-subject/.

6. http://springshare.com/libguides/.

7. At least two law libraries use LibGuides as the platform not just for research guides but also as the CMS for the entire website. These are the University of Kentucky College of Law Library, http://library.law.uky.edu/home, and Hamline University Law Library, http://lawlibrary.hamline.edu/.

8. Harvard Law School Library, "Tutorials & videos," http://www.law.harvard.edu/library/research/tutorials/index.html; Duke Law School, Goodson Law Library, "Research tutorials," http://law.duke.edu/lib/tutorials/.

9. Georgetown Law Library, "Tutorials," http://www.law.georgetown.edu/library/research/tutorials/.

10. In general, efforts to ensure good use of markup conventions (such as including a hierarchy of header elements) and a content-appropriate amount of linking and referencing are all the deliberate "search engine optimization" that should be required.

11. Other contributions that the library can make might also relate to publishing and Web development. Georgetown Law School publishes a database of faculty scholarship under the umbrella of its *library* website, as does Cleveland-Marshall, respectively http:// www.law.georgetown.edu/library/research/faculty-scholarship/index.cfm and https://www .law.csuohio.edu/facultystaff/publications. See Jan R. Babbit and Leslie Pardo, "The evolving nature of faculty publications," in *Public Services in Law Libraries* (Binghamton, N.Y.: Haworth Press, 2008). The numerous law schools who publish to institutional repositories on BePress's *Digital Commons* platform also manage that relationship through libraries.

12. Examples include Springshare's LibAnswers w/ LibChat, which supports the "Ask a Librarian" functionality on Harvard Law Library's website; Banckle Chat, which supports the "Ask a Law Librarian" function on the University of Michigan Law Library website; Olark, which support the "Ask a Librarian" interface on Yale Law Library's website; and LivePerson, which supports the Georgetown Law Library's "Chat with a Librarian" function. These services include a button or widget for display on the website, using techniques ranging from a simple "badge" image and link, to an embedded "iframe," to the inclusion of sophisticated interactive widgets through AJAX or JSON. On the back end, product offerings vary from purpose-built operator panels to integration with chat protocols such as Google Talk.

13. http://xkcd.com/773/.

14. Georgetown Law Library, for example, uses CommonSpot from Paper|Thin for its CMS, as does its law school. Note that in substance and UI these are an example of a library site that is functionally quite distinct from the law school site. CommonSpot is a "Saas, cloud based, Web content management system"; http://www.paperthin.com/.

15. Roger V. Skalbeck, *Law School Website Design Study 2009: A gallery of law school homepage screenshots and elements with data gathered February and March 2009*, Lulu, 2009. http://www.lulu.com/shop/roger-skalbeck/law-school-website-design-study-2009/paperback/product-5163077.html. For the most recent of several articles documenting the author's continued study of law school website technical practices, see also Roger Skalbeck, "The top 10 law school home pages of 2011," *Journal of Legal Metrics* 2 (2012): 25–52.

16. The Center for Computer-Assisted Legal Instruction, http://www.cali.org; http://www.teknoids.net/.

17. For a detailed analysis of using Google Analytics in the library setting, and setting library website performance goals, see Steven J. Turner, "Website statistics 2.0: Using Google Analytics to measure library website effectiveness," *Technical Services Quarterly* 27, 3 (2010): 261–278.

18. https://www.google.com/webmasters/tools.

19. Supporting browser-based uses of the smartphone over reliance on apps distributed through phone vendor closed gardens may also better align with libraries' interests in promoting an open, "generative" Internet. See, generally, Jonathan Zittrain, *The future of the Internet and how to stop it* (New Haven, Conn.: Yale University Press, 2009).

20. I can speak only for myself here, but websites that default to a "mobile version" on the iPad (which is quite competent at displaying regular, full-resolution websites) are a pet peeve of mine.

21. Even mobile versions of simple blog sites commonly omit the tools to navigate to particular archive displays by month, "category," author, or other such configuration—allowing nothing but a simple reverse-chronological scroll through the content and possibly a search window at the top of the page.

22. The slides (http://neasist.org/2012/04/24/going-mobile-slides-posted/) from the presentation "Going mobile: Library websites, services, and apps on mobile devices," *New England Chapter of the American Society for Information Science & Technology* (2012) offer several valuable ideas for mobile site development.

23. See, generally, Ethan Marcotte, *Responsive Web design*. A Book Apart, 2003.

24. http://librarybeta.princeton.edu/.

11

Web-Scale Discovery and Federated Search

Valeri Craigle

S. J. QUINNEY COLLEGE OF LAW LIBRARY, UNIVERSITY OF UTAH

In stark contrast to the library card catalogs of old, today's library search interfaces offer much more than one-dimensional, item-specific searching. Users are now engaged in a process of discovery in which they are empowered to control not only the sources of content being searched but also the context into which information is delivered, and the ways information is viewed in the search environment. By eliminating the barriers to information discovery, law libraries can position themselves as true partners in this process, defining their mission in new ways, and providing critical services in an ever-complex information ecosystem.

Discovery technologies are vital to a law library's core services. Users are driving the demand for a new generation of discovery tools through a now commonly held expectation—largely influenced by Google—that searches should include "everything" the library has to offer and should be displayed in a centralized location with a variety of options for organizing and contextualizing the information. Law library users are no exception, though by the very nature of the current legal information environment, legal researchers must have a certain tolerance for information silos such as Lexis and Westlaw.

As discovery technologies take hold in the legal information environment, law libraries need to understand the inner workings of these systems and the roles that must be fulfilled in order to meet the needs of their users. As discovery search services continue to make rapid progress, understanding how they work and having an awareness of the implications of deploying these services in law libraries will keep law librarians ahead of the curve.

FROM SEARCH TO DISCOVERY

Legal Research Beginnings

Legal research is unique from the research process of most other academic or professional disciplines. Practitioners must navigate a complex tangle of information scattered among a multitude of locations, formats, and iterations. Skill and experience are required to master the landscape. Core legal research in case law, statutory law, and other primary materials is generally conducted in the native environments of large and complex databases like Westlaw, Lexis, and Bloomberg Law. The use of print resources has dropped dramatically in the age of electronic databases, or what has historically been referred to in the legal research pedagogy as computer-assisted legal research (CALR).

CALR resulted in a sea change for the legal profession. Development of the earliest legal information databases began in the 1960s and 1970s out of a very simple need to more quickly and comprehensively search the exploding body of legal knowledge. There was just simply more law to research and it was growing at an exponential rate.

One of the first legal research databases developed was the "Horty Project." Known by some as the precursor to Lexis, it was an initiative started in 1959 by John F. Horty, a professor at the University of Pittsburgh. The purpose of the project was to compile Pennsylvania state statutes to facilitate legal research in healthcare law. In 1971, the OBAR (Ohio Bar Automated Research) Project launched to deliver the entire collection of Ohio state codes and court decisions. Soon after, in 1975, Westlaw came onto the scene with its famous electronic reporting system.[1]

By the early 1990s, Lexis and Westlaw were evolving into massive databases, developing large-scale business models and increasingly proprietary philosophies. The *pro se* movement spurred the evolution of online free legal research resources, even when the Internet was relatively new. The Legal Information Institute, started in 1992, offered a host of primary legal materials, both state and federal, and also offered a wealth of secondary sources, particularly in the areas of legal ethics and social security.

Today, there are too many free online legal resources to count. Google offers a search engine designed specifically for case law. FindLaw offers cases and codes, Supreme Court decisions, and lawyer directories. And many websites have begun offering free legal advice, or advice offered for a minimal fee. While legal practitioners and law schools continue to subscribe to the large, fee-based resources like Lexis, Westlaw, HeinOnline, and Bloomberg Law, the growing *pro se* phenomenon is driving the development of free online resources and self-representation collections in law libraries. As the economy continues to stagnate and the cost of legal resources remains prohibitive for much of the population, the growth of free and lower-cost research tools such as Fastcase and LoisLaw will only increase over time.

But free legal resources found on the Web do not offer the one thing that large database publishers do: the ability to organize legal information into a meaningful context. Legal research is not just a process of discovery but of contextualization among related resources and resources that provide deep analysis of a particular point of law. Serious researchers are looking for "good" law, or law that has received positive treatment in the courts over time, not just anything they find on a particular subject matter. The big legal databases offer this perspective and contextualization, but they charge a high price for it, and they are only available to a select few.

The Current Environment

With so many high-tech tools at our disposal, one would expect the legal research environment to be nimble, seamless—practically automated. In actuality, the tools we use are more numerous and complex than ever. Most primary legal sources reside in large, proprietary databases, accessible only to those of us willing or able to pay top dollar for high-end search algorithms and an endless array of tools and features.

The corpus of the law has grown in such size and scope that we need road maps to navigate the legal information environment. And as the size and complexity of legal information grows, so do the methods for finding these resources. For today's legal practitioner, knowing how to find very specific source material is a critical skill, particularly for new lawyers, who spend a large amount of time in their first few years of practice doing legal research.[2] As information professionals, how can we assist our patrons in navigating this vast and complex landscape of information and technology?

Discovery platforms and federated search are emerging as effective search solutions for legal researchers. As these systems gain recognition in the legal research community, law libraries must understand what products are available, how they function, and how to be proactive in the implementation and management of these systems.

DRAWING DISTINCTIONS

Federated Search

Federated search is a search technology that provides a one-box search experience for the user, similar to Google, that searches multiple library databases at the same time. In this way, patrons don't have to know whether they need to search HeinOnline or LLMC or Proquest Congressional to get their information needs met—they can just search once and find results.

The technology returns a results set and users can toggle between sources that were searched. The down side has to do with the fact that relevancy ranking is not really possible because each database uses its own algorithm for determining relevancy.

When a search is performed in a federated system, results are displayed in a unified interface in a Web browser. Federated searches do tend to be slow, however, as the system is required to search one database at a time with each search, returning results only when all of the databases in the federation have been queried. Because of this, federated search systems are known to frequently time out, particularly during times of high usage. As a result, libraries have largely abandoned federated search in favor of Web-scale discovery, a surprising development considering how popular federated search systems were just a few years ago. But federated search is still utilized in certain sectors, primarily by science and technology researchers for searching the deep Web.[3] And although federated search is no longer preferred as a discovery solution for libraries, it still offers a comprehensive, real-time search across multiple resources, without relying on preexisting indexes, which may not be updated on a daily basis.

Discovery Layers

A discovery layer is simply the public interface to library resources cooperatively selected by a library and discovery layer vendor for discovery in the system.[4] These resources may include anything purchased, licensed, indexed in the catalog, or living out on the Web. You could think of a discovery layer as a next-generation online public-access catalog (OPAC), only much more robust, with lots of search and browsing features, and some social media tools. It's likely that if you have done any searching in an academic library catalog in the past few years, you have used a discovery layer.

In discovery layers, researchers do not just search for information, they interact with it and can modify it to suit their needs, creating a personalized research experience. A medical researcher can limit search results to peer review articles only; an art history major can search for only audiovisual items, then limit the location of those materials to a digital collection; a law librarian can write a review for a resource he has found to be valuable for first-year legal research students.

One of the greatest advantages of a discovery layer is that it provides a unified view of all the items contained in the vast repository of library resources. Bibliographic and other descriptive information is gathered from all of these sources and placed in one location for searching, called the Central Index. Unlike a federated search, which sends out queries to all the different databases and then waits for all the results to come back, the discovery layers have preindexed the metadata for all of those sources instead. That way everything is in the same central database or index and can be quickly searched. Although there are a number of discovery layer platforms on the market, law libraries seem to prefer Encore as the platform of choice, according to an informal survey of technical services law librarians in 2010. Fifty-two percent of the librarians surveyed indicated that their libraries were either in process of implementing discovery layers or had already done so. Encore was by far the leading product, with WorldCat Local and Aquabrowser coming in second and third, respectively.[5]

Other discovery layer products include:

- Primo (Ex Libris)
- Enterprise (SirsiDynix)
- Ebsco Discovery Service (Ebsco)
- Summon (ProQuest)
- VuFind (Villanova University)
- Endeca (Oracle)

Web-Scale Services

In the past few years, there has been a lot of focus on more effectively managing user traffic to databases, particularly during times of heavy usage, which tends to slow down a Web service. Before Web-scale discovery services, heavy traffic would cause these systems to crash altogether. When a system is "Web-scale," it means that it is built on an architecture that balances and stabilizes a system when a lot of people are trying to access it at one time. When a Web-scale system detects an increase in traffic, it slows down the parts of the system that are not being used so that more resources can be allocated to the parts of the system that are being heavily used.[6] Web-scale services are a relatively new technological innovation, whose development gained traction in the early 2000s[7]. Web-scale services slow traffic just enough to keep the data flowing, then return to normal functioning when the loads are not so heavy. This results in consistent performance over time, ensuring a constant flow of data and few system failures. In many libraries today, Web-scale services form the backbone of search services, online bibliographic utilities, data management systems, and patron services. Indeed, Web-scale services are at the functional core of discovery layer platforms and online search engines.

INNER WORKINGS OF THE DISCOVERY SYSTEM PLATFORM

Technical Infrastructure

Discovery layer platforms contain several components, all working in concert to bring this new search experience to the user. Hardware, software, hosted services, and Web-scale services play a role in the proper functioning of the system as a whole. Each discovery layer vendor may have a unique approach in the deployment of these services. Some may require all components to be purchased and held locally, some may choose a hybrid solution of local hardware and some Web services, and others may deploy their services exclusively from the cloud.

In terms of hardware, vendors may ask a library to purchase a number of data servers to perform such functions as data backup, data testing and staging, and data harvesting. Often with discovery layer platforms, the central index is stored in the cloud to take advantage of Web scalability, so the servers that the library hosts are needed mostly to boost performance and provide extra data security. Some vendors will give libraries the option to have these data services hosted by them if they lack the budgets or personnel to administer the servers themselves.

Software components include the discovery layer interface itself; Web-based administrative interfaces for controlling the look, feel, and functionality of data; mobile applications; and statistical software for analyzing user behavior, collection use, and any other reporting needs a library might have. Law libraries may find great value in the reporting software when submitting their annual ABA or *U.S. News* collections statistics. Again, these services can all be deployed in the cloud or from a desktop client.

Ideally, all components would be deployed in the cloud to provide the scalability, flexibility, reliability, and seamless delivery required by the large data loads and numerous services necessary for discovery layer platforms. As these systems evolve, the need for them to be Web-scalable will only increase.

Information Gathering, Management, and Display

As previously mentioned, one of the greatest advantages of Web-scale architecture is its ability to quickly return search results by drawing from a preharvested central index. This index, compiled by the discovery platform vendor, contains resource metadata as well as authentication information for access to full-text materials. The metadata for databases, e-journals, and other publications are supplied by publishers who have entered into cooperative agreements with the library implementing the discovery layer platform. Metadata for digital collections, MARC metadata from the catalog, and metadata from external data sources must all be gathered along with publisher's metadata in the central index.

Once metadata from these various sources are placed in the central index, a series of customizations can be performed for the visual display of items in the discovery layer interface. Customizations can be made to features like browsing facets, social media tools, and even book icon images. Customizations can also be made for simplifying the display of items such as books that have multiple copies, locations, or formats. Permissions for performing these customizations are usually limited to server administrators. As patrons begin to interact with the system, adjustments may need to be made periodically to accommodate user preferences. For example, people may want more facets in a particular subject area, or requests might be made to display book chapter icons in the results list. For academic law libraries, campus-wide discovery layer implementations, which unify the resources of the entire campus, may make legal research more complex, adding to the mix of resources from hundreds of disciplines. Complicating things further is each party's definition of "relevancy." What is relevant in the biological sciences may not be relevant in the humanities, the arts, or the law. Who decides which results float to the top as more or less "relevant"? These issues must be resolved cooperatively among the campus libraries participating in the implementation, with the caveat that there may not be a perfect solution for all of the parties involved.

Search Results and Access

The first encounter with a discovery layer interface can be daunting. Users are often confronted with an overwhelming set of choices. Books, articles, audio files, digital photographs—almost any material type one can think of appears in the search results screen. Narrowing a search requires some skill, challenging the user to utilize a series of facets, filters, sorting options, and other advanced search features. Law library patrons must learn to master these tools in order to limit their results list to a manageable size. Law librarians will discover that their patrons, and perhaps themselves and their colleagues, require a bit of training to navigate the complexity of this new environment. Legal researchers may become frustrated by this new complexity and wish for the simplicity of the old catalog.

Indeed, it raises the question of whether it is beneficial for law libraries to participate in campus-wide implementations, as flooding the legal information landscape with materials that are irrelevant to legal researchers may be counterproductive. However, discovery platforms can be extremely beneficial for cross-disciplinary research and research involving secondary sources in the law. Materials that the researcher may never have thought of might appear. Environmental law researchers could discover water-rights maps from a digital collection, or an article in a scientific publication about the detrimental effects of pesticides. By expanding their horizons, discovery platforms open a new world to legal researchers.

For the patrons unaffiliated with the law library, and thus excluded from access to legal databases, a discovery layer search presents a litany of results for which they haven't the privileges to use. Some publishers prohibit any distribution of full text to nonlaw patrons, even if they come to the library personally to work with a librarian to gain access to these materials. Law libraries must decide what their policies are for making restricted information—such as that in Westlaw and Lexis—visible in the discovery layer. Sometimes it's a matter of choosing the lesser of two evils: whether it's better for patrons to know about a resource and not be able to access it, or whether it's better for them to be unaware of these resources entirely, so as not to cause frustration when they try to access these materials.

THE USER EXPERIENCE

Relevancy

Displaying search results according to relevancy ranking is one of the hallmarks of a discovery layer platform. Though the concept is not new, recent advances in the development of relevancy-ranking algorithms form the basis of search results display in next-generation catalogs. The underpinnings of relevancy algorithms are generally proprietary and are thus poorly understood. Factors such as currency, frequency, and proximity of search terms, as well as user preferences and behavior, are all part of the formula. Control of these algorithms is generally limited, but some vendors allow libraries to disable relevancy as the default display for search results.

Social Media

Have you ever wanted to post a book review, add a subject heading to a tag cloud, or "like" an article for your friends to see on Facebook? How about tweet a Supreme Court opinion, or select a novel for your Pinterest page? For the social media native, these might be perfectly natural activities, to be relished as part of the research process. And today's discovery layer platforms offer every social media tool you can imagine to indulge this new appetite for participation.

But one has to wonder if the incorporation of social media tools in the library catalog is simply a marketing ploy, or if it really is a serious attempt to involve researchers more in the discovery process. One could possibly view some of these activities as akin to metadata crowdsourcing. What is a tag cloud but a cluster of subject headings? Does "liking" an item increase its relevancy rankings? Will book reviews improve descriptive metadata? Will folksonomies replace the subject librarian? Only time will tell if these tools are nothing more than another set of superfluous features. But this is a time when patron-driven acquisitions and metadata crowdsourcing are becoming more accepted as legitimate tools to enhance library services. The Google experience has been a wake-up call for librarians, who can no longer think of themselves as the gatekeepers to information procurement and access. It will be interesting to observe the impact of social media tools on patron interaction in the discovery layer platform.

IMPLEMENTATION

Cost and Commitment

The cost and commitment of a discovery layer platform implementation is concomitant with the size and complexity of the library and the scope of its collections. Independently operating law libraries who are able to work with their current integrated library system (ILS) vendors and electronic resource publishers, and who have no additional digital collections, e-publishing services, or deep Web searching needs, can look forward to a relatively smooth process. Any costs incurred will likely be the vendor's time in loading the central index and working with librarians to develop the appropriate display options for the discovery layer interface.

That is the best-case scenario. Another scenario is an academic law library embarking on a campus-wide, multilibrary implementation. If this entails migration of the catalog and the purchase of a new suite of tools that are more compatible with the new discovery layer platform, the process can be costly and time consuming. In this scenario, libraries should plan on at least six months to a year for the catalog migration alone. Additional time and resources may be required to bring into compliance any external systems that will need to work with the discovery layer, such as an automatic retrieval system, or other systems that manage patron requests and circulation processes.

Understanding the full scope of the data to be harvested for the central index is paramount. A large digital collection, for example, may require additional servers or server partitions. Anticipate the need to purchase either additional servers, or bump up existing systems to handle these loads. During initial implementation, a significant amount of time is allocated to systems administration and the participation of technical services librarians. Trainings, meetings with the vendors, and hands-on working sessions to fine-tune customizations will monopolize staff time for the first six months.

Once the system is operational, downtimes for the catalog will need to be strategically planned. Cataloging lockouts of potentially several weeks may be required for loading the indexes and assessing data compatibility and integrity in the new system. It is during this time that unexpected issues occur. Data could be corrupt, peripheral systems might be incompatible, or server capacity may be deemed insufficient. All of these factors must be taken into account when planning for a discovery platform implementation of this scale.

Trends

Discovery layer platforms have not been around for very long, only really gaining traction since 2009. A conversation of how discovery layers were being used in law libraries was held during a presentation on discovery layer technologies given at AALL in Portland, Oregon. At that time, very few law libraries seemed to be participating in these initiatives.

An informal survey of technical services law librarians showed that a little over 50 percent of law libraries had implemented or were in the process of implementing discovery layers in their libraries.[8] Though the survey did not address the question of the types of libraries responding, the comments suggested a high number of academic libraries that were participating in campus-wide implementations. Law librarians try diligently to stay abreast of the latest technologies and want to offer their patrons the best services available. As discovery layer platforms gained traction, law librarians' interests were piqued by the prospect of simplifying the legal research process.

In the early days of implementation, most legal information sources were nowhere to be found in the knowledge bases of discovery layer platforms, a common complaint among academic law librarians. Partnerships between legal information publishers and discovery layer companies were slow to form. Today, the availability of these resources is not as much an issue as gaining full-text access to law library resources for patrons who are not affiliated with the law library.

This raises the question of whether it is prudent to enable these resources in a discovery layer at all. If the majority of the patrons who use the catalog can't get access to these materials, does this unduly frustrate the process? Alternatively, are legal researchers at large academic institutions frustrated by being flooded with resources that have nothing to do with the law? And does this feedback have an impact on other librarians' decisions to implement this new technology?

For solo law librarians or law firm librarians working in discovery platforms that manage materials specific to their organization, the problems of discipline-specific heterogeneity and resource access don't apply as much as they do for law librarians at large academic institutions. In fact, discovery solutions may be overkill for small libraries, who often subscribe to fewer databases and wouldn't necessarily need these sources to be visible in the catalog anyway. Allowing their patrons to "discover" resources that they can't get full-text access to may aggravate their patrons more than offer a good service If academic law librarians are going to spend a great deal of time, effort, and money to implement a discovery layer platform, they will want a quality product and rewarding experience for their patrons, not a product that leads to more complexity and confusion.

THE FUTURE

Predictions on the Cloud

In today's nimble technological environment, mobility and platform agnosticism are key. Using local hardware and software to run applications is a paradigm that has run its course. When the concept of cloud-based services started taking hold around 2011, libraries were quick to harness the power of this new technology. In fact, libraries were pioneers in the process, harnessing the power of the Online Computer Library Center (OCLC) collective in copy cataloging services. Libraries, like many organizations both in the private and public sectors, have begun to relinquish the acquisition and control of the hardware and software that formed the backbone of data management and dissemination. The cloud offers many benefits for managing the various functions required for information management. As with any system, there are disadvantages to be wary of, but these are far outweighed by the benefits of cloud-based services.

Scalability, resource aggregation, and the unified search interface have transformed library services. The vast amounts of data libraries manage today and the user loads library systems experience while patrons access this data 24/7 will keep library services in the cloud for decades to come. Data providers, even the most restrictive legal publishers, may feel pressure from their customers to conform to this new model, perhaps encouraging them to expose their metadata to cloud-based aggregators for harvesting to discovery platform providers, which will become ubiquitous in law libraries in the next decade.

As libraries move away from the idea of local control of resources and turn over the management of their systems to remote providers, they will continue to rely on the cloud to provide these services. All indications point to use of cloud-based computing well into the next decade and beyond.

Next-Generation Frameworks

The culture we live in today, though tech savvy and highly sophisticated, is fraught with information overload.[9] It is no wonder that finding the best formula for information discovery has become one of the hottest topics of our time. Information and data are now valuable commodities to be bought, sold, and traded. The value of data is compared to that of oil or gold. Personal data, the sine qua non of the information economy, is exchanged by governments, companies, and marketing bodies, unbeknownst to most Internet users as they readily opt in to share personal data from their smartphones, Facebook posts, and Web-browsing activities.

With this ocean of data and personal information to work with, next-generation search developers are positioning their products to engage the user in a process of sense-making in which the search experience moves beyond discovery to contextualization and personalization. Boolean searching may be less relevant, as searchers use more phrase-like, natural language search strategies, maybe even in the form of voice recognition, to condense multiple variables into one search. The new Facebook graph search exemplifies this process. With one phrase, the

Facebook Knowledge Base can be searched with multiple criteria such as people living in your town, who among those people like to ski, and who among *those* "like" a certain ski resort. Imagine phrase-searching in a library's discovery layer to include all articles with a certain author written in the past six months. Of course, we do this now in library catalogs, but one has to know how to do an advanced search and how to use facets and filters to obtain the desired results.

In the future, we may see library knowledge bases living in the cloud, along with library management workflows in acquisitions, cataloging, and circulation.[10] In the near future, we may see desktop clients become entirely virtual environments whose applications are Web based and data are stored in the cloud. OCLC is already making progress toward this end with its Worldshare Management Services (WMS).[11] Harnessing the power of the collective, OCLC WMS taps into aggregated bibliographic data in WorldCat, licensed data, vendor records, authority records, and local holdings information. Cataloging, acquisitions, and circulation workflows that libraries have traditionally managed from desktop clients are now done entirely in a centralized Web-based environment, eliminating the need for systems upgrades and maintenance.

In law libraries, the future may follow a different course. If anything, the role of law libraries as intermediaries in the search and discovery realm is becoming increasingly murky as large legal publishers and vendors offer more services native to their own databases. Legal information is becoming more proprietary, not less. This model stifles innovations in search and discovery within the broader context of the information economy. Legal information metadata are not exposed to Web-based search engines and the full text itself is locked behind license restrictions. These ideals run contrary to most data models today.

It remains to be seen whether consumers of legal information will drive either a continued dependency on legal information silos or a more open paradigm in which legal information will be incorporated into the larger information economy. The next generation of law students and practitioners will drive these decisions into the next decade.

Where Law Libraries Are Headed

Recently, a third-year law student asked me to post a list of the ten most frequently used databases to the law library website. As we talked about his request, I let him know that these databases were listed in several areas of our website as well as the website for the main campus library, and were also searchable in the catalog. Nonetheless he wanted just this list, in a place that he thought made sense to him on our website.

Most of us have experienced odd requests from attorneys, law faculty, and law students who demand the easiest possible customizations to their legal research experiences. I often have requests at the reference desk to find news articles that I find so easily with one search in Google that it leads me to worry about the most basic proficiencies of the community I serve.

The generations of legal professionals who have grown up using print will most likely not be on board with the newest technological innovations in legal research technology; however, their successors, our current law students and new lawyers, have a long way to go in becoming proficient in the newest technologies, much less expert users. The development of discovery layer technologies, federated search engines, and Web-scale services have all been driven in large part by consumer demand. Yet the legal community remains conservative in its views on making the legal research environment more open to the rest of the information economy. If anything, it seems we have become even more loyal to the largest legal information publishers, who are becoming increasingly proprietary and monopolized than ever before. In looking at the quality of these products and the features offered, it's easy to see why. Work folder sharing, litigation profiles, case mapping, and enhanced mobile apps are just a few of the extras offered in Lexis and Westlaw. Relevancy is geared specifically toward the law. Indexes are created by legal experts themselves. And students know they will be using these tools in practice.

Case law research and research in primary sources can be complex and time consuming. The importance of finding legitimate resources cannot be overstated. Discovery layers and Web-based searches are best left to researchers looking for secondary sources, an area of law that is rapidly expanding. Law libraries that are already using Innovative Interfaces products will find the implementation of the Encore discovery layer relatively simple. Academic law libraries, as part of a larger implementation, might want to weigh the costs and benefits of opening their catalog to the campus community at large. If they decide to do so, the costs of implementation and the training of their constituents must be taken into account.

In the end, all of us will need to become more tolerant of the vast array of information resources available in our discovery systems, and we must be open to learning how to navigate this new terrain. Legal researchers are becoming more interdisciplinary and tech savvy. As the next generations of law students and legal practitioners become part of our organizations and academies, the hope is that these digital natives will drive the further opening-up of the legal information environment.

FURTHER READING

General

Arewa, Olufunmilayo. 2006. "Open access in a closed universe: Lexis, Westlaw, law schools, and the legal information market." *Lewis & Clark Law Review* 10:797; Case Legal Studies Research Paper no. 06-03; Northwestern

Public Law Research Paper no. 07-02. Available at SSRN, http://ssrn.com/abstract=888321.

Breeding, Marshall. 2011. "Preparing for the long-term digital future of libraries." *Computers in Libraries* 31 (1): 24–26.

Childers, Scott. 2006. "An introduction to relevance-ranking systems." *Faculty Publications, UNL Libraries.* Paper 117. http://digitalcommons.unl.edu/cgi/viewcontent.cgi?article=1123&context=libraryscience.

Fagan, Jody Condit. 2011. "Discovery tools and information literacy." *Journal of Web Librarianship* 5 (3): 171–178.

Hoeppner, Athena. 2012. "The ins and outs of evaluating Web-scale discovery services." *Computer in Libraries* 32 (3).

OCLC WorldShare Management Services website. http://www.oclc.org/webscale/default.htm.

Singhal, Amit. "Giving you fresher, more recent search results." *Google Official Blog.* November 3, 2011. http://googleblog.blogspot.com/2011/11/giving-you-fresher-more-recent-search.html.

Weare, William H., Jr., Sue Toms, and Marshall Breeding. 2011. "Moving forward: The next-gen catalog and the new discovery tools." *Library Media Connection* 30 (3): 54–57.

Westrum, Anne-Lena. 2011. "The key to the future of the library catalog is openness." *Computers in Libraries* 31 (3): 10–14.

Web-Scale Discovery

Axford, Mary A. 2012. "Ultimate debate program on Web-scale discovery services: A report of the LITA Internet Resources and Services Interest Group meeting, American Library Association annual conference, New Orleans, June 2011." *Technical Services Quarterly* 29 (2): 142–147.

Barr, Jeff. April 2, 2009. "Amazon Web services: Building a 'Web-scale computing' architecture." Slideshow. http://www.slideshare.net/goodfriday/amazon-web-services-building-a-webscale-computing-architecture.

Elder, Carol, et al. 2011. "Single search? The promises and pitfalls of Web-scale discovery." *BCLA Browser: Linking the Library Landscape* 3 (3).

Gross, Julia, and Lutie Sheridan. 2011. "Web scale discovery: The user experience." *New Library World* 112 (5–6): 236–247. doi:10.1108/03074801111136275.

Howard, David, and Constance Wiebrands. 2011. "Culture shock: Librarians' response to Web scale search." *Edith Cowan University Research Online.* http://ro.ecu.edu.au/cgi/viewcontent.cgi?article=7208&context=ecuworks.

Meadow, Kelly, and James Meadow. 2012. "Search query quality and Web-scale discovery: A qualitative and quantitative analysis." *College & Undergraduate Libraries* 19 (2-4): 163–175.

Vaughan, Jason. 2012. "Investigations into library Web scale discovery services." *Information Technology and Libraries* 31 (1): 32–82. http://ejournals.bc.edu/ojs/index.php/ital/article/view/1916/1744.

Federated Search

Shokouhi, Milad, and Luo Si. 2011. *Federated search*. Hanover, Mass.: Now Publishers.

Discovery Layers

Becher, Melissa, and Kari Schmidt. 2011. "Taking discovery systems for a test drive." *Journal of Web Librarianship* 5 (3): 199–219.

Caplan, Priscilla. 2012. "On discovery tools, OPACs and the motion of library language." *Library Hi Tech* 30 (1): 108–115.

Craigle, Valeri. 2010. "Discovery layers in law libraries." *AALL Spectrum* 16 (3): 7–9.

Little, Geoffrey. 2012. "Thinking about discovery layers." *Journal of Academic Librarianship* 38 (6): 346–347.

Vaughan, Jason. 2011. "Chapter 1: Web scale discovery what and why?" *Library Technology Reports* 47 (1): 5–11.

———. 2011. *Web Scale Discovery Services. Library Technology Reports* 47, no. 1. Chicago: American Library Association, 2011. PDF e-book.

Yang, Sharon Q., and Melissa A. Hofmann. 2011. "Next generation or current generation? A study of the OPACs of 260 academic libraries in the USA and Canada." *Library Hi Tech* 29 (2): 266–300.

The Cloud

Gresham, Anne. "Next-generation automation: The cloud and OCLC's Web-scale Management Services." http://www.annegresham.com/mls/courses/534/automation/Gresham-Anne-Assignment2.docx.

Hartman, Robin R. 2012. "Life in the cloud: A WorldShare Management Services case study." *Journal of Web Librarianship* 6 (3): 176–185.

Wale, Carla P. 2011. "Cloudy with a chance of open source: Open source integrated library systems and cloud computing in academic law libraries." *Legal Reference Services Quarterly* 30 (4): 310–331.

History

William G. Harrington. 1985. "A brief history of computer-assisted legal research." *Law Library Journal* 77:543–56.

NOTES

1. William G. Harrington, "A brief history of computer-assisted legal research." *Law Library Journal* 77 (1985): 543–556.
2. Jody Condit Fagan, "Discovery tools and information literacy." *Journal of Web Librarianship* 5, 3 (2011): 171–178.
3. Milad Shokouhi and Luo Si, *Federated search* (Hanover, Mass.: Now Publishers, 2011).
4. Jason Vaughan, "Chapter 1: Web scale discovery what and why?" *Library Technology Reports* 47, 1 (2011): 5–11.
5. Valeri Craigle, "Discovery layers in law libraries." *AALL Spectrum* 16, 3 (2010): 7–9.
6. Jason Vaughan, "Investigations into library Web scale discovery services." *Information Technology and Libraries* 31, 1 (2012): 32–82. http://ejournals.bc.edu/ojs/index.php/ital/article/view/1916/1744.
7. Jeff Barr, "Amazon Web services: Building a 'Web-scale computing' architecture." April 2, 2009, slideshow. http://www.slideshare.net/goodfriday/amazon-web-services-building-a-webscale-computing-architecture.
8. Craigle, "Discovery Layers in Law Libraries."
9. Marshall Breeding, "Preparing for the long-term digital future of libraries." *Computers in Libraries* 31, 1 (2011): 24–26.
10. Mary A. Axford, "Ultimate debate program on Web-scale discovery services: A report of the LITA Internet Resources and Services Interest Group meeting, American Library Association Annual Conference, New Orleans, June 2011." *Technical Services Quarterly* 29, no. 2 (2012): 142–147.
11. Anne Gresham, "Next-generation automation: The cloud and OCLC's Web-scale Management Services." http://www.annegresham.com/mls/courses/534/automation/Gresham-Anne-Assignment2.docx.

12

The Cloud

Roger Vicarius Skalbeck
GEORGETOWN LAW CENTER

In our digital age, successful and responsive law libraries are choosing to carefully and consciously adopt cloud-based solutions for their operations. Decisions to host services, systems, and servers in the cloud can have significant benefits for law libraries, large or small. Admittedly, there is some amount of hype around cloud computing. However, there is a good degree of hope beyond the hype, where advantages and opportunities outweigh drawbacks in most circumstances.

In this chapter, we first define broad categories of cloud computing to differentiate types of solutions law librarians should understand. Following this are examples from each category, focusing on perceived benefits to using cloud-based solutions in law libraries.

CLOUD COMPUTING: BASIC DEFINITIONS AND BROAD BENEFITS

Cloud computing is fundamentally a way of taking advantage of computing resources, tools, platforms, or infrastructure not hosted by your library. It's a way of leveraging computing power outside your organization. From infrastructure to software to services, law libraries are turning to cloud-based systems and platforms to meet a wide range of needs. The move to cloud computing changes the technology paradigm for libraries away from one of acquiring and managing physical equipment. Instead, this marks a movement toward treating technology as a service or utility, just like power or air conditioning.

Cloud computing does not mean that servers exist somewhere in the ether or in a celestial location. The concept is meant to imply that computing resource locations and hardware specifics are amorphous, and that location is largely immaterial. Cloud computing tools still run on hardware that uses electricity and requires Internet connectivity. Servers and equipment are run by technology staff, but they are people you'll probably never meet in a location you'll probably never visit.

Cloud computing allows law libraries to provide and improve services that may otherwise be too costly to maintain or complicated to establish. For many law libraries, demands for technical staff and equipment need to be balanced against the needs of other departments or business units. For instance, a law firm library may have to share technology needs with its lawyers and support staff. If this library had to "wait in line" to get a new server installed or to coordinate software upgrades, innovation and progress could be delayed. Success is more likely when a library is able to integrate servers or services that require neither dedicated talent nor complex in-house equipment.

Very broadly stated, cloud computing encompasses any technology services and capabilities not directly managed by a law library or its parent organization. Worrying about a specific definition seems less important than understanding the great potential and possible pitfalls in cloud-based services and platforms.

To understand some fundamental concepts about cloud computing, let's start with a framework adapted from a document published by the National Institute of Standards and Technology (NIST).[1] It gives good context for thinking about opportunities for law libraries. Also, because cloud computing is used across many organizations and industries, having a standardized foundation ensures that terms are used in a broadly-accepted fashion.

> Cloud computing is a model for enabling ever-present, on-demand access to a shared pool of configurable computer devices (e.g., networks, servers, storage, applications, and services) that can be rapidly created and released with minimal management effort or interaction with the service's provider.

Technology can be expensive to manage with servers and staff in your organization. This overview shows some key features attractive to law libraries today. Cloud-based solutions are offered as shared resources configurable to meet organization needs. When more storage is needed or more computing power is required, it can be quickly added. In times of lower demand or to cease certain projects, allocated resources can be released partially or entirely.

Following is a view of some of the essential characteristics, as well as a brief summary of the three major types of cloud computing service models. These definitions are adapted from the NIST framework.

Essential Characteristics

- *On-demand self-service.* A library can add additional computing capabilities, such as server time and network storage, as needed automatically without requiring human interaction with each service provider.
- *Broad network access.* Because all systems reside in the cloud, they can be easily accessed by any Web-based device such as mobile phones, tablets, laptops, and workstations.
- *Resource pooling.* Cloud providers pool resources in order to serve multiple organizations simultaneously, with different physical and virtual resources dynamically assigned and reassigned according to their needs and demands. This results in a significant cost savings that is passed on to the customer.
- *Rapid elasticity.* Organizations can quickly add computing power, expand storage space, and increase other resources through cloud services. They can also scale back if they don't require as much at any given time.
- *Measured service.* Cloud services are metered much as utility services and customers only pay for what they use. Usage can be monitored, controlled, and reported, providing transparency for both the service provider and library.

These characteristics are especially useful to assess for law libraries seeking to configure servers to run applications directly. There is great potential in the area of cloud computing for libraries needing to host complex solutions. As is shown below, it's possible to configure robust server technologies and host large data collections in the cloud, configured with Web-based tools.

Service Models

- *Software as a Service (SaaS).* This provides libraries the ability to use a provider's applications running on a cloud infrastructure. Applications are accessible from client devices through either a Web browser (e.g., Web-based e-mail), or desktop software tools. The library doesn't control or manage underlying infrastructure such as the network, servers, operating systems, storage, or even individual application capabilities.
- *Platform as a Service (PaaS).* This provides libraries the ability to deploy library-created or acquired applications created using computer code, services, and tools supported by the cloud provider. The library does not manage or control the underlying cloud infrastructure such as the network, servers, operating systems, or storage, but has control over deployed applications and configuration settings.
- *Infrastructure as a Service (IaaS).* The library has the capability to provision processing, storage, networks, and other fundamental computing resources where the library is able to run software, which can include operating

systems and applications. The library does not manage or control the underlying cloud infrastructure but has control over operating systems, storage, and applications.

The critical element of these three models is found in recognizing them each as a *service*. No longer are libraries required to maintain servers, equipment, and in-house staff to provide content and services. Definitions of service type are important to understand the types of problems each might be suitable to solve. For instance, here are some examples of each service model in use in libraries today:

- *Software as a Service (SaaS):* Gmail, Google Apps, LibGuides, Microsoft SharePoint, Web-based statistical tools
- *Platform as a Service (PaaS):* Heroku, Windows Azure, Google App Engine, OCLC Web Services
- *Infrastructure as a Service (IaaS):* Amazon Web Services (such as EC2), Google Cloud Platform

In some cases, lines blur between each of these service models. For instance, a library may use Pinterest to host images related to its own organization, instead of developing tools to support these needs. In this case, Pinterest works like a SaaS solution for a library. However, Pinterest itself uses Amazon Web Services to host its service and content, so they're using more of an IaaS model. Choosing the "correct" definition or category for a cloud-based service for your library is relatively unimportant. The important thing is to recognize that *service* is at the core of each model and not hardware in your server room or technical staff in your break room.

The NIST definition of cloud computing describes four deployment models for cloud services: private, community, public, and hybrid. The important concept to understand here is that cloud-based services might be provided on shared systems or dedicated equipment exclusive to your organization. The impact this may have on your organization is in the area of service reliability and availability. Whatever deployment model a service uses, libraries will be well served by scrutinizing service level agreements, data security, and performance metrics.

Unremarkably, technology tools and infrastructure can support numerous operations and services in law libraries. Let's look at examples for some specific areas where cloud computing is helping libraries solve problems today and into the future.

CLOUD COMPUTING FOR NEW SERVICES (SAAS)

Many law libraries already rely on the Software as a Service (SaaS) model of cloud computing. When a library wants to introduce a new service or dramatically change the way an existing service is implemented, there are many innovative services to consider. For instance, a library may provide Web-based forms for users to submit research questions that can be tracked online. Alternately, a library may run a live chat service or have a way to track performance metrics, such interaction at a reference desk or another service point. Following are library-specific options for providing new service features that would often be too complex to build or develop in-house.

SaaS Solutions for the Reference Desk

- *Altarama* (www.altarama.com): Altarama provides Web-based tools for libraries to manage reference questions and track internal statistics. The forms can be integrated with an existing website, and all content is hosted on their servers. Their offerings include RefTracker (online forms and question banks), DeskStats (statistics tracking), as well as live chat tools. Corporate and academic law libraries are among their client base.
- *Compendium Library Services* (www.compendiumlib.com): With their DeskTracker and KnowledgeTracker tools, Compendium Library Services offers online tools to manage reference questions and to track desk statistics, aimed to be tailored for local needs and practices. Their clients include public and special libraries, as well as some ARL libraries.
- *EOS.Web Enterprise Reference Tracking* (www.eosintl.com/modules/reference-tracking/): Available as a premium module of the EOS Web-based integrated library system, the Reference Tracking service is meant to capture online reference requests and give libraries an option for building a knowledge base of topical questions and answers. The EOS.Web Legal tools (www.eosintl.com/eos-web/legal/) are aimed specifically at public and private law libraries, especially in the area of traditional library automation.
- *SpringShare* (www.springshare.com): They provide four popular solutions for common library services. LibGuides is for research guides. LibAnswers covers knowledge base and FAQ development and support. LibCal is a calendaring application that also works for online room booking and event registration. LibAnalytics is a platform for tracking library statistics, such as reference desk or circulation desk interactions.
- *Topaz Research* (www.topazresearch.com): Their Quatrove service is aimed primarily at law firm libraries, which provides online tools for managing reference transactions, integrating client billing details, capturing materials for a knowledge base, and providing analytics for reference transactions. This service is meant to be integrated into an intranet or other protected network space.

These are but a few examples of the kinds of Web-based cloud solutions being used for reference support and activity tracking. Live chat is another common Web-based service offered by many law libraries. Some providers focused exclusively on live chat include Libraryh3lp (www.libraryh3lp.com) for academic and nonprofit libraries, LivePerson (www.liveperson.com), Comm100 (livechat .comm100.com), and a host of options to integrate instant message tools with Web-based widgets.

In choosing tools to support services in the area of research and reference, it's critical to look at workflow of these activities to increase efficiency. Adding a chat widget or a submission form to a webpage can be trivial work. However, if a form only transmits a request to an e-mail inbox, that can end a service's utility. For instance, if requests are handled by multiple librarians, a submission form alone is little better than publishing an e-mail address or owning a physical mailbox. Useful features to consider are question management, storage, and categorization. Some libraries create interactive Q&A resources and knowledge base tools to help solve tomorrow's questions with today's answers.

A library that cares about the quantity and quality of services helped by these tools must think about how relevant data are captured by the system. Also, in metric-conscious operations, it is essential to have statistics to inform decisions about service and staffing.

Many law firms and even law schools are using Microsoft SharePoint to manage internal systems. SharePoint can be deployed on traditional servers installed within your organization, and SharePoint is also available through some providers as a cloud-based solution. However it is implemented, SharePoint should be considered for workflow management and internal system tools if your library has access to it.

Of course, library vendors themselves offer options for hosted or virtualized servers they manage. Marshall Breeding predicts that traditional ILS systems will go away, to be replaced by a cloud-based "library services platform."[2] For instance, instead of upgrading an ILS running on internal servers, Pepperdine University Libraries adopted the OCLC's cloud-based WorldShare Management Service, which covers services for acquisitions, cataloging, and discovery.[3] This particular project is still very new, but their list of member stories illustrates librarian views and reactions to this offering.[4]

Office Operations: E-mail, Documents, and Collaboration

Web-based e-mail and document storage solutions are an area where many businesses are taking advantage of cloud-based services. A great benefit with these is that they can be easily configured to access on computers, smartphones, and tablets, and in a variety of apps. Many universities and even municipalities have moved to services like Google Apps for Domains, which provides e-mail, calendar, and document management platforms. Microsoft also remains competitive in this area, with their SkyDrive service and online collaboration options in their latest Office Suite tools and cloud-based platforms.

- Google Apps for Business and Education (google.com/enterprise/apps/busi ness and google.com/enterprise/apps/education): Providing e-mail support, calendaring, and document collaboration, Google provides significant core services for libraries, businesses, and public sector organizations. Alone in the education sector, their customer list includes the University of Michigan (90,000 users), Northwestern University (24,000 users), as well as Nigeria's University of Benin (40,000 users) and England's University of Westminster (20,000 users).[5]
- *DropBox* (www.dropbox.com): DropBox was developed as an online storage solution for storing personal and shared documents, which the company sees as a "file system for the Internet."[6] The service integrates with many smartphone and tablet apps, and it also provides very good synchronization options with desktop computers. In early 2013, the company added features for photo management that may rival Facebook for social content sharing,[7] and they may move to more of a content-delivery model.[8]
- *Microsoft Live & SkyDrive* (skydrive.live.com): Between Windows 8 and the Windows Phone operating systems, Microsoft bundles connectivity with their cloud-based file storage system SkyDrive. Coupled with this, the latest version of the Microsoft Office suite integrates with cloud-based applications, and they are moving to a subscription model for their products.

For companies providing cloud-based file storage, the "table stakes" have turned into a limited amount of free storage, as well the ability to access data on multiple device types. Tools that support collaboration, publishing, security and data management are the next area of competition for these services. Some companies such as Box (www.box.com) have started providing enterprise-wide document hosting for colleges and universities, with 85 institutions signed up by early 2013.[9]

CLOUD COMPUTING FOR NEW INFRASTRUCTURE (IAAS)

Libraries should look closely at options to host the entire library infrastructure in the cloud on virtualized servers. This is one of the more radical departures from traditional library systems architecture. Whether it is for your organization's website, an institutional repository, the integrated library system, or all of the above, there could come a day when cloud hosting is the rule instead of the exception.

Some of the biggest players in this space include Amazon, Microsoft, and Google. Let's look at the types of services they offer to get a feel for whether, when, and how libraries might move to these models. Though these four providers differ in cost, experience, and price, they all provide one core approach to cloud computing. They offer systems infrastructure configured through a Web browser on massive servers and data centers running virtualized systems.

Amazon Web Services (aws.amazon.com)

Begun in 2006, Amazon offers a series of Web services, now offered under this umbrella name. Especially for newer Web-based companies, "Amazon Reshapes Computing"[10] as we have known it. Their offerings include:

- *Elastic Compute Cloud* (EC2): On this platform, you can configure servers in a number of configurations and install your own software on a very flexible server infrastructure. They offer multiple server types and dozens of database engines. A typical system setup uses what's called an Amazon Machine Image with software, operating system, and all settings preconfigured. This can be a simple WordPress site, a Drupal instance, or many complex configurations available through an online marketplace.
- *Simple Storage Service* (S3): This aspect of Amazon Web Services is focused on data storage and retrieval stored on redundant and extremely fast servers. Objects managed through S3 can range from 1 byte to 5 terabytes for each item. Through the AWS setup, you can have secure and customizable access keys to serve data and other objects for your application.

Google Cloud Platform (cloud.google.com)

Google developed their cloud service later than Amazon, so their products are not as well-developed in some areas. For instance, Google's platform for virtualized Linux services, called Google Compute Engine, was announced in the middle of 2012 and was still a limited preview in early 2013. Many of Google's cloud platform solutions provide limited free access to test the services, though you typically have to enable billing to cover costs when tests exceed free quotas.

Here's a quick overview of their other cloud platform products:

- *Google Cloud Storage*: Online storage and content delivery for vast amounts of data.
- *Google BigQuery*: Intended for working with big data sets, this cloud service offers a robust and powerful computation engine for running queries against extremely large data sets.

Beyond these, other companies, including Rackspace (www.rackspace.com), offer cloud storage and application solutions, which has the effect of driving down costs, though not to a level that might seem cheap for complex computing. Choosing to host entire portions of your server infrastructure in the cloud requires a degree of coordination and commitment. Two major contenders for systems libraries might put in the cloud can include a library's website or the integrated library system.

CLOUD COMPUTING FOR DEVELOPMENT (PAAS)

With PaaS, a computing platform is provided that supplies tools and a development environment to help companies and developers build, test, and deploy Web-based applications. Businesses do not need to invest in the infrastructure required for building Web and mobile applications but can rent the use of platforms such as Windows Azure, Google AppEngine, or Force.com in order to build their applications and get them up on the Web much faster. Because each platform is developed by the company providing it, systems are optimized for performance, speed, and reliability. One thing to be aware of however, is that the applications which are built using these provider's services are usually locked into that one platform.

- *Microsoft Windows Azure* (www.windowsazure.com): The Microsoft Azure service lets you build and deploy Web servers, virtual machines, data repositories, and mobile services. Azure excels at Microsoft branded products, such as SQL Server, Asp.NET, and Windows Server. They also offer all major non-Microsoft products, including MySQL, Linux, Apache, and other widely used technologies.
- *Google App Engine* (https://developers.google.com/appengine/): Built on the same platform as many of Google's own products, this is a set of tools that organizations can use to deploy applications using Java, Python, or Go with *application programming interfaces* (APIs) to interact with the system. It powers sites like the Khan Academy.
- *Google Cloud SQL* (https://developers.google.com/cloud-sql/): A solution to run MySQL databases in conjunction with App Engine applications on Google's cloud infrastructure.

For libraries, a major group active in this area is OCLC. For developers, they provide a series of Web services to interact with OCLC sources, such as querying the WorldCat database or resolving OpenURL entries. Bringing together many of these Web services is the OCLC WorldShare Platform,[11] which has three main components: a *framework* accessible through their Web services, *architecture* to enable library-developed and OCLC-created applications, and an *infrastructure* to support shared applications and tools across a community of library users.

BEST PRACTICES: INTEGRATION AND CONFIGURATION

When integrating cloud-based services into library operations, libraries should think carefully about how users experience these services, especially if they are new or are moved from one platform to another. Following is a list of configuration considerations and integration ideas to ensure compatibility for the library's long-term goals.

Prefer to Use Your Own Domain URL

When selecting a service to be run entirely on an external server, it is possible to configure a hosted or virtual service to use a URL that includes your organization's own domain name, instead of needing to adopt one within the vendor's commercial domain. At times there's an extra cost and some server settings to configure. Here are two example URLs. There's nothing wrong with a commercial domain, but consider the differences of these for the same hosted service:

- Jacob Burns Law Library at George Washington School of Law: http://law.gwu.libguides.com/
- Harvard Law Library: http://libguides.law.harvard.edu/

Prefer Flexible URLs

With many SaaS types of services, part of the URL path relates specifically to that service. Whether it's an institutional repository, your research guides, or even your catalog, you need a short string of text to reference in the service's URL. Having flexible URLs can give a broader range of options when a service's scope changes. If you host a class schedule at something like: www.mylibrary.org/classschedule or classes.mylibrary.org, what happens when you need to add nonclass events to this site?

Read the Terms of Service Carefully

When contracting for cloud-based services, you need to read the provider's terms of service carefully. It is important to look for any statements of who will own the data, and your rights to get the data back if a service contract ends.

Beware of Vendor Lock-in

When selecting a service, be conscious of how reliant you are on a particular vendor for this service. If a vendor's prices change, or if a competing service becomes available, consider how easily your library can change to something new. Many law firms and universities have been around for 100 years or more. It's unlikely any cloud service will last that long.

Create a Backup Plan

As part of disaster and recovery plans you should do across your library, be sure to include a plan specific to cloud-based services as part of this. For instance, what will you do if the Amazon Web Service goes down, or if there are service outages for your hosted research guides or live chat?

Consider Compliance

Law firms have an added need to be careful with client information and sensitive files. Part of this is a need to evaluate data integrity and service reliability. Beyond this, law firms need to investigate what can legally be stored in the cloud.

Develop a Data Storage Policy

Understand and review any laws governing precautions needed for sensitive information such as social security numbers, credit card information, and even names and e-mail addresses. Your library should have a policy for what is and is not allowed to be stored in the cloud, and you should review provider terms to see that they comply with your policy goals.

DIFFICULTIES AND DRAWBACKS IN CLOUD COMPUTING

Internet and Network Performance

Every cloud-computing project considered in this chapter assumes one important element: reliable Internet access. This is true for accessing virtualized servers as well as hosted solutions and integrated platforms and networks. Internet unavailability and service disruptions can cause major problems if core network services are not reliable.

Reliability

The first time any organization decides to undertake a cloud solution for its services, there is a degree of uncertainty to the reliability and speed of this service.

By all means, libraries should review service level agreements and talk to others who have adopted certain technologies. In the area of SaaS solutions, such as virtual reference or live chat, you can often find peer libraries online to evaluate performance directly.

However, with a cloud-based infrastructure, it might be difficult to find peer comparisons. In some circumstances, it may be difficult or impossible to know if a library's website or catalog is hosted virtually. In fact, it's only when a site performs poorly or goes offline that you will notice problems. Whenever possible, it's a good idea to ask service providers for names of similar clients who have adopted the service under consideration.

Privacy

When your data are stored on shared servers or in remote locations, it is critical to investigate policies and safeguards employed to protect these data. In some cases, there may be few concerns. It seems unlikely that a record of a library's holdings will be sensitive. However, circulation records for those items and account data for people using those services should be treated differently.

Data Portability

It is always important to have a backup plan for your data and applications. Irrespective of where applications are hosted, law libraries need records of service transactions, backups of unique data objects, and an archiving and retrieval strategy. With cloud services, having a backup onsite or in a mirrored location is useful. Beyond that, it is helpful to consider contingency plans in the event a service is discontinued or becomes too expensive to manage.

Whether contracting for a new service or configuring a system on a remote machine, always ask about rights and procedures for obtaining your own data. Beyond a blank promise of "trust us, we'll give you what you want," libraries should explore data formats and record structures. If items are stored in relational databases, you need to preserve those relationships in order for data to be portable from one system to another.

Private Library Considerations

A large number of law libraries exist in private organizations such as corporations or law firms. For these groups, all of the noted drawbacks and difficulties will apply. With some, such as a security and data privacy, additional considerations are needed, such as the geographic location of data that might not be stored on servers in your country. Beyond this, private organizations should evaluate firewall access issues, privacy of privileged communication, and any potential exposure problems with data transmission.

Private Channel Communications

Especially in the area of SaaS kinds of services, features may tend to favor creating resources that foster public or flexible community communication channels. Consider Facebook and Twitter. They both enable vibrant and flexible channels of communication, providing attractive models of interaction and information dissemination. Unremarkably, a law firm library would be unlikely to use Facebook for internal communications. Instead, similar models might be achieved with private channel communications, such as live chat authenticated through a firm's Active Directory server. Alternately, a product such as Yammer, owned by Microsoft, is meant for secure and internal enterprise communication.

Secure Data Transmission

Private law libraries must ensure that data are secure when submitted and transferred, as well as for storage more generally. For this, forms or tools that might convey confidential or sensitive information should be sent over secure connections. Any highly sensitive data transmission should also be encrypted if reasonably possible.

CONCLUSION

It is inevitable that law libraries will move to more cloud-based solutions for services and content access. With hosted services and platforms, many law libraries have adopted cloud solutions with great success. Competition between Amazon, Microsoft, Google, and other industry players should hopefully drive costs down. Also, lower-bandwidth operations are often economical already. Moving away from dependence on hardware ownership creates flexibility both for service performance as well as for business continuity and location independence. Cloud-based solutions can also create a more mobile workforce. Also, in the very near future, the idea that a law library will have a "cloudy future" may become a positive expression. Law libraries that are able to successfully leverage cloud infrastructure solutions should benefit from the flexibility and hardware independence it can create.

FURTHER READING

Black, Nicole, and American Bar Association. 2012. *Cloud computing for lawyers*. Chicago, Ill.: American Bar Association, Law Practice Management Section.

This is an overview of cloud computing topics for lawyers, written by lawyers, covering topics such as ethics, security, privacy, and implementation. There are good examples of virtual law practice as well as cloud technology tools used at law firms today.

Bradley, Sharon. 2012. "Ethics on the wing: Examination of opinions on electronic services and cloud computing." *Continuing Legal Education Presentations*. http://digitalcommons.law.uga.edu/cle/March_2012/Schedule2012/3/. This report covers state bar opinions in the area of electronic storage and access to client information. This includes a chronological summary of these opinions, as well as a checklist assessing cloud services and one's own competence to evaluate and monitor them.

Breeding, Marshall. 2011. "The systems librarian: A cloudy forecast for libraries." *Computers in Libraries* 31 (7). http://www.infotoday.com/cilmag/sep11/Breeding.shtml.

Breeding, Marshall, and Library and Information Technology Association (U.S.). 2012. Cloud computing for libraries.

Corrado, E. M., and H. Moulaison. 2012. "The library cloud." *Library Journal* 137 (4): 49–51.

Dula, M., L. Jacobsen, T. Ferguson, and R. Ross. 2012. "Implementing a new cloud computing library management service." *Computers in Libraries* 32 (1): 6–40. Discusses a project between Pepperdine University Libraries and OCLC. It discusses how Pepperdine University transferred its library management functions to the cloud using OCLC's Worldshare Management Services (WMS). It outlines the stages of adopting and deploying WMS, described as the first Web-scale, cooperative set of library management services.

Marchini, Renzo. 2010. *Cloud computing: A practical introduction to the legal issues*. London: BSI Print.

Murley, Diane. 2009. "Law libraries in the cloud." *Law Library Journal* (Spring): 249–254.

Wale, Carla P. October–December 2011. "Cloudy with a chance of open source: Open source integrated library systems and cloud computing in academic law libraries." *Legal Reference Services Quarterly*: 310–331.

Wang, Lizhe. 2012. *Cloud computing: Methodology, systems, and applications*. Boca Raton, Fla.: CRC Press. This is a well-researched collection of materials covering both technical implementation as well as organizational planning of successful cloud-based solutions.

NOTES

1. "The NIST definition of cloud computing," special publication 800-145, September 2011, http://csrc.nist.gov/publications/nistpubs/800-145/SP800-145.pdf.

2. Edward M. Corrado and Heather Lea Moulaison, "The library cloud pros and cons—The digital shift," *Digital Shift*, March 5, 2012.

3. M. Dula, L. Jacobsen, T. Ferguson, and R. Ross, "Implementing a new cloud computing library management service," *Computers in Libraries* 32, 1 (2012): 6–40.

4. "Member stories: OCLC WorldShare Management Services," http://www.oclc.org/us/en/webscale/success.htm.

5. "Google Apps for Education: Customer list," http://www.google.com/enterprise/apps/education/customers.html.

6. Jason Pontin, "Dropbox: Founder Drew Houston simplifies the cloud," *MIT Technology Review* (February 21, 2012), http://www.technologyreview.com/qa/426976/dropbox-founder-drew-houston-simplifies-the-cloud/.

7. Tom Simonite, "Dropbox photo upgrade threatens Facebook's idea of social," *MIT Technology Review* (January 31, 2013), http://www.technologyreview.com/view/510621/dropbox-photo-upgrade-threatens-facebooks-idea-of-social/.

8. Kim Eaton, "Dropbox reveals a future where more 'content' is shared, less files," *Fast Company*, January 31, 2013, http://www.fastcompany.com/3005311/dropbox-reveals-future-where-more-content-shared-less-files.

9. Liz Young, "BuckeyeCloud storage to cost Ohio State $250k yearly," *Lantern* (January 2013): 16.

10. Quentin Hardy, "Active in cloud, Amazon reshapes computing," *New York Times*, August 28, 2012, 27, http://www.nytimes.com/2012/08/28/technology/active-in-cloud-amazon-reshapes-computing.html.

11. OCLC WorldShare Platform, http://www.oclc.org/developer/platform.

13

Social Software

Marcia L. Dority Baker
UNIVERSITY OF NEBRASKA COLLEGE OF LAW, SCHMID LAW LIBRARY

> so-cial *adj.* 1. a. Living together in communities. b. Of or relating to communal living. c. Of or relating to society. 2. Living in an organized group of similar close aggregate.
>
> soft-ware *n.* The programs, routines, and symbolic languages that control the functioning of hardware and direct its operations.
>
> —*Webster's II New College Dictionary*, 2001

Libraries, and law libraries in particular, are in the supply and demand business. As the demand for information from our users has increased, law libraries are finding new ways to supply those information needs with technology. The many types of social software now available allow law libraries to customize how we meet these information requests from users in a rapidly changing world.

So, what is social software? According to Wikipedia—yes, Wikipedia, one of the oldest social software sites, founded in 2001—"Social Software" comprises the interactive tools used to communicate. *Webster's New College Dictionary* defines "Communication" as the exchange of information, messages, or ideas. Humans have been communicating in some form or another since the dawn of time; our most current adventure with twenty-first-century technology is a big step up from the running messengers of the ancient world to the telegraph machine of the nineteenth century. The common denominator is not the technology but the need to communicate with each other.

The phrase "social software" encompasses a number of technologies used for communication purposes. The spectrum includes social media sites, social collaboration opportunities, social cloud storage, and social organization—all options allowing users to communicate, share, and exchange ideas with others. As libraries of all types evolve as hubs of information services, social software offers many opportunities for law libraries to engage externally, with library users and for law librarians and staff to better communicate internally. While some social software sites such as Facebook or Twitter allow the law library to promote library information with interested "friends" and/or "followers," other social software like instant messaging (IM) help law libraries' workflow and productivity by providing law librarians instant contact no matter where in the building workspace is assigned.

Law libraries actively use social software to help users find answers, share information, and communicate with others. This chapter reviews what types of social software are being used by law libraries, as well as how law libraries are incorporating social software in the real world, the ever-changing digital realm, and the evolving third place of life. It also discusses current issues and concerns that both law libraries and law librarians face, and reviews best-practices advice from law librarians working in the field.

WHAT TYPES OF SOCIAL SOFTWARE ARE LAW LIBRARIES USING?

To determine what social software law libraries use, a SurveyMonkey survey, "Social Software in Law Libraries," was distributed to members of approximately eight American Association of Law Libraries (AALL) LISTSERVs in December 2012. The 101 total responses included law librarians from academic law libraries (55 responses), state law libraries (4 responses), firm libraries (23 responses), court law libraries (9 responses), county law libraries (7 responses), and others (3). The ten-question survey asked a variety of questions relating to the kind of law library, the types of social software being used, responsibility for the social software, policies regarding the social software, as well as advice and best-practice suggestions.

After reviewing the survey results, the list of social software used by academic, firm, state, court, and county law libraries is both broad in possibilities and deep in engagement. Social software is being used by law libraries for a variety of reasons: (1) specific communication purpose, (2) building online relationships with friends and/or interested library users, (3) as collaborative software and programs for sharing data, (4) online storage options (for cloud computing, see chapter 12), (5) mobile check-in features, and (6) as organizational tools. The sky is truly the limit.

Social Media

The most obvious social software options are also the most popular with users around the world. Social media sites such as Facebook, Twitter, and LinkedIn attract millions of users. While these sites are very social in nature—allowing members to share information and photos, brief ideas in 140 characters, or professional experiences—social media sites are also used by law libraries to connect with friends and followers, or by law librarians to build professional networks.

A number of law libraries (thirty-five survey responses in the "Social Software in Law Libraries" survey results) are using Twitter to engage library users with short and quick tweets to specific information requests. For instance, the Oklahoma City University Law Library is using Twitter (@OCULawLibrary) at the reference desk to interact with students. Questions received via Twitter for reference librarians that pertain to the Oklahoma City University campus, law school, or law library are answered throughout the day Monday through Friday. A larger number of law libraries, fifty-five responses to the survey, also created and maintain a law library Facebook page for interested Facebook users to "like." Per e-mail communication with a MU Law Library librarian in regards to using Facebook, the purpose of the University of Missouri Law Library's Facebook profile is to connect with students and alumni in an informal fashion. To do so, the library staff uses the MU Law Library's Harry Tate Facebook page to engage students; as an alumnus of the law school, Harry Tate comments on local events, provides library updates, and encourages students throughout the academic year.

The broad range of social media sites also includes YouTube, Google+, and wikis. Many academic libraries, including law libraries (nineteen survey responses), are using the official YouTube channel provided by their affiliated law college or institution to post informative library videos, research tutorials, and library guides for student orientation. While initially Google+ (five responses) has not performed as well as expected, it is still a viable option for law librarians to communicate with other librarians or users within a determined "circle," especially those law libraries using Google's Web-based services. Google+ is being used by groups within law libraries, such as Teknoids, to create space for collaborative exchange of ideas, allowing interested members of the community to discuss information during a "hangout" event.

Wikipedia is the largest and best-known example of a wiki. Law libraries use wikis (4 survey responses) in-house for library specific needs such as: an internal Internet (intranet), as a reference question wiki, and as a cataloging or a technical services how-to wiki. The Hamilton County (Ohio) Law Library uses an internal staff-only wiki to post documents for library policy and procedures, staff member contact information, vendor contact information, and notes from meetings or conferences. External use of wikis by law libraries can include a collaborative means to complete law library committee work, a library FAQ for the public and a dedicated space to share legal resources and links with library users. Wikis were one of the first collaborative social software options allowing multiple users the ability to create, edit, and share information with a large audience. While wikis and Wikipedia are still relevant in today's world, the evolution of technology has created new social communication opportunities.

Law librarians are not immune to trying new social media sites and communication options. A small number (four survey responses) of law libraries are using the popular Pinterest site to "pin" items the law library staff find of interest; this includes new law books, legal movies, law school or law library photos, local attractions, legal humor, and even banned book week titles. Law students are using social media to document their days and nights spent in the law library. Instagram photos are a creative way to show off the space where a large amount of time is spent studying and preparing during law school. The Instagram photos tagged #lawlibrary demonstrate the beauty found by users within the law library's walls, a visual way to communicate with one's social group. Law libraries are also using Flickr (two survey responses) to capture what's happening in the library, as a picture is worth a thousand words.

Internet radio is an interesting example of "old" technology evolving to fit a new platform. Blog Talk Radio provides Internet airtime for radio shows, as well as a chat room for show attendees to communicate in, and the option to add each radio show to iTunes for listeners to download at a later date. The University of Nebraska College of Law, Schmid Law Library, has used Blog Talk Radio to produce the *Law Librarian Conversations* online radio show since September 2009. This radio show focuses on topics and issues for law libraries in general. Several special shows have featured firm librarians discussing what summer associates should know, law library technology such as Google Scholar, and interviews with well-known law librarians, law vendors, and publishers. Each episode of the show, either live or downloaded, is heard by several hundred law librarians and interested listeners, one more social communication option for law libraries.

Blogs are the shorthand term for "Web log," an online discussion of a particular topic available via blogging software. Currently, WordPress is the most popular platform for bloggers, while the oldest platform is Blogger, which started in 1999 but was absorbed by Google in 2003. A new blogging trend is using Tumblr, a microblogging platform, to connect instantly with users in a fresh format. "Tumblogs" are a hybrid blog and social media site allowing users to post original content along with posts from blogs one subscribes to. Librarians have used blogs for a number of years to share information and ideas on librarianship, technology, programs, and emerging trends. Law libraries use blogs (fifty-three survey responses) to share library-specific information to students, to discuss new technology with firm lawyers, and to post association or chapter reports and events to members. While the law librarian can add new posts to a blog on a regular basis, the expectation of regular feedback from readers is not assumed. Instead readers can post comments on a specific post or share the post with friends on most social network sites.

LibGuides, a product from Springshare, is a Web-based application allowing libraries to create and share interactive subject guides on specific topics with users. A number of law libraries (there are approximately thirty-eight law libraries listed in the Higher Education section and two law libraries listed in the Special Libraries section on the LibGuides Community website) are using LibGuides for discreet projects: as subject guides for classes such as bankruptcy, immigration law, or property; as a resource guide for a seminar class; and as a research guide for state-specific information. Instead of law librarians creating paper bibliographies and handouts for library users, LibGuides provides a digital platform for law librarians to create interactive guides with links to appropriate databases, current headlines in the subject area, and relevant secondary resources available via the law library, in print or digital format.

Social bookmarking became popular in 2003 due to the URL organizing website Delicious (three survey responses), a user-friendly social bookmarking site. Delicious was acquired by Yahoo! and then AVOS system, but is still available for users who want to organize and share bookmarked websites across devices or computers. Xmarks is another bookmarking site allowing users to organize, access, and share bookmarks across devices. Social tagging has morphed from organizing bookmarks to organizing other types of content; some social media sites such as Facebook include the ability to tag people in photos. The Encore discovery service by Innovative Interfaces Inc. (III) allows community tagging by library users within the library catalog. Special collections or material for a specific need can be organized by standard tags to find the material more efficiently, especially for collections housed in multiple sites. One example of community tagging is the Ethics Center Collection at the University of Nebraska-Lincoln (UNL). Initially, the library staff at the University of Nebraska–Lincoln Libraries and Schmid Law Library tagged the ethics collection resources by discipline, and then allowed library users to continue tagging the material. Since this collection spans several library locations and two campuses, the tagging feature organized the material in a user-friendly fashion and helped interested users find relevant material within a particular field of study. Crowdsourcing is another way of social tagging; this provides ownership of information to library users and allows large projects to be completed without using the library's limited resources. Wikipedia is a great example of the power of harnessing crowdsourced information, while the New York Public Library (NYPL) has finished a number of digital collections projects with the help of crowdsourcing.

Social Communication

Social communication software includes the programs that allow for communication of ideas or information between users; options include texting and instant messaging (IM). The common denominator between social communication software is the sharing feature; typically one user is pushing information out but not necessarily expecting a response. For instance, many law libraries use IM (twenty-six survey responses) in the library for the convenience of users who need information but do not want find or speak directly to a law librarian or library staff member. Text or chat reference services or programs allow library users the ability to ask for reference help while located elsewhere; libraries of all types offer a "text-a-librarian" service and law libraries are no different. Law libraries can provide the communication option, but it's up the customer to use it. Unfortunately, the primary instant messaging service used by numerous academic and public libraries is no longer an option; Google acquired Meebo, a free and easy-to-use IM service in June 2012. As social software options are bought or absorbed by other social software giants, it can be a challenge for law libraries to maintain service features for users. It is imperative that law librarians keep up with technology, including social software options and emerging trends.

Social Online Storage

Social online storage is a convenience in our mobile society. The need to access and store information in a variety of formats or on multiple devices is essential to law librarians and information users. There are a number of services available to store and share information "in the cloud." Storing information in the cloud is when a third-party vendor hosts or stores digital content for a user; this service typically has a free basic account with a set limit for digital storage than can be increased for a fee. There is more discussion on third-party vendors later in this chapter. Cloud storage providers such as Box.net or Dropbox (seventeen survey responses) also allow account users to share and manage content with other users. Law librarians use cloud storage to work on projects between organizations or to back up professional content for access outside the building and to send large files for document delivery service. For instance, the chapters for this book were uploaded to individual Dropbox folders to share with the editor.

Organizational Tools

Organizational tools or social media management systems (SMMS) allow law librarians the ability to create and organize information for library users across multiple channels. SMMS tools are a one-stop shop for sharing information to multiple social media sites at once. Organizational tools software can also manage social media website statistics to help law librarians gauge the relevancy of what the law library is posting. Some SMMSs have free basic accounts while other organizational tools are for-hire services. Organizational tools such as HootSuite (eleven survey responses) and TweetDeck (four survey responses) provide users with a dashboard control panel to post and integrate content to a number of social media platforms such as Twitter, Facebook, LinkedIn, and Foursquare.

Librarians use TweetDeck because it allows users the ability to manage the multiple people they are following by organizing the feeds by type such as: librarians, marketing or communications, personal, and so forth. It has a robust search option to find Twitter profiles and allows users to communicate directly with other Twitter users within the TweetDeck application. The SMMS Hootsuite is popular with users as it provides a calendar feature to schedule future tweets in advance and will post to multiple social media sites at once, including Facebook. Law libraries encouraging several librarians to manage the library's online presence can use a SMMS site to schedule updates or to post information in tandem with the organization's calendar, being consistent on what is shared across channels. Organizational tools software helps law librarians with time management and creates a professional profile on the social media sites the law library subscribes to. It is important to review the law library information for content, grammar, and library voice before submitting or scheduling the post in TweetDeck or HootSuite, as it is sent to a number of platforms instantaneously. Instead of "Measure twice, cut once," librarians should practice "Read twice, post once."

Location-Aware Mobile Social Networks

The ability to "check in" at a location has created new location-aware mobile social networks. Currently, Foursquare is the most popular location-based social software (six survey responses), allowing users to check in at favorite places. It may not be an application used by law libraries but law librarians should be aware of Foursquare, as many students are using it to check in with friends at the library. As useful as identifying where one is currently located can be, there are other location-aware mobile social apps more productive to smartphone users. Organizations and universities can use Google Maps to create location-awareness applications for visitors to find and locate offices or campus buildings. The ability to layer relevant information with tags and identify locations of interest for future customers or students allows location-aware mobile social networks robust connectivity with users and instant feedback of services. While users may be annoyed at having to "okay" the use of their location with a mobile app on a mobile device, this permission allows the application to provide location-specific information to the device user. Several concerns about privacy and security of location awareness apps have increased discussion of the use of such apps, especially when a particular app broadcasts user location to those outside a defined network of friends or followers.

Social Collaboration Software

Librarians are organizers by trade; books, databases, technology, and library functions are just a few of the many things that law libraries catalog, collect, and streamline. Social collaboration software such as Google Drive (formerly Google Docs), Yammer, and Zoho assist law libraries in managing the internal work that allows law librarians to better assist library users. These Web-based "office suites" provide a free or low-cost option for libraries to operate in a virtual office instead of purchasing software to install to library computers. Law libraries are using the Google Calendar function to manage circulation and/or reference desk shifts or share work documents via Google Docs twenty-six survey responses). Zoho Creator functions as an in-house database for library staff, while Yammer (one survey response) provides libraries with private social networks for employees.

Social software will continue to morph as technology evolves and communication options change. What is here today may not be around in five years or even in the next decade but the need to communicate remains constant. Law librarians must follow technology and communication trends to appropriately interact with users but should not select all the available options. Part of being an information professional is the ability to pick and choose what is most suitable for your law library and primary users, and then implement the social software options to the best of the law library's ability.

WHY ARE LAW LIBRARIES USING SOCIAL MEDIA?

The primary reason "why" law libraries use social software is for communication purposes. As discussed throughout this chapter, the need to communicate with one another is part of being human. Law libraries use a number of channels, both analog and digital, to connect with and market to library users. The "Social Software in Law Libraries" survey found that 95 percent of law libraries use social software for communication purposes. Other motives for using social software include the desire to connect or engage with library users and law library marketing opportunities available via the social software channels. Law librarians are a curious group; the desire to play with new technology and to have an online presence are other reasons why law libraries use social software. If library users are on a particular social software site, then most likely a law librarian has tried it—either for personal interest or to scout out how the law library could engage users via that social network.

Understanding why and how law librarians use social software is important, especially since technology users can try new and free tools 24/7 or easily participate in the latest technology trend. But understanding *why* law libraries are, or should be, engaged with social software is essential. What is the goal for a project or the reason why law libraries should support staff time and budgets toward social software engagement? A basic game plan is necessary for law libraries to best harness the information the law library provides to the intended audience and on the appropriate platforms. When the *why* is understood, then the *how* is evident; if a law library wants to engage library users via social media, then it should join and use the social media sites its users are involved with. As exciting as jumping into social software options can be, setting boundaries and understanding why and how the law library should use social networks will create a more professional online presence and hopefully prevent future communication issues.

SOCIAL MEDIA POLICY FOR LAW LIBRARIES

Law libraries are actively using social media to engage library users who range from law students to practicing attorneys. The many social media sites that law libraries participate in is broad, but the most important part of using social media is understanding why and how the law library is dedicating energy toward this type of communication. Let's discuss what should be included in a law library social media policy. There will be a few items specific to the type of law library (e.g., academic, court, firm), but in general the following are suggested: (1) Define what social media means to the law library, on what platforms it will engage users, and how often the law library will post to these sites; (2) What are the goals for using social media and the strategy to achieve these goals? (3) What are the basic parameters for the law librarians to best manage the law library's online presence? What library information is appropriate to post/share or not, and what "voice" will the law library post in? (4) Who is responsible for the law library's social media presence, to whom does this person(s) report? (5) How often will the social media policy be reviewed?

First and foremost, why is the law library engaging in social media? As individuals, joining social media makes sense: One can share updates, photos, and location information with friends and followers, allowing for personal interaction in the digital world. Yet this personal interaction is different for organizations such as law libraries. While social media can help the law library fulfill its mission, how social media works must be understood so the digital content remains professional, useful, and pertinent to busy library users. Since most law libraries use social media to communicate with library users, what type of information does the library want to share? This could include general law school or law library information, highlights of library resources, timely information on local events or attractions, and links to relevant legal articles. This simple question should set the basic boundaries of a social media policy; once it is decided what content the law library will share, the guidelines will include how often and why library users may be interested in this information.

Social media is a bigger pond to play in, a place where law libraries can share information with users in the digital world; setting the guidelines before the law library gets in the water can prevent a tidal wave later on. According to the results of the "Social Software in Law Libraries" survey, over 90 percent of law libraries do not have a social media policy for staff. Although a formal policy may not be in place, law librarians understand the need to represent the law library in a professional manner. Several survey responses mentioned that a social media policy was in the works. Law libraries crafting policy can look to other law libraries or library associations for assistance; a number of blogs and library conference presentations have discussed how to create a social media policy as well. Examples of social media policies are available online at library and law association websites, or review the chapter resources list for suggestions.

As a law librarian, there are many places available to express what one does, to describe the benefits of using the law library, and to promote library services; social media options are one more avenue to share this information with a greater audience. That being said, it is necessary to have a common "voice" for posting law library information, especially if more than one law librarian is responsible for maintaining library content on social media sites. The "voice" is the professional tone the law librarians post information in. Some libraries require librarians to include their initials in each post to easily identify who is speaking for the library; others have librarians post under the law library's name as a corporate user. It is necessary for law librarians to separate personal posts from the law library's posts to maintain professionalism while online. If everyone involved at the law library is aware of why the library is using social media along with the rules and guidelines, then the wealth of library information posted for users will increase. It is also important that all library staff members, from administration to part-time employees, are informed of what social media sites the law library participates in and who is responsible for posting information; transparency within the physical law library is just as necessary as online.

While knowing the law library's communication chain-of-command is vital, so too is nimbly responding to changing technology formats and trends. A social media policy is only relevant to law librarians if timely and reviewed for necessary updates. Since law librarians must keep up with evolving technology and information needs, assessing the law library's social media policy should be added to the list of annual items to review. Other opportunities for evaluation include the hiring of new law librarians, administrative changes, or when major technology updates or migrations occur. It may be a good idea to review the law library's social media presence at regular meetings for the law librarians or staff who are responsible for social media, and with other communication departments within the institution or organization. Since social media can be viewed as a marketing tool, it is imperative that the law library use current branding, logos, and information to professionally communicate with users. Engaging with users online is the benefit of social media but the face-to-face involvement of library staff within the organization will help build the word-of-mouth support for digital library services.

Finally, it is necessary to discuss how to handle library users' comments or engagement with the law library. For instance, should the law library respond to students taking Instagram photos in the law library or tagging friends in photos posted by the law library on a social media site? If so, how will the library respond to online referrals or comments to posts? Law library staff members responsible for the library's online presence need guidelines outlining the appropriate way to best respond to good and/or bad comments and whom to report such comments to. Permission forms may be necessary for posting online photos of library users or dialog from students promoting library services; it is best to include these requirements in the social media policy. When crafting guidelines, it is important to discuss if the policy is good for the law library and staff, and how the social media policy may affect user's engagement with the library. In the long run, a social media policy will provide a professional path for law libraries to communicate with users, as well as precedent for the law library as a third place. Libraries exist in more than physical buildings and the digital environment; they are a safe place for library users and an area many law students spend a great deal of time while in law school. Libraries are more than computers or books; they are a place of learning, studying, engaging, and creating. Social media is a distinct platform to build relationships with library users, and a relevant social media policy allows for positive interactions.

PRIVACY CONCERNS FOR LAW LIBRARIES

As technology has created a new sense of sharing among friends and followers, it has also changed how individuals and organizations handle privacy issues. The bottom line is that anything posted online or created in the digital environment has the potential to be seen by unintended eyes. Law librarians must be aware of privacy issues in regard to the information posted externally on social software sites *and* internally for "library use only" needs.

First, read the privacy policy on all social networks the law library subscribes to. These privacy policies continually change, sometimes without warning! The law librarian(s) or staff members responsible for reviewing privacy policies need to be clearly defined, as does the frequency of the policy review. While reading the privacy policies, also review the privacy settings for all social software sites the library is using. Privacy settings change just as frequently as privacy policies.

Another important social software option to review is the permissions for mobile apps. When downloading new apps or when app updates are needed, it is important to check the permissions, privacy, and security settings; these are what allow outside users access to the app and its content. Depending on what type of app the law librarians or law library is using—for example, a social media Twitter app or a library task app for the catalog—some types of information will be more sensitive than other types but just as important. Use common sense when posting information online, as digital content is never truly private.

After reviewing the social software privacy settings for the sites the law library subscribes to, it is necessary for the law library to create and update its privacy policy for library content housed on university or firm servers, internal wikis, or on cloud-computing sites such as Google Drive. While it is important to keep communication to the public secure, it is just as important—or perhaps even more so—to keep the information created and used for library workflow secure as well. A law library privacy policy can include: (1) who has access to specific library functions in the integrated library system (ILS) (acquisitions, collections development, circulation, technical services, and databases administration), and the library website; (2) who has access to, and/or can share, department information stored on a wiki or in Google Drive; (3) what departments or law librarians are responsible for posting content on social software sites; (4) what library "voice" will be used to post content in or speak for the library; (5) who is ultimately responsible for questions, comments, complaints, and setting privacy policy for the law library. While many of these concerns can be answered by the law library organizational structure and staffing, law library privacy issues must be written, reviewed, and communicated to all law library staff.

Creating a privacy policy also provides boundaries for law library staff, especially for the law librarians engaging library users via social media sites. As law librarians monitor online chatter about the organization, the guidelines are in place for quick response to good and bad situations. The privacy policy should work in tandem with the law library's or parent institution's social media policy, not restrictive of communication efforts between the law library and library users but complementary and professional. Just as there should not be too many cooks in the kitchen, there should not be too many librarians independently posting social software content. Law libraries should delegate responsibility to the appropriate law librarian(s) for posting to social software channels and keeping up with online privacy and permission issues.

Social software has changed how people interact online and face-to-face. Let's review some advice from law librarians in the field, on what they wish they had known *before* using social software: (1) Make social media part of a bigger marketing plan, (2) Share the work, (3) Get all the professionals involved, (4) Plan ahead as using social software is more time consuming than expected, (5) Set goals, and (6) Always be professional.

THIRD-PARTY SITES, OR WHO OWNS MY DIGITAL CONTENT?

To fully engage in the digital world, users must share information. This could be a comment on a friend's Facebook status or work documents shared with a co-worker by way of Dropbox. Content created in the digital world is not necessarily private; to share information with others via social software, users must use a third-party site. Just as a law librarian's work e-mail is housed on a university or firm's server, the law library's posts on social media sites are potentially sitting on an unknown vendor's server.

Before using any type of social software, it is imperative to read the fine print! The Terms of Service (TOS) agreement includes the rules to use a service and any disclaimers relevant to that service. There may be a notice to users, dependent on the account a user or institution subscribes to (free or fee-based), and a schedule of changes or updates. The TOS is a basic contract for service between the user and provider. Be aware that some social media sites such as Facebook change the TOS and privacy settings when convenient for the company, often without notice or warning to users. Law libraries need to review what people, or outside entities, have access to content stored on a third-party site, as well the privacy settings and restrictions of use. Some information is more sensitive than other types of content; for example, Facebook or Twitter posts may be general library information for any audience, whereas the library policy and procedures for staff are not to be released to the public.

Law libraries also need a contingency plan for handling library data if a social software vendor is bought or absorbed by another company, or if the vendor quits providing service to its users. It is a good idea to be proactive in providing notice to law library users or clients of a third-party vendor, especially if a specific vendor is housing or providing backup of sensitive material. Being responsible for library users' data and information is a long-held practice in libraries; the digital realm of third-party sites should be navigated with professionalism.

BEST-PRACTICES SUGGESTIONS

Although technology and social software continue to evolve, the principles behind how to communicate with library users are fairly standard. Let's review a number of best-practices suggestions shared in the "Social Software in Law Libraries" survey by law librarians with technology experience in the field.

Have a Plan

It is a challenge to accomplish a new task or project without an action plan. Librarians are great at setting policy, trying new technology and programs, or embracing change in formats. Yet jumping on the newest technology trends bandwagon may not work well for every library or be necessary. Plan, document, and review why the law library is using particular social software sites to assess how functional these options are for the law library. Once a decision is made, it is best to provide sustainable energy into the social software by supporting staff endeavors to truly engage with users and master the technology.

Do What Works Best for Your Library

Just because the "cool kids" are participating in popular social software doesn't mean that site is appropriate for your law library. Law librarians who understand what their law library is about, who the target audience is, and how to best engage those users make good decisions on using social networks. For instance, using Google Drive may work well for tracking collection development choices among law librarians, or reference librarians may use a wiki to register students for a training session; these choices are made for functionality, not for popularity reasons. If it is not broken, sometimes it's best to leave it alone.

Be Where Your Users Are, Not Everywhere

While it's easy to sign up for free accounts on a variety of social software sites, it may not be in the best interest of the law library to be everywhere in the digital world. Research what social software sites the law library users are part of, and then determine if the law library should have a presence there as well. It is better to participate in a few social media sites very well than to use too many ineffectively. Go deep, not broad, in regard to social software engagement with library users.

Create the Time to Be Successful

The number of orphaned blogs, wikis, Twitter accounts, and stagnant Facebook profiles suggest libraries are willing to try new technology but do not consistently budget the time necessary to maintain success and engagement with users. When law librarians decide what social software to play with, it would be beneficial to create a contract for use, set a date to assess the product (annually), calendar when to review staff involvement, and determine if the law library should continue using the specific site. Social software can be time consuming for law library staff and librarians, especially when it is added to existing job duties. Learning how to use a new technology in a useful way and implementing it in the law library takes time and patience; both are essential to support the new endeavor.

Just the Library Facts

It's easy to get caught up in the social media hype of posting photos, sharing what one had for lunch, or ranting about poor service, but law librarians who are responsible for their law library's online presence must post responsibly. It is not appropriate to use the law library's platforms to discuss work issues, employee details, life events, or use vulgar, derogatory words and/or emotional language in postings. It behooves law libraries to create policies that set boundaries for employees who communicate with library users in the digital world. All posts or commentary must be professional, timely, relate to the law library's mission, and provide relevant information to library users.

Use Common Sense

Social software is one more area where law libraries can provide great service to users. When reviewing the options, approach the project with curiosity; identify and manage the risk involved, but do not try to control everything. Set good but flexible policy, and a review date to assess how things in the digital world are going for the law library. Remember, if it's supposed to be private, don't post it online.

Be Flexible

Life is busy; there are too many things, events, people, and commitments vying for attention. While having a plan of action is a great start to any project, especially social software, life happens; don't be frustrated if the number of followers does not increase as hoped or the law library staff are confused by "why" social media matters. Assign responsibility for communication and posting content among several librarians, if possible; collaborate on ideas; and don't forget that word-of-mouth marketing is best. Express to customers what the library is doing online, as promoting the social software cause face-to-face will build stronger relationships with users and open the door to other connections.

Have Fun

The attraction of social media is the ability to connect with people, share interests, and engage in digital conversation. The law library is probably the "happy place" for most law students and library users; providing relevant information and having fun will demonstrate to friends and followers what a great place the library is.

CONCLUSION

> And no, we don't know where it will lead. We just know there's something much bigger than any of us here.
>
> —Steve Jobs

Law libraries provide a distinct information service to users; where else can people turn for free assistance in finding the answer to a question? Social software gives law libraries and law librarians another communication avenue with library users. As technology continues to morph throughout all aspects of life, law librarians have numerous emerging social software options to help customers find information that matters and to build better relationships with library users. Law libraries using social software help fill the need for a third place between the real and digital worlds in which people live, share ideas, and communicate.

FURTHER READING

AALL and ILTA. N.d. "AALL/ILTA digital white paper: The new librarian." *AALL*. http://www.aallnet.org/main-menu/Publications/aall-ilta-white-paper.

Andrews, Lori. 2012. *I know who you are and I saw what you did: Social networks and the death of privacy.* New York: Free Press.

Barber, Peggy, and Linda Wallace. 2010. *Building a buzz: Libraries & word-of-mouth marketing*. Chicago, Ill.: American Library Association.

Bejune, Matthew, and Jana Ronan. 2008. *Social software in libraries*. Washington, D.C.: Association of Research Libraries.

Bradley, Anthony J., and Mark P. McDonald. 2011. *The social organization: How to use social media to tap the collective genius of your customers and employees.* Boston, Mass.: Harvard Business Review Press.

Carter, George, 2012. "Trending on Twitter: How Twitter can help #lawlibrarians." *AALL Spectrum* 17 (2): 7–8.

Chester, Simon, and Daniel Del Gobbo. January–February, 2012. "How to create a law firm social media policy." *LawPractice*. http://www.americanbar.org/publications/law_practice_magazine/2012/january_february/how-to-create-a-law-firm-social-media-policy.html.

Cofield, Melanie, and Kasia Solon. 2012. "Making the most of LibGuides in law libraries." *AALL Spectrum* 16 (5): 17–18.

Elefant, Carolyn, and Nicole Black. 2010. *Social media for lawyers: The next frontier.* Chicago, Ill.: American Bar Association.

Farrell, Elizabeth. 2010. "If you build it, will they come?" *AALL Spectrum* 14 (4): 8–9.

Friedmann, Ron, and Eleanor Windsor. July 4, 2010. "Law libraries transformed." *LLRX.com*. http://www.llrx.com/features/lawlibrariestransformed.htm.

Ginsberg, Debbie, and Meg Kribble. January 19, 2008. "Social networks for law librarians and law libraries, or how we stopped worrying and learned to love friending." *LLRX.com*, http://www.llrx.com/features/socialnetworks.htm.

Ginsberg, Deborah, Meg Kribble, and Bonne Shucka. 2009. "I want my Web 2.0." *AALL Spectrum* 13 (6): 28–29.

Jones, Kyle M. L., and Polly-Alida Farrington. 2012. *Learning from libraries that use WordPress*. Chicago, Ill.: American Library Association.

Kennedy, Dennis, and Allison C. Shields. 2012. *LinkedIn in one hour for lawyers.* Chicago, Ill.: American Bar Association.

Lambert, Greg. 2009. "Where do LISTERVS fit in a social media world?" *AALL Spectrum* 13 (8): 8–9, 13.

Lapachet, Jaye A.H., and Andrea Rubin. 2009. "What's on the library's mind: Facebook's place in the law library." *AALL Spectrum* 13 (9): 20–21.

Mendelson, B. J. 2012. *Social media is bullshit.* New York: St. Martin's Press.

Nelson, Sharon D., John W. Simek, and Michael C. Maschke. 2012/2013. *The solo and small firm legal technology guide*. Chicago, Ill.: American Bar Association.

Russell, Anna, and Carli Spina. 2012. "Law libraries linking data to mobile devices." *AALL Spectrum* 17 (3): 16–18.

Solomon, Laura. 2011. *Doing social media so it matters: A librarian's guide.* Chicago, Ill.: American Library Association.

———. 2013. *The librarian's nitty-gritty guide to social media.* Chicago, Ill.: American Library Association.

Springshare. N.d. "LibGuides community site." *LibGuides*. http://community.libguides.com.

Stratmann, Jo. January 18, 2011. "22 social media management tools—A wiki in development." *Business2Community*. http://www.freshnetworks.com/blog/2011/01/25-social-media-management-tools-list-wiki/.

Studwell, Bobbie. N.d. "Isn't it time to establish library-related social media best practice guidelines?" *AALL Spectrum Online*. http://www.aallnet.org/main-menu/Publications/spectrum/Spectrum-Online/social-media-guidelines.html.

Wiseman, Megan. 2012. "The law library as a place." *AALL Spectrum* 16 (4): 18–20.

Part III

REFERENCE SERVICES

14

Reference Services in a Law Library

Carol A. Watson
UNIVERSITY OF GEORGIA LAW LIBRARY

> Reference transactions are information consultations in which library staff recommend, interpret, evaluate, and/or use information resources to help others to meet particular information needs. Reference transactions do not include formal instruction or exchanges that provide assistance with locations, schedules, equipment, supplies, or policy statements.
>
> Reference work includes reference transactions and other activities that involve the creation, management, and assessment of information or research resources, tools, and services.
>
> —Guidelines, Reference and User Services Association

Reference services have always been at the heart of law libraries whether the law library is located in an academic institution, government agency, court, law firm, or even the corporate sector. Although the definition of *reference* continues to be relevant for today's law librarian, the scope of legal reference services has evolved tremendously. Technology has significantly enhanced the ability of law librarians to conduct reference work and contributed to the evolution of law library services.

Legal reference librarians are experiencing many of the same trends that general reference librarians in various types of libraries are reporting. For example:

- Foot traffic to the reference desk is declining but chat, e-mail, and other virtual reference transactions are increasingly popular. In particular, the demand for services via mobile devices is continuing to rise.

- Reference desks are often physically reconfigured to accommodate technology needs. Ergonomic and approachable desk designs with dual monitors for consulting are prevalent.
- Service points are being collapsed. It is not unusual for reference and IT to work at the same desk.
- Roving reference and embedded reference with wireless tablets and laptops are being implemented. These services meet the needs of researchers who might not otherwise approach the reference desk or make an appointment with a reference librarian.

Technology has also provided some unique opportunities for legal reference librarians:

- Legal research requests are often more interdisciplinary and extend beyond traditional legal resources. A reference librarian may be asked to locate an accident reconstruction expert, an authoritative citation for sociological research, or a twenty-year-old stock quote.
- In academia, empirical data analysis is a current scholarship trend. From recommending datasets to solving statistical software issues, academic legal reference librarians have become data experts.
- Competitive intelligence is also a particularly popular trend due to the increased availability of business information via the Internet. Some law firms have reference librarians exclusively assigned to client development.

Without a doubt, the information landscape is evolving at a lightning pace. There are substantial opportunities for expanding legal reference services. As legal researchers are confounded by information choices, legal reference services are more critical than ever.

REFERENCE TRANSACTIONS

Personal Interactions

For many years, law libraries have offered traditional reference services by staffing a reference desk for face-to-face or telephone questions. Legal reference librarians, like their counterparts in other types of libraries, are trained in the art of the reference interview. Learning to listen and ask relevant open-ended questions are skills that all reference librarians must master. Successful reference librarians clarify the deadline for answering the reference inquiry and determine the depth of information that is needed. Upon ascertaining the scope of the reference question, the legal reference librarian must be familiar with appropriate legal resources and formulate an effective search strategy. Finally, it is important to follow up with the patron to determine if the question has been sufficiently answered or if additional information is needed.

One of the main caveats of legal reference that still exists regardless of the influence of technology is the prohibition on librarians from providing legal advice to nonlaw patrons. Legal reference librarians must be cautious about engaging in the unauthorized practice of law. In general, it is permissible to recommend research resources, teach legal research strategies, and provide advice on constructing searches. However, a legal reference librarian should not interpret the law or provide a personal opinion regarding the patron's legal situation.

Reference Service Policies

Reference service policies help establish guidelines for library employees and ensure that all library users receive similar and consistent services. Libraries should determine to whom they will provide reference service and at what level. It is recommended that legal reference librarians work together to reach consensus on the major components of their library's reference services and draft a formal policy documenting the library's service. Ideally a reference service policy should reflect best practices and include input from appropriate stakeholders. The policy should be approved by the library's governance structure and be readily available to library users. When drafting a reference policy, some parameters that may be addressed include:

- Who will the library serve? For example, academic libraries must determine whether to serve the general public or perhaps the undergraduates from its institution.
- What are the limits of reference service? Is there a prohibition against providing legal advice? Is there a defined time limit specifying how much time a librarian will devote to answering a reference question?
- How is reference service prioritized? Do walk-in patrons receive service over telephone reference questions?
- How will privacy and confidentiality be protected?

Examples of Reference Policies

- Arizona State University Law Library: http://www.law.asu.edu/library/RossBlakleyLawLibrary/LibraryInformation/AskaLibrarian/ReferencePolicies.aspx
- University of Iowa Law Library: http://www.law.uiowa.edu/library/ref.php
- University of San Francisco Law Library: http://www.usfca.edu/law/library/policies/reference/

- University of Washington Gallagher Law Library: http://lib.law.washington.edu/ref/ref.html
- Wisconsin State Law Library: http://wilawlibrary.gov/about/reference.pdf

VIRTUAL REFERENCE

> Virtual reference is reference service initiated electronically, often in real-time, where patrons employ computers or other Internet technology to communicate with reference staff, without being physically present.
>
> —Guidelines, Reference and User Services Association

Reference librarians have eagerly exploited the benefits of technology and are using technology to "go where patrons are." Whether using chat reference, e-mail reference, or even video reference, the most successful legal reference librarians are fulfilling user expectations for immediate and virtual answers to their research questions. Today's library patrons are comfortable with electronic interactions, particularly using their mobile devices. Texting and e-mail are ubiquitous. However, when communicating virtually, librarians must keep in mind the nuances of electronic messages. Without face-to-face communication, it may be difficult to read nonverbal cues. While the librarian is typing or searching reference resources, a patron may not understand the slowness or lack of response. Indeed, digital natives communicate differently from digital immigrants. Be sensitive to the different styles of communication from your library patrons.

No matter what type of virtual reference is implemented, a policy governing basic issues should be developed first. When developing a virtual reference policy, be sure to address the following issues:

- Know the limits of your database agreements. Some database contracts prohibit the sharing of proprietary data beyond the defined user base in the licensing agreement.
- Even though the communication is online, all exchanges should be treated as private and confidential.
- Establish a reasonable time frame for responses. How often will messages be checked? What is the time frame for response?
- Clarify the depth of research that will be undertaken. What types of questions are permissible?
- Establish a protocol for dealing with difficult patrons.

Types of Virtual Reference

E-Mail/Web Form

One of the simplest and oldest types of virtual reference to implement is via e-mail or online Web form. E-mail is a format that most librarians are familiar and comfortable with. Although the possible slow or delayed response time and asynchronous communication can make e-mail seem to be an inefficient form of communication, it is still generally the standard form of communication in business environments. E-mail reference transactions lack the opportunity for clarification or give-and-take rapport that a face-to-face conversation provides. On the other hand, e-mail replies provide researchers with a record of the librarian's advice and can include screenshots or hyperlinks that can be saved for future recall. While most academic law libraries answer legal reference questions submitted via e-mail, reference librarians in law firms are likely to be heavy users of e-mail for reference transactions. In particular, in the firm environment, reference librarians have to be alert to incoming e-mails in order to respond to attorney requests promptly.

Chat/IM/Text/SMS

Chatting or texting offers library patrons a much faster response time than e-mail communication. This format is particularly useful for answering quick, fact-based questions. For the patron, chatting or texting provides quick access to a reference librarian, perhaps even from a different geographic location and at a time that is convenient for the patron. Users who might be reluctant to initiate face-to-face communication with a reference librarian may be more comfortable with a virtual transaction and thus more likely to seek assistance. In today's culture, texting and chatting is a normal mode of communication. Consequently, library users expect to be able to reach a reference librarian through the means of communication that they regularly use. Since texting is prevalent among young adults, chat reference is available and popular in the majority of academic law libraries.

For successful implementation, librarians should receive staff training. Otherwise, for example, if multiple questions arrive simultaneously, the reference librarian may find it unwieldy to balance more than one transaction (particularly if the librarian if also responsible for telephone and face-to-face reference transactions at the same time). Another hazard of chat reference is that occasionally a patron may disappear. Training librarians to remember that regular communication, perhaps every thirty seconds, can lessen the probability of a patron losing interest.

Several software options are available for managing chat/e-mail reference service. For a list of vendors and software, see http://www.ala.org/tools/libfactsheets/alalibraryfactsheet19#Vendors.

Videoconferencing/Skype/VOIP/Google Voice

Virtual reference services are not limited to text interactions. Recent technologies allow reference transactions to include audio and/or video as well. Examples of enhanced virtual reference services include:

- *Skype*: The Cleveland Law Library invites legal researchers to contact their reference department via Skype for reference transactions (http://www.clelaw.lib.oh.us/Reference/Reference.html). Taking the Skype concept one step further, Ohio University experimented with setting up a digital kiosk with an open Skype session to a reference librarian.[1] The kiosk was available for use by any passerby. Ultimately, the service was too time consuming, but it is an excellent example of how reference librarians are pushing the boundaries and offering innovative reference services.
- *Google Talk*: In addition to Yahoo! Messenger, AOL Instant Messenger, and MSN/Windows Live Messenger, American University Law Library has expanded its virtual reference services by also including the option for patrons to contact them via Google Talk (http://library.wcl.american.edu/ask.cfm). Google Talk is a voice-over IP (VOIP) service that permits patrons to talk with reference librarians using their computer's microphone and speakers.

Social Media

Many legal reference departments are integrating their services into their library's Facebook page and other social media outlets. In addition to publicizing research resources, search strategies, and special events, often libraries include a link to their chat reference services in their social media messages. For example, the Utah State Law Library Facebook page includes a link to its instant messaging chat reference (http://www.facebook.com/UtahStateLawLibrary). In essence, such librarians are "embedding" themselves into Facebook to be available to answer chat reference questions. There are many challenges of using social media such as Facebook for reference services, such as ensuring that a reference librarian provides timely responses to inquiries, training librarians about the nuances of various social media tools, and potentially increased staffing needs. Additionally, librarians must walk the fine line promoting library services without invading the personal digital space of library patrons. In addition to Facebook, some law libraries integrate Twitter into reference services. For example, the LA Law Library uses its Twitter account to post recent reference questions presented to the reference desk. Some of the reference questions mentioned on the LA Law Library Twitter feed include:

- "How do you prevent access to a sealed court record?"
- "Can a personal name be trademarked?"
- "Where can I find the text of a Los Angeles city ordinance?"

Artificial Intelligence

One of the most cutting-edge examples of virtual reference that might seem like a scene from a science fiction movie is the development of artificial intelligence logic to answer reference questions without human intervention. Artificial intelligence programs can provide consistent answers to frequently asked questions with the additional benefit of 24/7 access. In addition to serving as a unique and creative marketing tool, artificial intelligence reference can fill basic staffing voids created by budget cuts. Although human reference librarians will likely still be needed for quite some time, at a minimum, artificial intelligence systems have the potential to serve as triage for reference services by answering basic questions and referring more complex inquiries to human librarians. As proof of the possibilities for artificial intelligence, the University of Nebraska–Lincoln Library has developed "Pixel," a chatbot that guides students through the library's website and databases.[2] See: http://pixel.unl.edu/.

Checklist of Virtual Reference Best Practices

- Develop a script for online services, especially the introduction.
- Ask open-ended questions.
- Send a resource for the patron to review early in the process.
- Ask the patron, "Am I on the right track?"
- Sign off by summarizing the session.

Sample Virtual Reference Service Policies

- Catholic University Chat Policies: http://www.law.edu/library/AskALibrarian.cfm
- Georgia State University Law Library Chat Reference Policy: http://law.gsu.edu/library/2047.html#lawchat
- Guidelines for E-mail Reference Service at the Maryland State Law Library: http://www.lawlib.state.md.us/aboutus/policies/EmailReferenceGuidelinesMSLL.pdf

THE REFERENCE DESK AND BEYOND

The Reference Desk

Traditional reference—whether face-to-face or virtual—generally takes place at a centrally located reference desk. Continual staffing of a physical reference desk for "just in case" questions can be time consuming and labor intensive. It may prevent librarians from engaging in other reference activities such as in-depth instruction or embedded services. Also, a noisy reference desk may not be the best environment for answering in-depth or detailed reference questions.

With staff reductions due to shrinking budgets, increased demand for online course support, and more requests for instructional services, reference librarians have developed a number of creative solutions such as combined service desks, tiered reference, reference by appointment, roving reference, and embedded reference to meet user needs.

Combined Services Desks

Some libraries have consolidated their reference, technology, and circulation desks to a single service desk. The main advantage of single service desks is the elimination of confusion that library users often experience when they are confronted with multiple service desks. Instead, librarians can focus on providing the best possible customer experience at one point of contact.

Although the opportunity for collaboration between departments can be exciting, it can also be challenging to combine the functions of various service desks. The merger requires significant staff cooperation and flexibility. New terminology, skill differences, and changed expectations are all hurdles that must be overcome.

Tiered Reference

Tiered reference is based upon the concept of triage medical treatment. Libraries staff their reference desks with paraprofessionals or even students. The first-tier staff are trained to answer basic or recurring reference questions. Advanced or in-depth questions are referred to second-tier professional librarians. This model frees up the librarian's time to focus on more complicated questions. This model has been developed in response to the trend of reference requests becoming more in-depth and interdisciplinary. Most users are answering basic questions on their own. One of the main challenges for this type of model is training first-tier reference staff on when to refer the question to an appropriate librarian.

Reference by Appointment

Some law libraries have eliminated their reference desk. These libraries are influenced by the fact that ringing telephones, steady foot traffic, and lack of privacy can inhibit serious research conversations. Instead, reference librarians are available for individualized research consultations by appointment. Individual appointments allow for uninterrupted research consultations. More time can be spent on task and the librarian can prepare for the interaction in advance. Some libraries provide weekly open office hours in addition to being available for appointments. It is essential to effectively market research assistance by appointment to increase awareness that individual service is available.

Roving Reference

Roving reference is defined as conducting reference outside of the reference desk. Reference librarians move about an area where patrons are located and offer assistance using tablets, netbooks, or laptops. Roving reference librarians help patrons at their point of need. Furthermore, some patrons have library anxiety. It is theorized that librarians are less intimidating and more approachable when they are not behind a desk. Sometimes researchers are hesitant to interrupt a librarian but are content to consult away from the barrier of the reference desk.

Roving librarians must be sensitive to the possibility that a library user may not want to be interrupted. Roving reference librarians will be best received if the library markets that the service is available and in operation. The reference librarian should wear a name tag and spend time in areas where library patrons are located, such as public computers, computer labs, or soft seating. Use welcoming languages such as, "Are you finding what you are looking for?" or "Let me know if I can assist you."

Embedded Reference

Embedded reference services range from setting up a reference desk in a different or temporary location to purposeful collaborations between librarians and teaching faculty where the librarian is fully integrated into a course, virtual or real. Embedded librarianship is based upon the concept of embedded journalists in Iraq. In academia, studies show, student performance is improved when an embedded librarian participates in a class or clinic. Faculty endorsement of librarians is very influential with students and encourages students to reach out for library services. The University of the District of Columbia Law Library is an example of a law library that has implemented embedded librarian services.[3]

In the law firm setting, reference librarians have successfully embedded within practice groups. Embedding librarians within a practice area provides the opportunity for reference librarians to become actively engaged with attorneys, paralegals, marketing personnel, and support staff on a daily basis. It enables reference librarians to become more proactive in providing service. Cindy Adams, at the Atlanta law firm branch of McKenna Long & Aldridge LLC, provides practical advice on establishing an embedded reference librarian service.[4]

Personal Librarian Model

A popular trend that began in undergraduate and medical libraries that has been experimented with in law libraries is to assign each incoming student an individual librarian. Yale University Library has been a leader in implementing the personal librarian model.[5] Personal librarians contact their assigned students throughout the year with news of library tools, services, and research strategies. Students are encouraged to work with their personal librarian on library research-related issues. Most of the librarian's contact with students is online either via texting or e-mail.

CREATING AND MANAGING RESEARCH RESOURCES AND SERVICES

Research Guides

In addition to answering questions from library users, reference librarians connect researchers to relevant information by creating research guides, tutorials, and other instructional resources. One of the most popular and traditional methods for guiding library users is the library research guide. Guides can be print or electronic. The definition of a research guide is a list of resources to assist on specific topics designed to help with research needs. The list of resources may include books, journals, periodical articles, databases, and websites, as well as any other pertinent resources.

Research guides can be particularly helpful for researchers who are at a "point of need." They can significantly shorten and simplify the research process. Some librarians also create research guides to support their presentation or instruction sessions. Research guides are handy takeaways for students to use after they may have forgotten some of the resources covered in a class setting. On the other hand, be forewarned: research guides are time consuming to create. They need to be regularly updated. Also, research shows that research guides are not frequently used by patrons.[6]

Research guides are primarily online and are included as part of a library's website. They can be created using in-house content management systems or Web-editing software such as Adobe Contribute. Springshare's LibGuides product (http://springshare.com/libguides/) has become increasingly popular among reference libraries. LibGuide's templates are easy to customize without programming skills. It is simple to add widgets and features to LibGuides such as RSS feeds, chat reference widgets, and images.

A checklist of best practices for creating research guides includes:

- Don't overwhelm. Curate the selected resources.
- Keep text to a minimum. Don't overwrite.

- Clearly describe the subject and purpose of the research guide.
- Know your patrons and their expectations.
- Use resources specifically available at your library.

Tutorials

In addition to telling patrons how to do something, reference librarians need to "show" patrons how to do something. Video or screencasting tools are useful for providing how-to instructions at a patron's point of need. Video tutorials record the mouse movements from the computer screen and include a narrative audio track. For example, tutorials can demonstrate how to use a software program or navigate a research resource.

Online tutorials are particularly useful for guiding a researcher virtually and have become increasingly popular with the advent of distance-learning initiatives. They can be viewed and even repeatedly watched at any time of the day on the researcher's individual schedule. An added benefit is that online tutorials can be especially helpful for visual and auditory learners. Screencasts are more ideally suited for discrete defined tasks rather than general tasks such as developing a research agenda.

Reference librarians have become experts at using screencasting tools that record computer screen and mouse movements combined with audio narration to produce videos that are useful for demonstrating how to use a particular database or computer resource. There are free screencasting tools. Jing is one of the most popular free screencasting tools. The free software platforms have limited capabilities such as no editing tools, but are a good option for experimenting with screencasting. They generally lack interactivity and do not provide quizzing options. Try creating a screencast of a PowerPoint presentation to get a feel for screencasting.

Two of the most popular commercial software tools for creating screencasts are TechSmith Camtasia and Adobe Captivate. Both programs work well for creating professional-quality screencasts that capture keyboard, screen, and mouse input integrated with audio and video. They both offer quiz options as well. However, each platform has distinct advantages. Camtasia is well suited for straightforward narrated demonstrations. Captivate offers more advanced options and flexibility as well as sophisticated editing options. Bottom line: Camtasia is best for "show me" demonstrations; Captivate is best for "you try" tutorials.

Camtasia Summary

- Limited quiz features—multiple choice, fill in the blank
- Limited hot spots

- Can output in different video formats, including Flash, QuickTime, and Windows Media Video
- Affordable— approximately $300; nonprofits and educational institutions should check for possible discount pricing

Captivate Summary

- Multiple quiz features—multiple choice, fill in the blank, true/false, matching, short answer
- More options for interactivity and complex demonstrations; branching logic available
- Limited range of output formats but can export directly to YouTube with a single keystroke
- Integrates with other Adobe products
- Steeper learning curve
- Expensive—approximately $800

Checklist of Best Practices for Creating Screencasts

- Prepare a script or list of talking points.
- Keep videos focused. Have one goal and one answer.
- Limit screen casts to three minutes or less.
- Don't forget to use close captioning for hearing-impaired viewers.
- Practice before recording.

Examples of Online Tutorials

- Harvard Law Library: http://www.law.harvard.edu/library/research/tutorials/index.html
- Georgetown Law Library: http://www.law.georgetown.edu/library/research/tutorials/
- Suffolk Law Library: http://www.law.suffolk.edu/library/services/tutorials/index.cfm
- William and Mary Law Library: http://law.wm.edu/library/research/tutorials/index.php

Podcasting

Podcasts are audio or video files that are downloaded through a series although video podcasts may be referred to as vodcasts. Because podcasts are available via RSS, users can subscribe to a podcast series and new podcasts will be automatically downloaded. The user does not have to seek out or search for the latest podcast or audio recording. Once downloaded to an MP3 player, podcasts offer the advantage of being able to be played anytime and anywhere (or even more than once).

Another advantage of podcasts is ease of creation. The main investment for podcasting is staff time. It takes time to create thoughtful, relevant, and creative podcasts that can compete with entertainment podcasts and will interest researchers. Podcasts can generally be created with free software and a computer microphone—a relatively inexpensive piece of hardware. Audacity (http://audacity.sourceforge.net/) is one of the most popular free software platforms. The CALI Classcaster tool has also been frequently used in academia to manage blogs and podcasts: http://classcaster.net/.

Many types of podcasts are available, including TedTalks, NPR programs, music, stories, news programs—the list goes on and on. Reference librarians have successfully used podcasts for creating content such as research tips, interviews, building tours, book talks, and special event promotions.

Podcasting examples include:

- One of the most popular law librarian podcast series is Law Librarian Conversations (http://www.blogtalkradio.com/lawlibcon).
- Visit http://www.libsuccess.org/Podcasting to view of a list of libraries currently offering podcasts.

Document Delivery

Many reference librarians assist in locating materials for document delivery services offered by their libraries. Typically, document delivery is defined as copying materials (whether or not they are owned by the library) on behalf of and at the request of a library user. Document delivery is increasingly important for law firm libraries since most firms no longer maintain significant collections of library materials but instead rely upon obtaining materials quickly from various suppliers. Whether materials are in print or digital format, they are subject to U.S copyright law restrictions. Consequently, reference librarians must become familiar with copyright law as applied in a library setting. In particular, with the increasing popularity of distance learning, reference librarians must be savvy about providing digital materials to remote students within copyright restrictions.

In addition to providing document delivery services, reference librarians must be familiar with where to obtain document delivery services for their clients. Often, a law firm librarian will be expected to quickly locate a reliable and inexpensive source for obtaining documents such as court filings, annual reports, patents, journal articles, and so on. Document delivery sources range from commercial ventures that physically retrieve copies of court or government documents to online publishers who permit direct, one-time purchases of digital materials. Law firm reference librarians must know their document delivery suppliers well in order to be successful. Many major academic libraries such as Vanderbilt (http://law.vanderbilt.edu/library/document-delivery/index.aspx) and William and Mary (http://law.wm.edu/library/services/interlibraryloan/delivery/index.php) provide document delivery services. There are also commercial suppliers of documents as well. Examples of popular commercial suppliers include Reprints Desk (http://www2.reprintsdesk.com/), Information Express (http://www.ieonline.com/), and Infotrieve (http://www.infotrieve.com.

ASSESSING REFERENCE SERVICES

It is essential for reference librarians to use library assessment tools to determine the needs of their users and to measure how well they are meeting the needs of their community. Assessment provides concrete information for libraries to use when planning future services and evaluating the effectiveness of current services. Some libraries designate a specific librarian to be responsible for assessing all types of library services. For example, at the Duke Law Library, the digital initiative librarian is responsible for library services assessment.

Types of Assessment

Statistics

There are many types of library assessment tools available ranging from focus groups to conducting an in-depth study using ARL's LibQual+ survey instrument. However, assessment tools generally fall into either qualitative or quantitative categories. One of the most traditional types of quantitative measures of reference service is to examine the types and volume of reference transactions. There are many digital tracking tools available for tallying the number and types of reference transactions such as Compendium Library Services Desk Tracker (http://www.desktracker.com) and Knowledge Tracker (http://www.knowledgetrackerlib.com), as well as Gimlet (http://http://gimlet.us). Most libraries establish an agreed-upon definition of reference transactions and generally do not include directional questions. Some libraries further distinguish between quick and extensive reference questions.

Studying the number of reference transactions helps inform decisions about reference desk staffing, hours of service, and trends such as whether questions are primarily being submitted virtually or in person. These measurable statistics can also be compared with peer libraries. Note that reference transactions can be more difficult to track as the types of reference transactions such as chat and embedded expand. It can also be tricky to distinguish reference questions at consolidated service desks. With the addition of newly created resources such as screencasts and other tutorials that anticipate questions, how many reference questions are being diverted from the reference desk? Before putting too much effort into and focus upon tracking traditional reference statistics, a cost-benefit analysis should be carefully considered.

Surveys

Conducting user satisfaction surveys is popular means of qualitative assessment in law libraries. Many free and paid online services such as Survey Monkey (http://www.surveymonkey.com/), Zoomerang (http://www.zoomerang.com/), and Qualtrics (https://www.qualtrics.com/) are available to create and analyze survey results. ARL's LibQUAL+ (http://www.libqual.org) is a standardized survey used by many nonlaw academic libraries that is sometimes used by law libraries. Surveys are a quick and economical method for identifying gaps in services or resources. It is possible to reach a large number of survey participants. Consequently, general trends and themes can be easily identified.

The downside of using surveys to assess reference services is that designing well-worded surveys and accurately analyzing results requires advanced skills that many librarians must learn more about.[7] In an academic institution, reference librarians may need to obtain the approval of institutional review boards before distributing a survey since human subjects are involved. Also, users sometimes complain of survey fatigue if they are subjected to too many customer satisfaction surveys.

Checklist of Best Practices for Creating an Online Survey

- Define the goals of the survey.
- Identify the appropriate audience.
- Decide whether to use open-ended or close-ended questions.
- Keep the survey simple and brief. Limit the number of questions.
- Pretest the survey before administering.
- Consider offering incentives to survey participants.

Focus Groups

Focus groups are also useful for obtaining qualitative feedback about library reference services. Focus groups are in-depth interviews or guided conversations that are typically one to two hours long with seven to twelve participants. Focus group discussions provide the opportunity for the moderator to ask follow-up questions and to clarify ambiguous answers. They are a particularly useful discovery tool for identifying gaps in service and awareness of library services. Most focus groups are video-recorded, creating an opportunity to observe nonverbal behavior as well. Focus groups do not require an elaborate setup, however. They can be a simple and quick method for reference librarians to obtain immediate feedback about services and resources.

The quality of focus group results depends upon the qualifications of the moderator, the make-up of the participant pool, and how well drafted the focus group questions are. The focus group conversation can be influenced by the vagrancies of group dynamics. One participant can dominate the conversation. Sometimes participants state opinions they think the facilitator wants to hear. Also, because the focus group is small, it may be inaccurate to generalize based upon feedback from the participants. Careful planning and preparation for the focus group conversation is required to overcome these obstacles.

Checklist of Best Practices for Conducting a Focus Group

- Define the objective of the focus group.
- Brainstorm potential questions.
- Choose a comfortable setting for hosting the focus group.
- Draft a script and pretest the text.
- Select a neutral facilitator.
- An audio and/or video of the focus group discussion should be recorded.
- After the interview, transcribe results and write a summary report.

FURTHER READING

Reference Transactions

Estes, Mark E. 2007. "Reference services in law firm environments." *Legal Reference Services Quarterly* 26:91.

Kern, M. Kathleen. 2009. *Virtual reference best practices: Tailoring services to your library* Chicago: ALA Editions.

Lotts, Megan, and Stephanie Graves. 2011."Using the iPad for reference services: Librarians go mobile." *College and Research Libraries News* 72:217. http://crln.acrl.org/content/72/4/217.full.pdf+html.

Moraisa, Yasmin, and Sara Sampson. 2010. "A content analysis of chat transcripts in the Georgetown Law Library." *Legal Reference Services Quarterly* 29:165.

Nann, John B. 2010. "Personal librarians: The answer to increasing patron contact may be simpler than we think." *AALL Spectrum* 14:20.
Sachs, Dianna E., Edward J. Eckel, and Kathleen A. Langan. 2011. "Striking a balance: Effective use of Facebook in an academic library." *Internet Reference Services Quarterly* 16:1.
Selby, Courtney 2007. "The evolution of the reference interview." *Legal Reference Services Quarterly* 26:35.

Creating and Managing Research Resources and Services

Notess, Greg R. 2012. *Screencasting for libraries*. Chicago: ALA TechSource.
Sampson, Sara, Catherine M. Dunn, and Morgan Stoddard. "Preparing strong research guides: Lessons from the Georgetown Law Library." Presentation at the 2012 annual meeting of the American Association of Law Libraries, Philadelphia, Pennsylvania, July 16, 2012. http://www.aallnet.org/sis/ripssis/pdfs/2012-StrongResearchGuides.pdf.

NOTES

1. Char Booth, "Developing Skype-based reference services," *Internet Reference Services Quarterly* 147 (2008), http://www.informaworld.com/10.1080/10875300802103684.
2. DeeAnn Allison, "Chatbots in the library: Is it time?" *Faculty Publications, UNL Libraries,* Paper 280 (2011), http://digitalcommons.unl.edu/libraryscience/280.
3. Vicenç Feliú and Helen Frazer, "Embedded librarians: Teaching legal research as a lawyering skill," *Journal of Legal Education* 61 (2012): 540.
4. Cindy Adams, "Law firm librarians: Out of sight, out of mind," *3 Geeks and a Law Blog*, January 28, 2013.
5. Yale Personal Librarian Program, January 28, 2013, http://www.library.yale.edu/pl/.
6. Dana Ouelette, "Subject guides in academic libraries: A user-centered study of uses and perceptions," *Canadian Journal of Information and Library Science* 35 (2011): 436.
7. James M. Donovan, "Back away from the Survey Monkey! Optimize survey results with an honest assessment of methodology," *AALL Spectrum* 14 (2009): 20.

15

Introduction to Legal Research

Rhea Ballard-Thrower
HOWARD UNIVERSITY LAW LIBRARY

You jump onto the train as the doors are about to close. As you begin to stroll to the back of the car, your cell phone rings. The administrative assistant calls to inform you that your employer's major grant provider will be at the headquarters in a couple of hours. The grant provider is concerned that a recent Court decision will affect how the grant is handled. Before the train reaches the next station, you check Scotusblog for the latest news on the U.S. Supreme Court. As you peruse the website, you get a text message from your supervisor. She wants to make sure you are aware of grant provider's visit and that she needs your research assistance before grant provider arrives. After you text that you are looking at the specific case, you pull out your tablet and begin to research. You smile to yourself. So many things in legal research have changed and yet, so many things have stayed the same.

Legal research is simply a process whereby you use electronic and print resources to answer legal questions. The documents you retrieve are analyzed in relation to the facts given and a response is presented to the one who posed the question. Legal research has three main purposes: one, to help you solve a problem; two, to help you prevent a problem; and three, to help you understand what laws may be applicable to your issue. The process of legal research is so extensive that textbooks have been dedicated to the topic. However, this chapter attempts to provide an overview of the necessary steps for answering a research request in the digital age.

HISTORY OF AMERICAN LEGAL RESEARCH

Legal research has a long and detailed history. The American legal system owes much to our British heritage. As a former British colony, America adopted English common law. Common law is the practices and customs developed and administered by the king's courts.[1] With the advent of a judicial system, common law evolved from community practices to judge and made law passed from generation to generation. The judges drew their decisions from existing principles of law that represented the values and attitudes of the English people.[2] A unique aspect of the common law is the adherence to precedent. Precedent is when a judge uses a prior decision to determine the outcome of a later case involving similar facts or issues.[3] The American colonists embraced the English common law to become the basis of American common law.[4]

SOURCES OF LAW

In 1787, our founding fathers convened to draft a new constitution for the United States. In October of that year, Alexander Hamilton, James Madison, and John Jay began to publish a series of articles in support of the Constitution's ratification. Those articles, using the pseudonym *Publius*, are known as the *Federalist Papers*. They provided a counterargument for those opposed to the Constitution's ratification. After much discussion, debate, and compromise, advocates garnered enough support for the new Constitution. One year after members convened, New Hampshire became the last state needed to ratify the new Constitution.[5] For 225 years, the U.S. Constitution has provided the basis of our government structure. It is considered the preeminent document of American legal research materials. Yet, it was a book published 39 years prior to the Articles Congress that is credited with influencing the Constitution's drafters. In 1748, Charles de Montesquieu published *The Spirit of Laws*. Although not the concept's originator, Montesquieu is credited with promoting the idea that liberty depends on a separation of governmental powers.

The U.S. Constitution is organized into eight distinctive parts. After the Preamble, seven articles follow to describe the structure of our government. The significant arrangement of the Constitution is its separation of powers doctrine. James Madison forcibly stated that, "the accumulation of all powers legislative executive and judicial in the same hands whether of one a few or many whether hereditary self appointed or elective may justly be pronounced the very definition of tyranny."[6] The first three articles of the Constitution provide the parameters of the three branches of government—the legislative, executive, and judicial. Article I, Section 1 of the Constitution states, "All legislative powers herein granted shall be vested in a Congress of the United States, which shall consist of a Senate and House of Representatives." As the longest of the articles, Article I describes congressional membership, how federal laws are created, and the powers of Congress. Article II, Section 1 of the Constitution states, "The executive power shall be vested in a President of the United States of America." Article II describes who can qualify to run for president, the Electoral College, and the powers of the presidency. Article III, Section 1 of the Constitution states, "The judicial power of the United States, shall be vested in one Supreme Court, and in such inferior Courts as the Congress may from time to time ordain and establish." Article III provides the scope of the federal judicial system.

The three branches along with the Constitution are referred to as the *four main sources of law*. The first main source of law is the Constitution. The documents and materials produced by the legislative, executive, and judicial branches form that basis of the materials used for legal research. The second main source of law is the legislative branch that produces the statutes or laws. The third main source of law is the federal executive branch. The federal executive branch includes the President, Vice President, the cabinet, and the federal agencies. The executive branch produces executive orders and regulations.[7] Executive orders are directives the president issues to federal government agencies or officials to accomplish certain tasks.[8] President George Washington issued the country's first executive orders. Because the executive branch implements and enforces the laws, the majority of the federal regulatory agencies are considered part of the executive branch. Although an extension of the executive branch, it is the legislative branch that creates the agencies to develop rules and regulations to administer the laws. Agencies serve a dual function. They have a legislative purpose whereby they promulgate or create administrative orders and rules. Moreover, they also have a quasi-judicial role whereby they conduct hearings and render judgments. After the Constitution and the legislative and executive branches, the final or fourth main source of law is the judicial branch, because it produces the opinions or cases.

TYPES OF LEGAL AUTHORITY

Legal research is a process in which you locate authoritative materials to answer a legal question. The facts of the circumstance are reviewed in light of documents located during the research process. Unique to legal research is the use of primary authority. You can use Wikipedia to learn about the recent U.S. Supreme Court case. However, you cannot use Wikipedia as an authoritative source for how the Supreme Court interpreted the law.

To answer a legal question, a legal researcher must use *primary resources*. A primary legal authority is the verbatim text of a document that is legally binding, meaning that it must be followed. Examples of primary authority are the U.S. Constitution, statutes, regulations, executive orders, cases, and treaties. All of the aforementioned documents have the force of law and are legally binding. The U.S. Constitution and U.S. Supreme Court decisions are mandatory, primary authorities that must be followed by all. Thus, if Grant Provider wants to know if he can continue to distribute funds as it has done previously, Grant Provider would have to locate primary authority. In particular, Grant Provider may research U.S. Supreme Court judicial decisions for an answer.

When performing legal research, you will also find documents that are *persuasive authority*. Persuasive authority documents are nonbinding and do not have to be followed. Persuasive authority can be primary authority (statutes, regulations, cases, etc.), and at the same time still persuasive. For example, a judicial opinion from the Supreme Court of Georgia is primary and mandatory for Georgians. However, that same case is only persuasive in Florida. Statutes from other states, however, are only primary authority. They are not even considered persuasive authority in other states.

The American Court System

The American court system is organized in a hierarchical manner. Imagine a pyramid divided into thirds. Each level has a specific jurisdiction. Jurisdiction is the power a court has to hear a case and decide the issues.[9] The bottom and largest third of the pyramid represents the trial level. The trial courts are where litigation usually begins. They are the lowest level of the court system. The trial level is where the facts in the case are decided. Typically, the trials at this level are jury trials. The trial court hears witnesses and views evidence. A jury then decides the outcome of the case.[10] For the majority of trial level cases, the outcome of a trial is typically *unpublished.* The jury declares either guilt or innocence and does not render an opinion. Due to being the lowest level of the court pyramid, trial court decisions do not have precedential rank over the higher courts.

The second level of the American court system is the *intermediate appellate.* This level is usually referred to as the court of appeals. However, for some state jurisdictions, that may not be the nomenclature. The intermediate appellate section of the pyramid is smaller than the trial level. It is different from the trial level, because attorneys write briefs and argue the cases before a panel of three or more judges. The judges listen to the argument and rule based on the written record from the trial. Appellate judges are bound by the facts determined at trial. They review the decision of the trial court to determine if the law was fairly applied to the facts. One of the judges will then write an opinion on behalf of the court. Judges who disagree can write a dissenting opinion. Appellate-level decisions can become precedential legal argument used for future cases. Opinions of an intermediate appellate-level court are binding on the courts within its jurisdiction. However, not all intermediate appellate level decisions are published.

The third and highest level of the court system is the *court of last resort.* This court is usually referred to as the supreme court. However, for some state jurisdictions, the court of last resort is not called the supreme court.[11] This section of the court pyramid is the smallest. The court of last resort is an appellate court that hears appeals from the trial or intermediate appellate courts. The judges on the court of last resort review the trial court record and exhibits. As an appellate court, it can review questions of law. It also determines if the law was properly applied to the facts. Decisions from the court of last resort are binding on all lower courts in its jurisdiction. Court of last resort opinions are published.

The Federal Courts

The *District Court* is the trial court level of the federal system. There are ninety-four federal judicial districts with at least one district in each state. There are also a few specialty courts: for example, the United States Tax Court. Federal district courts are referred to by their geographic location: for example, the U.S. District Court for the Northern District of Georgia.

The intermediate appellate level of the federal court system is the *United States Court of Appeals*, and the individual courts are often referred to as circuit courts. The United States is organized into thirteen circuits, which are comprised of eleven numbered circuits, the United States Court of Appeals for the District of Columbia, and the United States Court of Appeals for the Federal Circuit.

The U.S. Supreme Court is the court of last resort for the federal system. As an appellate court, the Supreme Court determines each year which cases it will hear. It is within the Court's power to determine whether a state or the federal government has violated rights granted by the U.S. Constitution. Judgments by the Supreme Court are binding on all American courts.

The State Courts

The United States has two parallel court systems. The federal and state systems are independent of each other. Yet, under specified circumstances, they can interact.[12] Like the United States Constitution's arrangement of the federal court system in Article III, the state constitutions established the state courts. Unlike federal judges, who are appointed, state court judges are either elected or appointed, depending upon the laws of each state. Therefore, each state judicial system is unique. Typically, like the federal system, state courts are composed of three levels: trial, intermediate appellate, and the court of last resort.[13] State courts are usually organized by county. Trial state courts often have names that indicate the type of legal matters that are brought before that court—for example, probate court or family court. State courts have jurisdiction in their respective states over disputes that are not exclusively federal. Decisions by the court of last resort are published, while intermediate appellate-level opinions are selectively published.

COURT REPORTS

When an appellate level case is decided, it is made available on the court's official website and published in a book called a reporter.[14] Reporters are a set of books that contain judicial opinions from a particular jurisdiction.[15] Cases in the reporters are arranged chronologically. The governmental authority either prints or authorizes the publication of official reports. Commercial vendors publish unofficial reporters.

For the federal court system, the U.S. Supreme Court cases are available at http://www.supremecourt.gov/opinions/opinions.aspx. They are also published in print in the official *United States Reports* and in two unofficial reporters: the *United States Supreme Court Reporter* and the *United States Supreme Court Reports, Lawyers' Edition*. Federal appellate and trial court decisions are available at http://www.uscourts.gov/Court_Locator/CourtWebsites.aspx. They are also published in print in two commercial publications: the *Federal Reporter* (courts of appeal) and the *Federal Supplement* (district courts). Because the U.S. government neither prints courts of appeal cases nor district court cases, the *Federal Reporter* and the *Federal Supplement* are treated as official sources.

Many states no longer print their own state reports. However, some have authorized a commercial publisher to publish the appellate decisions and designate it as the official state reporter. State courts have also made their decisions available via their official court websites. For example, Nevada Supreme Court decisions are available at http://www.nevadajudiciary.us/index.php/advance-opinions. State appellate decisions are also commercially published in regional reporters. A regional reporter contains decisions of more than one state. The regional reporters are the *Atlantic Reporter*, *North Eastern Reporter*, *North Western Reporter*, *Pacific Reporter*, *South Eastern Reporter*, *Southern Reporter*, and *South Western Reporter*.

DEVELOP A RESEARCH PLAN

A research plan is a guide you create to help your research be more efficient, accurate, and complete. You want to be mindful that quality legal research is efficient, cost-effective, and thorough. Thus, the research plan is an essential guide map.

The Legal Research Process

Locating legal materials to answer a legal question is called the *process of legal research.* Once you have a basic understanding of the American legal system, you can begin the legal research process. The process or steps you take will depend on the type of research needed. There are different types of legal research. The most common types are retrieval, general inquiry, and fact-based. *Retrieval research* is when you are asked to find a particular document. For example, Grant Provider wants a copy of a recent Supreme Court decision. *General inquiry* legal research does not involve a specific set of facts but is a systematic investigation of a topic. For example, Grant Provider would like to know if other Grant Providers in the area have selected the same grant recipients. Finally, the legal research process can be *fact based.* Typically, you will be given a number of facts and then asked to research a legal question. For example, in accordance with federal regulations, Grant Provider awarded several grants to recipients. However, a recent Court decision has ruled the federal regulation invalid. Thus, Grant Provider is concerned how the recent Court decision will affect how their grant is awarded.

You can take the following steps below. However, always remember that legal research is not a linear process. You may have to alternate between research steps and sometimes even repeat them.

1. Facts are reviewed and legal question is posed. Often you will refer to the facts throughout the research process. You may also need to review the legal question.
2. Determine whether federal or state law applies. Refer to the facts.
3. Research the issue using your research plan. Refer to the facts.
4. Check for updates to the law.[16]

Gather Preliminary Information

The research plan involves four steps: one, gather preliminary information; two, identify potential search terms; three, identify the research resources you plan to use; and four, locate the resources. To begin your legal research, you must gather as much preliminary information as you can. Conduct a reference interview with the requestor (the person who gave you the assignment). The best way to accomplish this task is to use what are called the *reporter questions*: who, what, when, where, why, and how.[17] However, as a law librarian, you need to convert those terms into legal terminology. The first question, *Who,* refers to the parties involved. The second question, *What*, refers to the legal issue. The third question, *When*, refers to the statute of limitations.[18] The fourth question, *Where*, refers to venue.[19] *Why* did the requester give you this particular assignment, is the fifth question. And, the final question, *How*, refers to how the requester wants to receive the information from you. For example, does the requestor want the answer in an e-mail, via text, or a written report to the client?

Generate Search Terms

Once you have completed the reporter questions, you can then use the answers to generate potential search terms. For example, the answer to the *Who* question is much more than just the plaintiff's and the defendant's names. This question can refer to whether the party is a child, the government, a corporation, or a grant provider. It is significant to understand which parties are involved, since their status may implicate certain laws or regulations. For the *Why* question, a possible answer could be to demonstrate to Grant Provider that your organization keeps aware of any cases or regulations that could adversely affect Grant Provider's business. The answers to the reporter questions become your potential search terms. Answering the who, what, when, where, why, and how questions will provide the search terms you need to proceed with the research process.

Identify Resources

Once you have created a list of search terms, you can then identify which legal research resources you may want to use. Legal resources can be either one of three types: primary, secondary, or finding aids. *Primary sources* are the rule of law. They are authoritative but can be mandatory and/or persuasive. Judicial opinions, treaties, executive orders, regulations, and laws are all primary resources. They can be cited in court documents as a declaration of the rule. For example, a United States district court must follow the decisions of its court of appeals. The appellate court opinion would be primary and mandatory. A district court in a different jurisdiction would not be required to follow that appellate court's ruling. However, that same appellate case in a different jurisdiction would be primary, but persuasive.

Although primary resources have the force of law and are cited in court documents, *secondary sources* will prove to be more valuable at the beginning of your research. Secondary sources are materials that help you understand primary resources. A law review article is an excellent example of a secondary resource. For example, you use http://www.gpo.gov/fdsys/[20] to locate an applicable regulation for Grant Provider's issue. However, when you retrieve the regulation, you find it is over twenty-five pages long. Thus, you use Google Scholar to locate a law review article that provides a detailed explanation of the regulation. If you are unfamiliar with a legal issue or topic, it is recommended that you do not start your research by trying to locate primary sources. Rather, begin your search process with a secondary source.

Finding aids are the last type of legal resource. Finding aids are sources that help you locate primary and secondary sources. In the history of legal publishing, it is finding aids that made legal research possible. Without the indices, directories, tables, digests, and so forth, primary information would be almost impossible to find. Imagine trying to locate information on the Internet without a search engine like Google or Yahoo!. Although sometimes ridiculed for its lack of authority, Wikipedia can prove to be a helpful finding aid. When a concept or term is unfamiliar, perform a quick search on Wikipedia. However, you cannot rely on the Wikipedia explanation as authority. When using Wikipedia, always refer to the references at the bottom of the entry. Oftentimes, the references are great secondary sources and sometimes primary as well.

Once you have used the reporter questions to generate search terms, be sure to review the facts and issues again. Another look will help you determine the primary sources you will need to complete the requestor's assignment. Will you need laws, regulations, cases, or all three? After you have identified the sources, you will then be able to determine the best way to locate the materials you need.

Locate the Resources

The final step for your research plan is to locate the resources. The process of legal research has developed from print to proprietary electronic (like Lexis and Westlaw) to now free Internet research. Using the Internet is the most cost-effective way to locate primary sources. However, before using websites for legal research, you want to make sure the websites are reliable. For example, http://www.congress.gov is an excellent source for federal legislative documents. In addition, court decisions can be located for free on official court websites, like http://www.supremecourt.gov/opinions/opinions.aspx. After you locate the resources and find the information that you think will answer the requestor's inquiry, you will want to update your research to be complete. Because laws can be amended, superseded, overruled, or modified in many ways, updating is needed to make sure you have the most current information.

STATUTORY RESEARCH: A BRIEF OVERVIEW

As the United States is a common law country, often your legal research will involve case decisions. However, there are times when the legal question will concern a federal or state statute. After a law is passed by both the House and Senate and signed by the president (or governor), the law is given a number. Federal laws are designed as public laws or private laws. Public laws are applicable to all of society, while private laws affect an individual, family, or particular group (like veterans). For example, United States Public Law 92-3 is the third law enacted by the 92nd Congress. Because a session of Congress is every two years, public laws are also referred to as *session laws*. State statutes have similar designations, but are often referred to as *Acts*.

To locate federal laws via the Internet, you would use http://www.gpo.gov/fdsys/browse/collection.action?collectionCode=PLAW. In print, you would use the public law number to locate federal laws in the *United States Statutes at Large*.[21] If you do not have a public law number but rather a subject, use the *United States Code*. The *United States Code* is available at http://www.gpo.gov/fdsys/browse/collectionUScode.action?collectionCode=USCODE. The *United States Code* is the cumulative resource for public laws (private laws are not included). It is organized into fifty-one titles (or subjects).

To locate state laws, you would use the official state website. For example, California state laws are available via the Internet at http://www.ca.gov/about/government/state/lawsandregs.html. The state websites will have the session laws and the state code.

REGULATORY RESEARCH: A BRIEF OVERVIEW

In addition to judicial opinions and statutes, oftentimes when performing legal research you will find the need to locate and review a federal or state regulation. Administrative agencies, using their statutory directive, promulgate regulations that are primary authority. Agencies can issue rulings, conduct hearings, and adjudicate disputes. Both federal and state governments have regulatory bodies.

To locate federal regulations, use the U.S. Government Printing Office's website. Federal regulations are located in the *Federal Register* and the *Code of Federal Regulations*. The *Federal Register*, http://www.gpo.gov/fdsys/browse/collection.action?collectionCode=FR, is the daily compilation of agency notices, proposed and final rules and regulations, as well as presidential documents (executive orders and proclamations). The *Code of Federal Regulations*, http://www.gpo.gov/fdsys/browse/collection.action?collectionCode=FR, is the annual compilation of general and permanent rules that were previously published in the *Federal Register*. The *Code of Federal Regulations* is organized into fifty-one titles (or subjects).

State administrative regulations are also located on state official websites. For example, the *Maryland Register* (the daily compilation of Maryland agency notices, rules, and regulations) is available at http://www.dsd.state.md.us/MD Register/mdregister.aspx.

LEGAL CITATIONS

Like other disciplines, legal materials have a unique identification system or style. Legal citations are used to locate documents in the various legal resources. The *Bluebook*: *A Uniform System of Citation* is the legal citation style manual for the United States. The law review staffs at Harvard, Columbia, the University of Pennsylvania, and the *Yale Law Journal* compile the *Bluebook*. In addition to the *Bluebook*, there are other style manuals including those that are applicable for certain jurisdictions.[22] The following examples refer to the *Bluebook*.

A case citation will typically look like this: *Gideon v. Wainwright*, 372 U.S. 335 (1963). The citation will include the case name, reporter volume number, abbreviated reporter name, the beginning page number of the opinion, court (if not the court of last resort), and year the case was decided. In the example above, the case name is *Gideon v. Wainwright*; the reporter volume number is 372; the reporter name is the *United States Reports* (abbreviated *U.S.*); the reporter page number is 335; and the year is 1963.

A federal statutory citation looks like this: 20 U.S.C. § 1681 (2006). The citation will include the title number, *United States Code* abbreviation, section symbol, section number of the statute, and the year of the most current compilation of the *United States Code*.[23] In the example above, the title number is *20*; the abbreviation is *U.S.C.*; the section number is *1681*; and the most current compilation of the *United States Code* is *2006*. Similar to court reporters, there are two unofficial compilations for the federal laws: the *United States Code Annotated* and the *United States Code Service*. Both publications use the same citation format as for the *United States Code*.

A federal regulatory citation in the *Federal Register* looks like this: Importation of Fruits and Vegetables, 60 Fed. Reg. 50,379 (Sept. 29, 1995) (to be codified at 7 C.F.R. pt. 300). The citation will include the name of the regulation, the volume number, the *Federal Register* abbreviation, the beginning page number for the regulation or other document (like a notice of hearing), the publication date, and where the document is to be codified in the *Code of Federal Regulations*. In the example above, the name of the regulation is the *Importation of Fruits and Vegetables*; the title is *60*; the *Federal Register* abbreviation is *Fed. Reg.*; the beginning page number for the regulation is *50,379*; the publication date is *Sept. 29, 1995*; and the regulation will be codified in the *Code of Federal Regulations* at *7 C.F.R. pt. 300*.

A federal regulatory citation in the *Code of Federal Regulations* looks like this: National School Lunch Program, 7 C.F.R. § 210 (2012). The citation will include the name of the regulation, the title number, the *Code of Federal*

Regulation abbreviation, the section symbol, the specific section cited, and the edition year of the *Code of Federal Regulations*. In the example above, the name of the regulation is the *National School Lunch Program*; the title number is *7*; the *Code of Federal Regulations* abbreviation is *C.F.R.*; the specific section cited is *§ 210*; and the edition is *2012*.

Each state has it own citation system for statutes and regulations. Thus, to properly cite those documents, you need to refer to the official state website for a particular state or locate a research guide from one of the law school libraries located in that state.

THE BEST WAY FOR YOU

There is no perfect way to perform legal research. After you have gathered the preliminary information, identified potential search terms, identified the resources, located resources, and updated your research, you will have completed the process. Possibly, you will reach the deadline or other criteria given for completion. However, more likely, you will begin to see the same authoritative standard repeated in various sources. For example, you determine that the court uses a five-point test to determine if a law is unconstitutional. During your research, you locate a law review article, which analyzes the five-point test. You also locate a number of appellate cases that apply the same five-point standard. If you believe you have located the authoritative standard for your question, it is then when you should consider stopping your research.

How you start the research process and how you end it depend on what is comfortable for you. Over time you will develop your own methodologies for locating legal information and your research skills will improve. As long as you use techniques that provide you with authoritative materials that answer the issue posed by the requestor, you will be successful.

FURTHER READING

A century of lawmaking for a new nation. http://memory.loc.gov/ammem/amlaw/lwabout.html. Website created by the Library of Congress that consists of a linked set of published congressional records from the Continental Congress through the 43rd Congress, 1774–1875.

Administrative office of the U.S. Courts on behalf of the federal judiciary. http://www.uscourts.gov/Home.aspx.

Barkan, Steven M., Roy M. Mersky, and Donald J. Dunn. 2009. *Legal research illustrated*, 9th ed. New York: Foundation Press.

Harvard Law Review Association. 2010. *The bluebook: A uniform system of citation*, 19th ed., Cambridge: Gannett House.

Ilbert, Courtenay. 1904. *Montesquieu*. London: Clarendon Press.

Kunz, Christina L., Deborah A. Schmedemann, Ann Bateson, and Mehmet Konar-Steenberg. 2012. *The process of legal research authorities and options,* 8th ed. New York: Aspen.

McNeese, Tim. 2001. *The U.S. Constitution.* St. Louis: Milliken.

Olson, Kent. 2009. *Principles of legal research, successor to How to Find the Law Concise Hornbook 9th edition.* St. Paul: Thomson Reuters.

Sloan, Amy E. 2012. *Basic legal research tools and strategies*, 5th ed. New York: Aspen.

The United States Government Manual. http://www.gpo.gov/fdsys/browse/collection.action?collectionCode=GOVMAN. The official handbook of the federal government that provides information on the federal agencies.

SUGGESTED LEGAL RESEARCH PLAN WORKSHEET

1. Requestor's Name: ______________________________
2. Date Requested: ______________________________
3. Date Needed: ______________________________
4. Billing Account/Client Number: ______________________________
5. Brief Synopsis of the Facts or Question Asked:

   ____

   ____

   ____
6. Reporter Questions

   Who: ______________________________

   What: ______________________________

   When:

   Where: ______________________________

   Why: ______________________________

   How: ______________________________
7. Possible Search Terms ______________________________
8. Identify Resources

   Primary Mandatory ______________________________

   Primary Persuasive ______________________________

   Secondary ______________________________

   Finding Aids ______________________________
9. Locating Resources

   Websites ______________________________
10. Updating (Dates) ______________________________
11. Date Research given to Requestor: ______________________________

NOTES

1. Bryan A. Garner, *Black's law dictionary*, 9th ed. (St. Paul, Minn.: Thomson Reuters, 2009), 314.

2. Lawrence M. Friedman, *A history of American law*, 2nd ed. (New York: Simon and Schuster, 1985), 21.

3. Garner, *Black's law dictionary*, 1295.

4. Ibid., 313.

5. Robert L. Maddex, *The U.S. Constitution A to Z*, 2nd ed. (Washington, D.C.: CQ Press, 2008), 122.

6. "The particular structure of the new government and the distribution of power among its different parts," *The federalist papers*, no. 47, from the *New York Packet*, Friday, February 1, 1788.

7. At the state level, the executive branch would be the governor.

8. Charles H. Koch Jr., 3 *Administrative Law and Practice* § 7:31 (3rd ed.).

9. Garner, *Black's law dictionary*, 927.

10. A bench trial does not have a jury. Therefore, a judge decides the facts, law of the case, and its outcome.

11. In New York, the court of last resort is the court of appeals, while the trial level is the Supreme Court.

12. For example, the federal court has jurisdiction over cases in which two or more states are a party.

13. Some states, like New Hampshire, do not have an intermediate appellate court.

14. Typically, federal and state trial court opinions are not published. However, selective federal district court opinions are located in the *Federal Supplement*. In addition, not all intermediate appellate level cases are published.

15. The term *reporter* was originally used to refer to the people who compiled and published the court opinions. However, in the United States, the term is also used to indicate the set of published court decision. Thus, the terms *reporter* and *reports* are used interchangeably. Lawrence M. Friedman, *A history of American law*, 3rd ed. (New York: Touchstone, 2005), 241–242.

16. The three major legal database systems provide updating services are *KeyCite* (Westlaw.com and WestlawNext), *Shepard's* (Lexis.com and Lexis Advance), and *BCite* (Bloomberg Law).

17. David M. Malone, Peter T. Hoffman, and Anthony J. Bocchino, *Effective deposition: Techniques and strategies that work*, 3rd ed. (Chicago: National Institute for Trail Advocacy, 2007), 99.

18. The statute of limitations is a law that establishes a time limit for when a lawsuit can be brought. The idea is to encourage people to commence litigation within a reasonable time. Garner, *Black's law dictionary,* 1546.

19. Venue is the place where the lawsuit can commence. Charles Alan Wright, Arthur R. Miller, and Edward H. Cooper, *Federal practice and procedure*, vol. 14D, *Jurisdiction 3d §380* (St. Paul: Thomson/West, 2007).

20. The United States Government Printing Office's official website.

21. Private laws are published in the *United States Statutes at Large*. You can locate both public and private laws by their *Statutes at Large* citation. For example, the *No Child Left Behind Act of 2001* can be found at 115 Stat. 1425.

22. For example, the *ALWD citation manual: A professional system of citation*; *Texas rules of form*; and *The University of Chicago manual of legal citation.*

23. The *United States Code* is published every six years with annual cumulative supplements (or updates).

16

Online Information Sources

Sarah K. C. Mauldin
SMITH, GAMBRELL & RUSSELL, LLP

The modern law library, whether it is in a law firm, law school, court, or agency, is, in general, growing physically smaller while the number of available information sources and the demands of researchers increase with every passing year. Law libraries are also being asked to provide more and more information while the budgets shrink and prices of online sources rise. The job of a librarian involved in collection development in any law library setting is to provide the best information, in the fastest and most accessible way, to patrons at the reference desk, at home, in another office, on another campus, or traveling anywhere in the world.

ABUNDANCE OF SOURCES

For years the legal marketplace for electronic resources has been dominated by Westlaw and Lexis. Both provide primary law as well as secondary sources and news. Sometimes these sources are specific to a platform because the print resource is published by the same company or the publisher has made an exclusive deal to provide specific content on only one platform. The entry of BloombergLaw into the market has shaken up the landscape. Not only did BloombergLaw acquire BNA and all of its content, it has also positioned itself as a provider of primary law resources and exclusive content under a seat license or a firm-wide unlimited usage model. CCH/Wolters Kluwer has also entered the arena with Lois Law for primary and some secondary sources and with IntelliConnect, with libraries of exclusive content and tools that provide dynamic content based on user input, also on a seat license or a firm-wide unlimited usage model.

Every year there are also new entrants into the legal information field. Some of these are current awareness tools that push content like new court filings and breaking news to patrons based on a topic or on specific search terms defined by the end user or a librarian intermediary. Others are specialized databases designed for a very specific practice area or field of knowledge. These may be based on print books, journals, and primary law that have been packaged into a single platform that serves as a one-stop shop for legal researchers in a specialized practice area like tax, intellectual property, or international arbitration. These very specialized sources are often expensive and of limited or no use to attorneys and faculty outside of the particular practice area covered.

COST CONSTRAINTS

The digital law librarian must be able to evaluate resources to determine which are most cost effective for the institution, if it is possible to use information from a single provider, or if it is necessary to purchase from multiple vendors to get the right mix of information sources to the right people just in time. The librarian must also be able to provide training on an ad hoc basis to meet an immediate information need and scheduled training to patrons on the most effective way to use the many sources available and on how to determine which online information source is the best tool for the task at hand. The librarian also must be constantly on the lookout for new resources that can replace a current source, that provide better value for the dollars spent, or that provide a single platform that gives patrons the information they need and want in the most efficient, cost-effective manner possible. This balancing act can be difficult but also makes the library and the librarians who staff it an indispensable part of the law firm, law school, court, or agency it serves.

DUPLICATION OF SOURCES

A major issue, especially in law firm libraries, is the duplication of material in print and in various online sources. Nearly every electronic resource found in a law library, except the most specialized databases, provides duplication of materials, mostly primary law such as cases and statutes. Each resource treats this primary material differently, with some sources, like Westlaw, Lexis, and BloombergLaw adding editorial content to make their offering unique. Other sources, including free or low-cost online services like LoisLaw, CaseMaker, LII, and Fastcase, provide the straight text of cases and statutes without editorial enhancement. Some of these services do often add value via algorithms that provide lists of cases that have cited the case or statute being researched or other unique features.

It is vital that librarians monitor their electronic and print collections to minimize duplication. This may mean canceling print subscriptions or limiting fee-based research services to one primary vendor and finding workarounds to gain access to materials not available in the organization's primary research services. Cutting materials will almost always be a disappointment to one or more library stakeholders. These cuts are necessary to minimize cases where primary legal material that is available for free or nearly free with few, if any, editorial enhancements is also included in a fee-based service. In this situation, the library may be paying twice or perhaps three times for the same primary materials by subscribing to multiple online providers.

FORCED MOVES TO ONLINE SOURCES

Occasionally, there is no choice for a library but to move from a print resource to one that is online. These moves may be dictated by publishers that cease publication of a print resource and begin offering a source in an online-only format. A move may also come about owing to the shrinkage of the physical library space, making it necessary to discard print materials and replace them with comparable electronic resources. Cost constraints may also play a role in the move from print to electronic, often because it is cost prohibitive to keep multiple sets of a print title in different locations, while online resources can provide users in multiple locations with equivalent access to materials at a similar price. Moves to electronic resources can also be forced by end users who may prefer how a resource is presented online or who want access from anywhere at any time. This kind of move can be difficult unless all stakeholders are happy with the decision and the price is right. There are some materials in a law library that are simply easier or more intuitive to use in print than they are online, so the decision to change formats must be considered carefully and all pros and cons taken into account.

FEE-BASED SOURCES

General Purpose Legal Research

Law libraries most often have access to one or more of the large, well-known fee-based legal research services that like Westlaw or Lexis. Newer entrants into the general purpose legal research arena include Lois Law from Wolters Kluwer and BloombergLaw from Bloomberg BNA. All of these services provide access to primary legal resources like cases, statutes, and regulations, as well as specialized secondary resources and value-added editorial content, often limited to a single provider, (e.g., Matthew Bender and American Lawyer Media titles are currently only available on Lexis/Lexis Advance, and the West Key Number System and the accompanying Headnotes are exclusive to Westlaw/WestlawNext). For more information about researching in these and other fee-based general purpose legal research databases, please see chapter 17, Major Legal Databases and How to Search Them.

Law firm, agency, and public law libraries often have a flat fee or special offer contract with one or more of these providers and gain access to nonpreferred services on a pay-as-you-go basis or have no access at all. These contracts often include a selection of secondary resources or specialized resources like dockets and public records databases, depending upon the needs of the library's parent organization. Materials outside the contract can often be accessed for an additional fee. Law school libraries are more likely to have access to a wider selection of fee-based resources for teaching purposes. A well-rounded law librarian may be an expert searcher in one or more of these databases, but is also able to learn and use any of the fee-based services, as entities may change providers or a new employment setting may have a different preferred service. Content exclusivity is also subject to change as the major players regularly make agreements for different databases to be available exclusively on one or another search platform.

CaseMaker and Fastcase are special cases within the primary law market. Access to both databases is available for purchase by law libraries, but the most common way that attorneys gain access is as a benefit of membership in a state bar association. The current breakdown as of January 22, 2013, of bar associations and other membership organizations between the two vendors is shown in table 16.1. Both CaseMaker and Fastcase provide access to the text of cases (which have no copyright protection as government works), statutes (which require that all copyrighted editorial enhancements from codifiers be stripped out), and some secondary sources like bar journals. Neither provides the level of editorial enhancement that is available in other fee-based services, but each provides a valuable resource for attorneys who do not have, or choose not to use, fee-based general purpose legal research providers. These are often solo practitioners, small firms, and attorneys or firms that focus on one very specific area of law. Attorneys in these kinds of private practices or agencies are unlikely to employ a librarian but may use the services of a court, county, or membership law library.

Specialized Legal Research

Sometimes a general purpose fee-based service is more than is needed for particular attorneys or practice areas. Because of this, there are many specialized databases covering a slice of the law. These include RIA Checkpoint, standalone libraries on CCH IntelliConnect, BNA Practice Centers, and many others.

Table 16.1. Bar Association and Legal Membership Organizations Subscribing to Fastcase and Casemaker as a Member Benefit[1]

Fastcase	Casemaker
State Bar of Arizona	Alabama State Bar
Arkansas Bar Association	Alaska Bar Association
Cincinnati Law Library	Colorado Bar Association
Cleveland Metropolitan Bar Association	Connecticut Bar Association
D.C. Bar	Idaho State Bar
The Florida Bar	Indiana State Bar Association
State Bar of Georgia	Kansas Bar Association
Hawaii State Bar Association	Kentucky Bar Association
Illinois State Bar Association	Maine Bar Association
Iowa State Bar Association	Massachusetts Bar Association
Kern County Bar Association	State Bar of Michigan
LA Law Library	The Mississippi Bar
Louisiana Bar Association	Nebraska State Bar Association
Maryland Bar Association	New Hampshire State Bar Association
Minnesota State Bar Association	State Bar Association of North Dakota
The Missouri Bar	Ohio State Bar Association
State Bar of Nevada	Rhode Island Bar Association
New Jersey State Bar Association	Santa Clara County Bar Association
State Bar of New Mexico	South Carolina Bar
North Carolina Bar Association	State Bar of Texas
Oklahoma State Bar Association	Utah State Bar
Oregon Sate Bar	Vermont Bar Association
San Fernando Valley Bar Association	Washington State Bar Association
State Bar of South Dakota	Wyoming State Bar
State Bar of Wisconsin	
Tennessee State Bar	
Virginia State Bar	
West Virginia State Bar	

[1]*Source*: Greg Lambert, "Texas Bar launches the new casemaker platform for its members," *Three Geeks and a Law Blog*, January 22, 2013, www.geeklawblog.com.

RIA Checkpoint

RIA Checkpoint is a flat fee service that provides access to tax-related information through a series of libraries, including federal, state and local, estate planning, and employee benefits. Depending upon the mix of sources chosen, Checkpoint can serve as a one-stop shop for most of the research needs of a tax or trust and estate planning attorney. Checkpoint includes primary law (American Federal Tax Reports, Tax Court opinions, Treasury Regulations, the Internal Revenue code, IRS publications and forms, legislative documents, etc.) and also provides access to separately purchased treatises and journals from RIA, WG&L, and EBIA. Checkpoint can be set up to send alerts on new developments in the areas it covers. Much of the information available on Checkpoint is also available on Westlaw and WestlawNext, as they are all Thomson Reuters products. Checkpoint is accessed via a username and password, and the librarian is often the administrator for passwords, creating, assigning access, and deleting users. Checkpoint is sold in bands of one user, twenty users, thirty-five users, and up.

The Bloomberg/BNA Practice Centers

The Bloomberg/BNA Practice Centers are slices of materials also available in BloombergLaw. Common Practice Centers include Tax and Accounting, Intellectual Property, and Employee Benefits, as well as many others. The services provide access to primary law and secondary sources produced by BNA's editors and reporters. Each Practice Center can serve as a starting point for attorneys practicing in specific areas of law. The centers include cases, statutes, and regulations, as well as secondary source material, including journals and commentary. The Practice Centers can also be set up to provide alerts on new developments in the particular area the center covers. Practice Centers may be IP authenticated or accessed via a username and password and are sold for individual users or with firm-wide IP authenticated access.

CCH-IntelliConnect

CCH-IntelliConnect is another platform that includes specialized materials and is sold for firm-wide access so that the platform appears the same to users no matter which libraries or publications have been purchased. IntelliConnect libraries include publications from CCH, Wolters Kluwer, and Aspen Publishers, and can include nearly everything that is available in print from tax materials to business franchising to intellectual property to construction law and many more. CCH provides daily alerting features as well as SmartCharts, which are dynamically created charts of laws or other legal materials that respond to parameters supplied by the end user.

Each of the proceeding database services closely mimics the print titles on which it is based and can be relatively easy to introduce to end users. Librarians must keep in mind that each of the libraries and centers available from these publishers includes the same issues with duplicative primary law content that exist with general purpose search platforms. This means that it is imperative to carefully consider cost, user experience, and features before choosing to make the switch from print publications to their electronic analogs.

SINGLE-TOPIC RESOURCES

Some legal practices require very specialized, single-topic resources. Examples of these include databases for performing advanced searches of Securities and Exchange Commission (SEC) filings, databases used by a single practice group like Thompson for employee benefits, A. S. Pratt for banking and finance, Practical Law Company (PLC) for examples of documents used in transactional practice, and many others. Some, like Thompson and A. S. Pratt, are based on the publisher's print products, while PLC is an online-only resource with a print journal, and the SEC databases are overlaid on the EDGAR platform.

Examples of fee-based SEC databases include Accelus (owned by Thomson Reuters), Morningstar (formerly 10K Wizard), and Knowledge Mosaic (recently acquired by LexisNexis). These databases are useful for finding the detailed information that U.S. public companies and some other entities are required to file. These are all available on EDGAR, but the database allows access to the bare filings, which are only searchable by company name and form type. The fee-based resources add the ability to search by keyword, limit searches to particular form types or exhibit types, limit by other identifiers like SIC or NAICS code, provide snippets of the search terms in context, and download results in various formats. An organization that practices commercial law needs one of these services for finding exemplar documents and performing precise searches for company data and filings.

Practical Law Company (recently acquired by Thomson Reuters) is somewhat similar to the SEC databases in that it provides example documents for attorneys working in commercial and other transactional practices. PLC is different in that its examples are created by attorney-editors and are created to include all possible clauses of an agreement or other document, with commentary, instead of the negotiated documents found in a search of SEC filings. PLC also provides a glossary of common business terms and "know-how" articles for the benefit of attorneys who may be new to practice or unfamiliar with a particular concept or deal type.

There are numerous other single-topic databases that provide information or guidance on particular areas of law. These include databases of jury verdicts, information on international arbitration, information on judges, attorneys, expert witnesses, and arbitrators, or very specific materials like documents recorded with county clerks and trial court dockets.

CURRENT AWARENESS

Attorneys thrive on information and knowing what is happening with a client or industry before anyone else, including the client. This is where current awareness services become incredibly useful. Current awareness tools include customizable clipping services available for most databases in resources like Westlaw and Lexis. They can also be very specialized and targeted newsletters of news or case filings provided by a team of reporters and editors like Law360 (recently acquired by LexisNexis), Courthouse News Service (cases filed in specific geographic areas), or the Tan Sheet (FDA news). Other current awareness tools include legal newspapers, which may be local (county legal organs), statewide, or national in scope. Attorneys also use newspapers like the *New York Times*, *Wall Street Journal*, and others to keep abreast of industry trends. In the digital library, the librarian may act as a curator of this information, using software to create customized newsletters for different practice groups or people, or the librarian may simply offer the information sources to users to do with as they wish.

FREE OR LOW-COST SOURCES

Primary Law

The Internet teems with free or very low-cost primary legal information in the form of cases, statutes, regulations, and other materials. Common sources for this information include Findlaw, the Public Library of Law (PLOL), Cornell's Legal Information Institute (LII), Thomas, FDSys, PACER, and other governmental websites. In general, these sources are current and provide correct information, but they also provide few, if any, editorial enhancements and usually carry a disclaimer that the information they contain is "as is" and may contain errors, so it can be a good informational source but should not be relied upon.

Findlaw, PLOL, and LII all provide access to primary law that includes state and federal cases and statutes, usually from the highest court in a jurisdiction. However, none of these sites claim to include everything available. Findlaw is a service of Thomson Reuters and includes both a public site and a legal professional site. The public site includes links to find lawyers, explanations of common legal issues, and commentary on current events in law. The legal professional site includes U.S. Supreme Court opinions, many federal and state court opinions, and links to laws and regulations. Public Library of Law is provided by Fastcase and has similar coverage and features to the Findlaw's professional site. If material is not available for free, the searcher is invited to purchase access to Fastcase. LII is a nonprofit site run by the Cornell Law School. It also includes many links to primary law and includes features like the text of the Uniform Commercial Code. While these sites have many resources available, it must be remembered that neither they, nor any other free sites, are fully comprehensive and current. If the outcome of a case or issue depends on the absolute accuracy of the research, it is best to use a fee-based service with a high level of editorial control, like Westlaw, Lexis, or BloombergLaw.

Government sites are also excellent sources for free and low-cost primary law. FDSys is the new federal site that replaces GPO Access and Congress.gov as the source for authenticated materials from the U.S. government. These materials include the text and history of congressional bills, the *Congressional Record*, the text of the Statutes at Large, and the official United States Code, as well as the current and past editions of the *Code of Federal Regulations* and the *Federal Register*, including the issue for the current day. All documents on FDSys are electronically authenticated.

PACER provides access to federal trial, bankruptcy, and appellate court dockets for all courts that participate. It can be searched by individual court or by using the PACER Case Locator. Access to PACER is password protected and searches can be done for ten cents per page with a maximum of $3.00 per discrete document and no charge for court opinions.

Case materials are available for most courts back to the late 1990s, but there are exceptions. In some cases, even though the case filings may appear to be available, they are not accessible because they have been sealed by a court order. Anyone can sign up for a PACER account. Accounts are billed quarterly and any balance below $10.00 is not billed. A free source for many PACER documents can be found at www.recapthelaw.com. RECAP is a browser add-on, currently available for Firefox and Chrome, which uploads PACER documents to an archive as the user downloads them and alerts the user if a requested document is available for free in the RECAP archive.

The websites of states and their agencies, federal agencies, and local governmental entities like counties and municipalities are often treasure troves of excellent information. They often include links to statutes, ordinances, and regulations of the governmental entity and can also be useful for finding contacts and other information about the work of the agency or governmental entity. Examples of these sites include IRS.gov, which provides federal tax information for the public and practitioners, including links to forms and instructs from current and previous years; and Georgia.gov, which includes links to state agencies and counties and municipalities like atlantaga.gov or fultoncountyga.gov, where a researcher can find links to city and county departments as well as the governing ordinances. The same caveat applies to these websites as to any other site on the Internet. Just because the information is posted does not mean that it is authentic, current, or correct, so a researcher should always check an authoritative source if a case or issue depends on absolute accuracy.

Secondary Sources

Free and low-cost secondary law sources are much more difficult to find online than primary law. Most of what is published is in institutional repositories or on services like SSRN and JStor. These may or may not have full text available online, and they mostly consist of law review articles and white papers that will soon or have been recently published. Some are available as abstracts and may be available in full text from a law school or public law library, or by going directly to the author. Depending upon what the facts and circumstances of your research are, there may be reasons that the attorney, professor, or other patron would prefer that a librarian not contact authors for confidentiality and strategy reasons.

An excellent source for secondary legal materials is often databases available through the public library or university. Many articles, especially those on non-law-related disciplines like accounting or education law, are available in databases commonly available for use within the subscribing library or via remote access with a library card. SSRN and JStor articles, which are not freely available on the Web, can often be obtained in this way. Most of the secondary sources that a researcher might need to find for a patron are not legal documents, but instead are writings by expert witnesses or materials requested by patrons for researching and writing law-related articles and books.

Another useful source is Google Scholar and Google Books. These services provide access to cases, articles, and books, but are not complete or comprehensive. They are, however, a fine starting place if the researcher knows little about a topic or is looking for leads on materials that might be best used as print resources that can be obtained through interlibrary loan.

USING SEARCH ENGINES EFFECTIVELY

A search engine is a program or set of programs that allows for the indexing and retrieval of information from a database. This database could be your e-mail inbox, a law firm's document management system, or the entirety of the Internet. Internet search engines like Bing, Google, and Live Search create databases with the use of crawlers, sometimes known as bots or spiders. Crawlers are programs that use algorithms to visit millions of webpages every day to find changes and to produce an index of pages, pulling out keywords for fast retrieval of information.

Search engines use a variety of highly sophisticated algorithms to determine the relevance and ranking of results from any given search. This may be based on the number of sites that link to a particular webpage. It may also rely on the frequency of particular keywords or how much a particular website owner has paid to a search engine company to show its sites at the top of a result list. Some search engines make sponsored links obvious in a result list while others make it harder to know if a result is near the top of the list because it is popular, particularly relevant, or paid for.

Some search engines, like Dogpile, Kayak, and Metacrawler, are meta-search engines. This means that the engine aggregates results from multiple sources into a single result list. Kayak is a meta-search engine that aggregates a very specific subset of results, in this case travel information. Other meta-engines bring together results from the most popular search engines.

It is imperative that searchers try more than one search engine or use a meta-search engine to be—although not sure that everything has been found—at least sure that the majority of rocks have been overturned for information. LexisWeb is another option for finding legal secondary sources. It is a free service that includes only sites that have been vetted for quality by Lexis editors. Although this limits the number of available hits, it also improves the precision of a Web search for specifically legal materials.

Another option for searching freely available resources is OAIster. It is a collaboration of OCLC and the University of Michigan, and it is freely accessible on OCLC at http://oaister.worldcat.org/. The collection was harvested via the Open Archives Initiative Protocol for Metadata Harvesting. The materials available on OAIster can be especially valuable when searching for information on nonlegal topics that are required by attorneys. The most likely reasons that a legal researcher might require these kinds of materials is for researching the backgrounds of expert witnesses or arbitrators, searching for cases of infringing works, or for searching for information on nonlegal topics that may be of use in a lawsuit or other action. Available materials in OAIster include:

- Digitized (scanned) books, journal articles, newspapers, manuscripts, and more
- Digital text
- Audio files (wav, mp3)
- Video files (mp4, QuickTime)
- Photographic images (jpeg, tiff, gif)
- Datasets (downloadable statistical information)
- Theses and research papers

RESEARCH ON PEOPLE, COMPANIES, AND SO FORTH—FEE-BASED RESOURCES

Subscriber Sources

What is in the sources you already subscribe to? When looking for public records on the Internet, it is important to think of the resources that the institution already subscribes to, like Westlaw and Lexis. Westlaw has a service called People Map that synthesizes available public records into a report on the person or persons searched. Lexis's SmartLinx product produces a very similar report. These fee-based services also include numerous databases of standalone public record information, including databases of professional licenses, secretary of state filings, court dockets, utility records, and a host of other sources that go into the completed People Map or SmartLinx report.

Company Information

For researchers seeking information on companies and other entities, there are numerous resources that the organization may already subscribe to, including Hoover's, Dun & Bradstreet, SkyMinder, and many other niche players. Hoover's provides what it calls Company Capsules on public and private companies. These capsules include a summary of the entity written by an editor as well as information on employees, financials, and recent news.

Dun & Bradstreet provides several different reports and is somewhat like a credit-reporting agency for businesses, public and private. D&B reports vary in depth of treatment based on many variables, including how forthcoming the entity in question is when a Dun & Bradstreet investigator calls or visits. One very useful report from D&B is the Corporate Family Tree that makes it somewhat easier to determine how divisions and subsidiaries of a parent corporation are arranged. This can be especially important in a law firm environment by making it possible to avoid conflicts brought on by suing a subsidiary of a company the firm already represents and also in determining if a potential client is a good credit risk before the firm accepts the engagement.

SkyMinder is another resource to which an organization may already subscribe. It produces reports that are a kind of hybrid of Hoover's and D&B, but the focus is on companies outside the United States. These reports are often difficult or impossible to find without the use of a service like SkyMinder.

People Information

People information is often highly prized in a law firm setting. It can be used to find out more about an adverse party and their assets or can provide information needed to perfect service of process on a party to a lawsuit. Currently, the two most common services are Accurint (from LexisNexis) and TLO, which is a relatively new entrant to the legal marketplace. Both services work similarly and provide reports that include addresses, telephone numbers, possible employers, possible criminal records, cars and real property owned by a subject, and much more. Entities must be credentialed to use any public records service and must have valid reasons for searching under both the Gramm-Leach-Bliley Act and the Driver's Privacy Protection Act. *Any public records database must be used only for legitimate research purposes*, and there are numerous safeguards in place to keep private information private. These databases only have information for the United States, as other countries have even more stringent privacy laws than are in place in the United States. It is also important to remember that public records information is only as good as the provider. Many records provided by governments or other entities are incomplete or have typographical errors that may provide incorrect or misleading information. It is a best practice to check another source before assuming that all of the information found on a person is true and correct.

RESEARCH ON PEOPLE, COMPANIES, ETC.— FREE OR LOW-COST SOURCES

Many public records are available on the Internet in databases that are either free to access or that have nominal fees. PACER and other court websites can be treasure troves of information on people. However, outside of the federal PACER system, state courts vary widely in the information available. Some dockets are not searchable at all, while others are behind pay-walls, and still others are freely available on the Web with access to not only the dockets but also the documents filed. These resources are valuable for finding background on people and companies and their history of litigation.

County assessor and clerk/recorder sites can also vary widely in coverage, but can be helpful in determining ownership of property or chain of title. Nearly all assessor records are available online throughout the United States. These records include current ownership of property, information on property taxes, and may even include photographs and sketches of property. Occasionally assessor records are not found on the assessor's website but are instead found in a county or municipality's Geographical Information System (GIS). County clerks and recorders have custody of all documents required to be filed for real estate and other transactions. Some clerks' sites have information freely available online while others use a pay-wall model. The information in recorded documents is only available online since the mid-1990s to early 2000s, but is available from the clerk or recorder's office in print into the nineteenth century. Recorded documents can be a wonderful resource for patrons involved in quiet title actions, genealogists researching family history, attorneys involved in real estate deals, and for many other research reasons.

As noted earlier, databases from public and university libraries can also be a huge boon to the researcher on a shoestring budget. Some libraries have access to databases like ReferenceUSA, which provides information on addresses and telephone numbers for individuals and businesses throughout the United States. Libraries may also subscribe to databases like Hoover's that can provide detailed company information for free.

Another useful tool is the Social Security Death Index (SSDI). It is a listing of all persons for whom the Social Security Administration has a record of death. This was once freely available on Ancestry.com but now requires a free account.

When searching for information on public companies and people who are insiders, are highly compensated, or who own large blocks of stock in public companies, EDGAR from the SEC is an excellent tool. Although it is more difficult to search EDGAR without one of the overlay services discussed earlier in this chapter, it is still a useful tool for free company research.

Secretary of state websites are also great tools for finding information on individuals and companies. All corporations, partnerships, and limited liability entities must be registered with the secretary of state (SOS) to legally do business in a state. The SOS corporations database includes information on officers, addresses, registered agents, and sometimes full business filings for free for all business registered with a state. The coverage is highly variable from state to state. Some states use a simple Web interface for searching, while others require that a searcher log in to a legacy system through the Internet. Texas and Delaware are two states that require searchers to have an account or pay for access to search, but this is an exception to the general rule.

DECIDING WHAT YOU NEED—MAKING HARD CHOICES

The modern law librarian in the digital age, no matter what organization he or she works for, often has to make tough choices about electronic resources. A major driver of these decisions is budget. Is there enough money to acquire all of the electronic resources that users want and need? Are users willing to give up print and move to electronic resources voluntarily, or is the change being forced by space constraints and vendor changes? Is the information available in the online resource unique or does it duplicate other online or print resources already owned by the library? Will the resource be used often enough to justify the cost? How will usage be measured? What kind of licensing terms are required? Will this electronic resource provide equal access to everyone in the organization that needs it?

The most likely issues for most organizations are budget, space, and user acceptance. If there is no money to purchase the resource, then it is pointless to begin the process. Using electronic resources to replace large sets can be a great strategy if space is at a premium, as it usually is in law firm and court libraries. However, if users are unwilling or unable to learn new online resources to replace existing print resources, the battle may also be lost.

Another major issue to consider when determining whether to switch to an electronic resource is parity of access. Will every user that needs to use it have access? Is the resource priced by seat, by simultaneous log-in, by location, entity-wide, or in some other way? In my experience, I have moved two print resources to electronic specifically for access reasons. My firm began in Atlanta but now has offices in multiple states with attorneys practicing in tax and trusts and estates in three of the offices. We had CCH's *Standard Federal Tax Reporter* and *Federal Estate and Gift Tax Reporter* in print in Atlanta, so anyone from New York or Jacksonville who might want to use the print resource would have to call Atlanta to have access to the material. The cost for moving to firm-wide access for both of these titles was reasonable and the attorneys involved agreed that the format change would be a workable solution. While some users were not completely thrilled with the move to a new electronic platform, all have adjusted and are generally happy with how the situation is now. This only worked because the electronic resources were priced for a firm-wide subscription. If they had been priced by office or user, the solution might not have been able to work at all, as purchasing the right number of seats or for three offices might have been cost prohibitive and difficult to deal with when attorneys come or go from the firm. Other pricing models may also be difficult to implement because the license for an electronic resource may be restrictive as to location or users and the use that the library would like to put the resource to might be a violation of the user license.

When making the decision to switch from print to electronic access, it is also necessary to make sure that the resource is really required for teaching, access, or practice purposes. Many electronic resources can easily be considered "nice to have" rather than "required to have." The specter of the budget is always looming over the law library, so it is essential to make a business case for any move that might improve access or provide new information sources but will raise the annual cost of the library budget. Especially in law firms, it is sometimes difficult to justify new materials as the library is considered a cost center and each penny spent on information resources is a penny that isn't profit for the partnership. Academic and court libraries face many similar pressures to provide more and better service with an ever-increasing array of resources, rising prices, and dwindling budgets.

Librarians in the new digital library must also be prepared to defend spending yearly. Many organizations are now using a zero-based budgeting model, so just because an expenditure was allowable in the prior fiscal year, it might not be feasible in the current budget cycle. One way to defend the use of electronic resources is to keep usage statistics. Some vendors, like West and Lexis, make this very simple, but others either cannot or will not provide these statistics. One way around this issue is to use a monitoring tool like OneLog, Research Monitor, or Lookup Precision. These tools not only track usage statistics, they can also be used as a validator of client matter numbers and as a handy tool for storing passwords for various databases securely.

One of the most important questions to ask, and that is likely to be asked of the librarian is, "Are we as an organization getting value for the dollars spent on library resources?" Having usage statistics can make this justification easier. In law firms, this question may also be answered by how much of the cost of electronic resources is able to be passed through to clients, and how much of the pass-through charges are actually paid by clients and how much is written off before the client even sees the bill. This is not a possibility in academic or court libraries, so these law libraries must attempt to justify the budget with usage statistics alone.

CONCLUSION

Law librarians in the digital age are faced with an overwhelming array of choices when it comes to electronic resources for research, whether fee based or free. The librarian must balance the needs of students, attorneys, professors, and other users against the available budget and space when making the difficult decisions about what to keep in print and what to move to an electronic platform. Some choices are dictated by vendors ceasing publication of a print product while others may be dictated by an accrediting body like the American Bar Association for law schools, or by the geographic makeup of the organization, or by user needs and preferences. The modern law library must be nimble and prepared for the inevitable changes it will face.

FURTHER READING

Gil, Paul. 2012. "The 10 best search engines of 2012." http://netforbeginners.about.com/od/navigatingthenet/tp/top_10_search_engines_for_beginners.htm. This is a regularly updated site that provides a list of popular search engines for all purpose searches. The list is based on reader suggestions and looks at the key features: relevant results, uncluttered interface, and helpful options for broadening or tightening a search.

Incisive Interactive Marketing LLC. N.d. *Search Engine Watch.* www.searchenginewatch.com. This site is full of up-to-date information on the business and technology of search, including lots of information on search engine optimization (SEO).

Leavitt, Carole A., and Mark E. Rosch. 2011. *The cybersleuth's guide to the Internet*, 11th ed. Rio Rancho, N.M.: Internet for Lawyers. This is one of the most used guides to finding legal information fast and free on the Internet. While the resource is excellent, it is imperative to keep the most updated version at hand because the Internet changes extremely fast and what was a great resource yesterday may not even exist tomorrow.

MacLeod, Don. 2012. *How to find out anything.* New York: Prentice Hall Press. This slim volume provides up-to-date instructions on using Google to its best advantage as well as providing a good grounding in the basics of search strategy and a great list of resources you may not have considered, including the public library.

Notess, Greg. 2012. *Search engine showdown.* www.searchengineshowdown.com. This site, while beginning to show its age, is a great resource for finding information on the most popular search engines, including reviews, features, blog posts, and analysis. The Feature Chart, while somewhat dated, gives a concise picture of how major search engines work. As with anything online, take this site with a grain of salt because of the fluidity of the Internet.

The Search Engine List. www.thesearchenginelist.com. Another site that is slightly out of date but provides a long list of search engines grouped by type, with a description of major features. The list can also be seen as an alphabetical index.

Zimmerman, Andrew. *Zimmerman's legal research guide.* http://law.lexisnexis.com/infopro/zimmermans. This comprehensive encyclopedia of information for legal researchers is produced by Andrew Zimmerman, a law librarian at a firm in Washington, D.C. Although it is hosted on the Lexis-Nexis Information Professional website, it is agnostic as to platform and treats all legal publishers equally. It is an excellent resource for finding out what resources are available in print and online on a particular legal topic.

17

Major Legal Databases and How to Search Them

Theodora Belniak
SUNY AT BUFFALO LAW LIBRARY

WHY USE LEGAL DATABASES?

The Internet has opened up a world of information, but finding a specific document is a bit like diving into an ocean to find a particular type of seashell. In a search engine like Google or Bing, crafting an exacting query to locate all cases involved with the lemon law in the 20th century, for example, would be difficult and would likely yield far too many unrelated results.

Using legal databases is more like diving into a swimming pool: Each has visible boundaries and manageable sets of topically—or thematically—arranged data that a researcher may search systematically. In many databases, a researcher has the flexibility to define a time span, related subjects, or keywords, making a search like the one above easy to execute.

In addition to finding primary sources like case law more easily, many legal databases offer access to in-depth secondary sources like treatises, online looseleafs, legal encyclopedias, law journals and reviews, and editorial commentary. These materials enrich and assist the research process by providing guidance to both the novice and the expert.

Cost Considerations

The legal databases discussed in this chapter are proprietary and are available for a fee. The pricing structure for each database varies based on the organization that contracts with the vendor, but pricing typically is based on the level of access and the number of transactions or queries run within the database. With this in mind, the importance of well-framed, well-constructed, and nonrepetitive search queries take on special weight.

Finding the Right Database for the Job

There are a lot of legal databases available. Some reach for comprehensiveness and others focus on niche areas of law. Table 17.1a is not exhaustive but represents the databases discussed in this chapter and provides a sketch of their respective contents.

Although not discussed in this chapter, there are many other databases that may address a particular research need. Federal legislative materials, including hearings and reports, may be found in Proquest Congressional. Law reviews, journals, and books are indexed in the Index to Legal Periodicals and Books, the Index to Foreign Legal Periodicals, Index to Legal Periodicals Retrospective, and LegalTrac. Historical legal materials such as historical state codes, legal treatises, Supreme Court records and briefs, and materials from historical trials are available through Gale's Making of Modern Law modules. International law materials are also available through the Making of Modern Law modules, AccessUN, and the Foreign Law Guide. There are also other low-cost comprehensive databases, such as Fastcase.

Specialized Language, Terms of Art

Within the law, there are many practice areas with specialized vocabulary and terms of art unique to those areas. While researching using legal databases, take note of any language or phrase that is used frequently in court opinions or other primary legal sources and integrate that language into future search queries. As an example, in the practice of administrative law, the phrase "arbitrary and capricious" has a legal meaning, defined through case law and statutory law, that is beyond its plain-language meaning; understanding this legal definition is a necessary step in researching any issue in administrative law and will inform construction of future searches.

To help you with these definitions, there is a myriad of legal dictionaries available. From *Black's Law Dictionary* in print or its free online counterpart at http://thelawdictionary.org/ to dictionaries written for a particular practice area such as the *New Palgrave Dictionary of Economics and the Law*, there are many sources you can consult to help narrow your terms.

Constructing Searches in Legal Databases

Construction

Every legal database has its own tricks and quirks, but a well-constructed search will reduce time spent on reviewing results. Before logging onto a legal database, take a few moments to define the core issue. If there are terms of art or phrases,

Table 17.1a. Databases and Their Contents

Where should I look for . . .	Consult any of the following:
primary legal authority, like cases, statutes, and regulations?	• Bloomberg Law • HeinOnline (historical) • LexisNexis • Lexis Advance • LLMC Digital (historical) • Loislaw • Westlaw Classic • WestlawNext
treatises?	• Bloomberg Law • HeinOnline (historical) • LexisNexis • Lexis Advance • Loislaw • LLMC Digital (historical) • Westlaw Classic • WestlawNext
current legal news?	• Bloomberg Law • LexisNexis • Lexis Advance • Westlaw Classic • WestlawNext
practice-oriented forms?	• Bloomberg Law • LexisNexis • Lexis Advance • Westlaw Classic • WestlawNext
law reviews and journals?	• Bloomberg Law • HeinOnline • LexisNexis • Lexis Advance • Westlaw Classic • WestlawNext
current court documents, similar to PACER?	• Bloomberg Law
international materials?	• Bloomberg Law • LexisNexis • Lexis Advance • LLMC Digital (historical) • Westlaw Classic • WestlawNext

time periods, jurisdictions, or specific judges or attorneys that are relevant, be sure that your query incorporates them.

Also before beginning, think about where you are in the research process. If you are beginning and seeking the basics, your query should be more general in scope and should yield a lot of results. If you are working on a particular issue within an area of law, are familiar with the terms of art, and are only interested in the last ten years of litigation, your query should be more tightly written and should yield fewer results.

Types of Searches

- *Natural Language Searching.* If you've used a search engine like Google or Bing, you are familiar with natural language searches. Phrases like "How do I make a peanut butter and jelly sandwich?" and "law libraries in western New York" mimic everyday speech and writing patterns. Many legal databases offer natural language searching as an option. For example, the phrase "arbitrary and capricious case law in New York" will gather anything related to any of the words in the phrase. Novice legal researchers are usually more comfortable with this option because it yields high numbers of results. Despite the high yield, many of the results may be irrelevant because of the inclusiveness of the search.
- *Terms and Connectors Searching.* A generic phrase, "terms and connectors," describes vendor-specific conventions for complex search-query construction. Terms and connectors use symbols, directives, and truncations to craft a query. Terms and connectors are described in more detail for specific databases below, but, as an example, the natural language search example above, "arbitrary and capricious case law in New York," would be broken down into components. First, you will select a relevant jurisdiction, New York, and the relevant court system at the state level. After doing so, using the phrase feature of the terms and connectors conventions and placing the words "arbitrary and capricious" in quotes, your results will only include New York state cases that include that exact phrase. Legal researchers who are familiar with their topic will most often use this method, as it generates fewer and more focused results.
- *Navigation Assistance.* Legal databases are not always intuitive to the first-time user, and even expert legal researchers run into problematic searching issues occasionally. Take advantage of free training and guidance provided by vendors to hone your research skills. Most legal databases come equipped with a plethora of self-help options such as tutorials, live chat, free phone assistance, explanatory documentation, and webinars. If you notice a small circle containing an "i" or a "?" next to a resource or database term, you can

use this to find out more information about that particular item. If you are in a firm or library setting, many legal databases will send representatives to provide in-person training as well.

COMPREHENSIVE LEGAL DATABASES

Background

Comprehensive legal databases contain a variety of resources and are considered indispensable for attorneys and legal researchers. They provide access to case law, statutes and regulations, administrative proceedings, news, periodicals, international sources, secondary sources, and more. Many users think of and use these databases as "one-stop shopping."

The two well-known first-generation databases of this type are LexisNexis and Westlaw Classic, which both started in the late 1970s. A more recent competitor, Bloomberg Law, entered the legal research market with a comparable product in the past five years.

New iterations of LexisNexis and Westlaw—Lexis Advance and WestlawNext—respectively, were introduced in 2011 and 2010. Bloomberg Law and the new iterations of WestlawNext and Lexis Advanced are substantially different from their predecessors in respect to points of access, structure, and default settings, and are discussed in a separate section.

Searching First-Generation Comprehensive Databases

Origins of LexisNexis and Westlaw Classic

In the 1970s, LexisNexis and Westlaw Classic offered dedicated terminals connected via dial-up. These terminals represented closed universes of databases and resources organized by each vendor. The scope of these universes was defined by the purchasing institution, with more materials being added as time went on. Now with Web-based systems, LexisNexis and Westlaw Classic are accessible anywhere with a user name and password, and their respective contents continue to grow. Table 17.1a outlines the depth of offerings if an institution provides access to the full package in each database.

A Basic Search

The complexity and depth of information in LexisNexis and Westlaw Classic are such that there are multiple ways to search and to arrive at similar results. However, there are a few basics that must be tackled in any search. Table 17.1 is a basic outline of a search and includes overarching questions that you should keep in mind for each step.

Table 17.1. Sample Research Outline

Step	Task	Questions to Keep in Mind
1	Define your issue	• What terms are crucial for your search? • Are you looking for material that supports or contradicts a stance? • Are you just beginning your research? Or are you verifying past research?
2	Choose a source or sources within which to search	• What is your jurisdiction? • Are you interested in primary or secondary sources?
3	Construct your query	• What is your date range? • Are there key phrases or terms? • Are you looking for a particular judge, attorney, docket number, law, or citation? • Are your key phrases and terms likely to appear in the same sentence or same paragraph?
4	Run your query and review the results	• If no results, is there a way to broaden your search terms? • If too many results, is there a phrase or word you can use to narrow the list?
5	Save relevant results	• What system are you using to organize your research findings? • How will you find results when synthesizing your research?
6	Update relevant results	• Before using this result in your research, is it still good law? • Are there more recent cases, statutes, or secondary sources that have been published since you ran your original search?

Selecting a Source

In the classic iterations, you must select a source within which to search. Both LexisNexis and Westlaw Classic boast thousands of sources from which to choose; LexisNexis refers to each set of related documents as a source and Westlaw Classic uses the word "database" to describe each set. If you do not know where to look, it can be difficult to unearth the correct source.

Thankfully, each service provides a directory. In LexisNexis, the "Find a Source" tab opens to a search box, an alphabetical listing, and a link to the online directory. These directories are used by both novice and expert legal researchers to discover specific databases. In Westlaw Classic, on the top of the page is a menu option labeled "Directory." Clicking on this label will open up the directory, which reflects the organizational structure of Westlaw Classic.

As an example, if you are interested in the Mexican Civil Code in Westlaw Classic, you would begin by clicking on "Directory"; then you would click on:

1. International/Worldwide Materials
2. North America
3. Individual Country Materials
4. Mexico
5. Legislation and Regulations
6. Mexican Civil Code Annotated

Searching for this same resource in LexisNexis, you would open the "Find a Source" tab and type the title into the search box. Around three hundred results are returned, with the appropriate resource at the top of the list.

Using Terms and Connectors

Terms and connectors define the relationship of one search term to another and allow you to narrow your query before running a search in a database. Terms and connectors for LexisNexis and Westlaw Classic are listed in table 17.2.

As an example, you are interested in whether a personal damages claim might be mitigated due to a failure to wear a seat belt in an automobile in the state of Mississippi. If you were to enter this as a natural language search in the Mississippi case law source, you would likely return no results with an error message about the use of nonspecific language. A terms and connectors search in that same source might look like this:

- *In LexisNexis: automobile or car or vehicle and "seat belt" w/5 failure and damage! w/p mitigat!*
- *In Westlaw Classic: automobile car vehicle & "seat belt" /5 failure & damage! /p mitigat!*

When in the correct source, this search query directs the system to look for Mississippi state cases that contain:

- the words automobile, car, and/or vehicle
- and the exact phrase "seat belt" within five words of the word failure
- and the word damage, damages, or damaged within the same paragraph as the word mitigation, mitigate, mitigated, or mitigates

This search results in fewer than fifty cases, which may then be reviewed and narrowed further.

Table 17.2. Terms and Connectors

Connector	LexisNexis Symbol	Westlaw Classic Symbol	Sample Search	Outcome
AND	and	& and	warrant & arrest (WLC) warrant and arrest (LN)	Returns all documents containing warrant and arrest in the same document.
OR	or	A space or	attorney counselor (WLC) attorney or counselor (LN)	Returns all documents containing either attorney or counselor.
PHRASE	" "	" "	"article 78 proceeding"	Returns all documents containing the exact phrase "article 78 proceeding."
BUT NOT	AND NOT	%	vehicle bike % motorcycle	Returns all documents containing the word vehicle or bike, but not motorcycle.
Numerical restriction	W/n, /n	/n	"article 78" /5 judge	Returns all documents containing the exact phrase "article 78" within five words of judge.
Grammatical restriction	W/p, /p W/s, /s	/p /s	"long arm" /p insurance "long arm" /s jurisdiction	Returns all documents containing the exact phrase "long arm" in the same paragraph as jurisdiction. Returns all documents containing the exact phrase "long arm" in the same sentence as jurisdiction.
Root expansion	!	!	archit!	Returns all documents containing words with a root of 'archit-', e.g., architect, architects, architecture.
Word variation	*	*	organi*ation	Returns all documents containing possible spelling variations, e.g., organization or organisation.
Hierarchical searching	()	()	(car or automobile) and accident	Prioritizes search for items in parenthetical, then applies remainder.

Field Searching

Most documents in LexisNexis and Westlaw Classic have been divided into internal fields (also known as segments) by editors. The searchable fields correspond to a specific portion of a document, such as its title, case citation, synopsis, or the date. Availability of certain fields varies based on the type of document being viewed. As an example, a statutes database will not have a searchable field for a judge's name because no judge was involved in the passing of the statute. LexisNexis and Westlaw Classic fields are defined in table 17.3.

Table 17.3. Commonly Used Fields in LexisNexis and Westlaw Classic

Westlaw Classic Field Abbreviation	LexisNexis Field Abbreviation	What's Being Searched in the Document
AT()	COUNSEL()	Attorney(s) listed in case opinion.
CA()	CITE()	Contents of the citation field for a statute.
CON()	CONCUR()	Any concurring judges' opinions.
CO()	COURT()	Court of decision.
DN()	NUMBER()	Docket number of case.
JU()	JUDGES()	Judge that penned the opinion.
RE()	REFERENCES()	References used.
SY()	SUMMARY()	Editor-provided summary.
TE()	OPINION()	Text of main body of document.
TI()	n/a	Title of document.
YE()	n/a	Year of document.

You may use field searching in combination with terms and connectors to add another level of specificity to your search.

Searching a Topical Index

In addition to organizing documents into specific sources, such as New York case law or law reviews and journals, LexisNexis and Westlaw Classic also organize documents topically.

This organization represents a complex system of topics, divided into more discrete topical areas, and is a powerful tool for finding related documents on the same topic. Each document reviewed by staff editors is assigned one or many topics, depending on the number of topics covered in the document. A case dealing with adverse possession may also touch on civil procedure, property rights, or personal property and will be tagged appropriately to make retrieval under any of those labeled topics easier.

This topical organization feeds into the headnotes found in both databases. Headnotes are found at the beginning of many documents and provide links to other documents that are related by topic to the document being viewed.

- *LexisNexis Topical Index.* To locate a relevant topic, find the "Search by Topic or Headnote" box, and select "View More." A list of topics will appear, and you may then drill down into related subtopics. For example, if you were interested in a topic in education law, you will click on that initial topic, which will then open into a tree hierarchy. You can then drill down into your hierarchy until finding a topic on point. If you have a relevant case in hand, using the topical organization is straightforward. Open the case in LexisNexis and review the headnotes. Once you find a headnote that is on the topic you would like to investigate further, click on "More like this headnote" to move to other cases and documents tagged with that same topic.
- *Westlaw Classic Key Number System.* Westlaw Classic uses its Key Number System to organize documents by topic. A complete list of Key Numbers is available in the uppermost menu of Westlaw Classic. Table 17.4 is an outline of a small portion of the Key Numbers on adverse possession.

If searching by Key Number, you will discover cases and documents across all jurisdictions within the United States that are related to the Key Number of interest.

Table 17.4. Portion of Westlaw's Adverse Possession Key Numbers

ADVERSE POSSESSION
I. NATURE AND REQUISITES, k1–k95
  (A) ACQUISITION OF RIGHTS BY PRESCRIPTION IN GENERAL, k1–k13
  (B) ACTUAL POSSESSION, k14–k27
    k14 Necessity
    k15 Entry on land
    k16 Acts of ownership in general
      (1) In general
      (2) Marking boundaries and entry to make survey
      (3) Wild lands
    k17 Use and occupation

Imagine that you have been asked to research adverse possession, and you have been given a Pennsylvania case that discusses the issue in a way favorable to the legal facts of your ongoing case. It is labeled with the Key Number of "k16(3): Acts of Ownership in General, Wild lands." However, the Pennsylvania case is merely persuasive, and you need California cases that discuss the topic and establish precedent in the state. With the Key Number system in place, you can search all cases and documents labeled with the relevant topic, and narrow your results by jurisdiction to find California cases.

Evaluating and Updating Search Results

Documents in LexisNexis and Westlaw Classic have additional editorial content that will aid you in evaluating and updating your research. For many documents, each database provides critical information about:

- Documents that cite to the viewed document, known as citing references
- Citations to related documents
- Related topics and topical summaries, known as headnotes
- Historical notes
- The current disposition of the case or statute

Much of this evaluative information is communicated through the use of symbols. The most common symbols you might run across in a document in both databases are listed in table 17.5.

Table 17.5. Symbols Used by Shepard's and KeyCite

LexisNexis Symbol	Westlaw Classic Symbol	Legend
Red Hexagon	Red Flag	At least one point of law has been overruled
Yellow Triangle	Yellow Flag	Some negative history
Blue Circle with White A	Blue H	Document has some history
Blue Circle with White I	Green C	Document has citing references
n/a	Purple Quotation Marks	Citing document directly quotes to current document
Red Circle and Exclamation Point	n/a	Negative case treatment for the statute
Yellow Square with Black Q	n/a	Validity questioned by citing references
Green Diamond with White +	n/a	Positive treatment is indicated

Updating your search results is an important step in synthesizing your research. Before using a case or statute to bolster an argument or satisfy a request, you must ascertain whether the resource is still "good law," meaning that it has not been overturned by the legislature or by the courts.

LexisNexis uses Shepard's to update a citation, and the process of updating is known as Shepardizing. When you Shepardize a case or statute, a summary is provided that will indicate the status of the document with a symbol from table 17.5 and will provide you with any citations to the document.

Westlaw Classic uses KeyCite to update a citation. When KeyCiting a case or statute, a summary provides much the same information as in the Shepard's summary. In addition, KeyCite provides a full history of the case represented graphically and links to petitions, briefs, and filings related to the case or statute.

Shepard's and KeyCite offer alert services for the legal researcher to help automate the updating process and reduce duplicative work. Alerts will notify you of any changes to the case or statute in which you are interested, and may be set up by clicking on "Save as Shepard's Alert" or "Monitor with KeyCite Alert."

Searching Second-Generation Comprehensive Databases

Origins of the Second-Generation Databases

The popularity of single search boxes in Internet browsers has played a large part in the reimagining of comprehensive legal databases. Users expect to be able to search all content with one search, and database providers are reworking interfaces to meet this expectation.

Although the content remains the same as what is available in the first-generation systems, the method by which a user searches and reviews results is dramatically different in the new versions of LexisNexis and Westlaw and in Bloomberg Law. Each system offers a unique presentation, but the commonalities across the new comprehensive databases are that there is one search box into which a query may be entered and that a search is run through the entire contents of the database. The following sections explain the search strategy of the second-generation systems.

A Basic Search

Bloomberg Law, WestlawNext, and Lexis Advance represent a new approach to legal research. A basic search in these systems is outlined in table 17.6.

The research outline is truncated in the second-generation systems; the importance of locating a source within which to search and of framing an exacting query is de-emphasized. In many ways, the order of operations has been inverted: a search using whatever terms you might think are relevant defaults to natural

Table 17.6. Sample Research Outline for Second-Generation Databases

Step	Task	Questions to Keep in Mind
1	Define your issue	• What terms are crucial for your search?
2	Run your query and review your results	• If no results, is there a way to broaden your search terms? • If too many results, is there a phrase or word you can use to narrow the list?
3	Drill down into faceted results	• Are you looking for a particular jurisdiction, year, or type of document?
4	Save relevant results	• What system are you using to organize your research findings? • How will you find results when synthesizing your research?
5	Update relevant results	• Before using this result in your research, is it still good law? • Are there more recent cases, statutes, or secondary sources that have been published since you ran your original search?

language searching within the system, and the entire universe of documents is searched. The results are returned and faceted, and these facets incorporate many of the parameters you might have defined in Westlaw Classic or LexisNexis. However, you will still need to sort through the documents for relevancy and to evaluate, organize, and update your results.

Default Settings

Bloomberg Law, Lexis Advance, and WestlawNext have pared down default user interfaces. The functionality of the first-generation systems remains, but much of it is below the surface. It is important to be aware of the default settings of each system. Table 17.7 lists these defaults when searching.

Table 17.7. Default Settings

Default Setting	Bloomberg Law	Lexis Advance	WestlawNext	Able to Change?
Search Language	Natural	Natural	Natural	Yes
Searchable Universe	United States law	All of LexisNexis	All state and federal	Yes
Results List	Faceted	Faceted	Faceted	No

Evaluating and Updating Search Results

The second-generation systems organize results differently. After running your search, you can further sort results using the facet options on the left side toolbar of each system. In general, results will be broken into facets such as case law, statutes, secondary materials such as law reviews or treatises, and regulations. Notice that the faceting aspects in step 3 of table 17.6 is very similar to the selection that happens in step 2 in table 17.1. The second-generation systems allow you to consider the complete universe of related documents before deciding on a source type and narrowing your focus.

Lexis Advance and WestlawNext utilize the same symbols found in table 17.5 to help evaluate search results. Bloomberg Law uses BCite Indicators, outlined in table 17.8.

Table 17.8. BCite Indicators

BCite Indicator Symbol	Legend
Green Square with White Plus Symbol	Courts have used this opinion as precedent.
Blue Square with White Checkmark	Decision has been distinguished by other case law.
Yellow Square with White Pyramid	Opinion has been criticized without being overruled or has been modified or amended.
Orange Square with White Dot	Opinion has been superseded or is obsolete because of a statute, rule, or regulation.
Red Square with White Dash	Opinion has been reversed or vacated in full.
Gray Square with White X	Opinion is under review.

Updating your results in Bloomberg Law, Lexis Advanced, and WestlawNext is very similar to updating in the first-generation systems. Lexis Advanced and WestlawNext continue to utilize Shepard's and KeyCite, respectively. In Lexis Advanced, a "Shepardize" button is in the top right toolbar. In WestlawNext, KeyCite features are located in the top toolbar above a document. Bloomberg Law uses BCite, available in a tab to the right of the document tab while the document in question is in view.

SPECIALIZED LEGAL DATABASES

In contrast to the comprehensive databases discussed above, there are legal databases that serve specialized purposes. They may be devoted to historical legal resources, foreign and international law, legal periodicals, secondary sources within practice areas, and more. The databases discussed below represent a sample of these specialized legal databases and give you a hint as to the depth, complexity, and variations of legal resources you may encounter in your research.

HeinOnline

HeinOnline's Specialized Collections

HeinOnline began in 2000 with a single collection consisting of twenty-five periodicals that related to the law. Currently, it has expanded its law review and journal collection to cover every accredited law school's journals, and hosts over thirty collections on different topics.

In contrast to other legal databases that convert documents to a basic text format, HeinOnline generates and maintains full-text PDFs of documents, replicating a page exactly as it would be found in a print volume. These PDFs are scanned using optical character recognition (OCR) software, and are full-text searchable. The digital replication of the myriad charts, graphs, and mathematical formulae contained in many publications is invaluable to a legal researcher.

Organizational Structure

HeinOnline is organized into modules known as libraries that represent a set of documents. These libraries may be added a la carte to a subscription, and, in some cases, content overlaps from one library to another.

In general, there are two types of libraries:

- *Organizationally based.* Usually represents the entire run of a serial or periodical related to an organization or group. For example, U.S. Congressional Documents library includes documents related to the debates and journals produced by Congress. The entire *Congressional Record* (1874–present) has been digitized. In addition, HeinOnline digitized the *Congressional Record*'s predecessors, *Debates and Proceedings in the Congress of the United States* (1789–1824), the *Register of Debates in Congress* (1824–1837), and the *Congressional Globe* (1833–1873), and included them in the same library for research continuity if searching by topic or keyword.
- *Subject based.* Attempts to capture the most relevant monographs, articles, and data on a particular subject. For example, the Foreign and International Law Resources Database includes titles such as *African Yearbook of International Law*, *American Society of International Law Proceedings*, *Dutch Penal Code*, *Religion & Human Rights: An International Journal*, and others.

After selection, HeinOnline opens to the library, which has many or all of the following characteristics:

- The library title in bold above a search box, and a bread-crumb trail from the main page to the current page above the library title
- A citation navigator, including *Bluebook* citations in the left column
- A "Browse by" feature below the search box, offering sorting by publication title, state, country, subject, most cited, or by popular legislation or topic related to that library's contents

The variable nature of the libraries in HeinOnline makes it difficult to generalize about searching within each. Table 17.9 highlights some of the different libraries and demonstrates the variety of tools available across libraries.

Searching HeinOnline

HeinOnline offers various methods of searching its libraries. First, there are two meta-searches that stretch across all subscribed libraries on the subscription main page: (1) a directory search allows searching for a publication by author, title, ISBN, series, subject, or publisher, and (2) a general search allows searching across all subscribed libraries for a word or phrase.

Table 17.9. HeinOnline Libraries

Library	Field Search?	Sortable Titles?	Citation Search?	Browse Options?	Drill Down Into Title?
New York Legal Research	No	Yes	No	Yes	Yes
U.S. Federal Legislative History	No	Yes	No	Yes	Yes
Session Laws	No	No	Yes	No	Yes
U.S. Statutes at Large	No	n/a	Yes	Yes	No
World Trials	No	Yes	No	Yes	Yes
Taxation & Economic Reform in America Parts I & II, 1781–2012	Yes	Yes	No	Yes	Yes
Code of Federal Regulations	No	n/a	Yes	Yes	No

Second, within some of its libraries, HeinOnline supports more complex searching through the field and advanced search options. Paired with Boolean operators, these search tools can help you craft productive research queries. Popular HeinOnline operators are outlined in table 17.10. Note that all Boolean operators must be in all capitals.

Evaluating Search Results

Another feature of HeinOnline is ScholarCheck, indicated by a blue circle with a red checkmark symbol in your results list. ScholarCheck is available in some of

Table 17.10. HeinOnline Operators

Symbol	Example	Results
:	title: "Constitutional law"	Returns documents that contain the phrase "constitutional law" in the title field.
AND &&	constitutional AND Miranda constitutional&&Miranda	Returns documents that contain the words constitutional and Miranda.
OR	arbitration OR litigation	Returns documents that contain the word arbitration or the word litigation or both.
+	+Internet rights	Returns documents that contain the word Internet and may contain the word rights.
NOT - !	"aspiration" NOT "inspiration" "aspiration"-"inspiration" "aspiration"!"inspiration"	Returns documents containing the word aspiration but not documents containing the word inspiration.
date:[YYYY TO YYYY]	Date:[1909 TO 2012]	Returns documents dated between 1909 and 2012.
?	organi?ation	Returns documents containing the word organization or organisation.
*	architect*	Returns documents containing the word architect, architects, architectural, architecture, et al.
~n	"brown field"~5	Returns documents containing the word brown within five words of the word field.

the libraries, and it tracks the number of times an article or statute is cited and provides links to those citing articles.

ScholarCheck allows you to sort search results by most to least cited and is a tool to help a researcher find related documents that may not have appeared in the original search. For example, if you are searching for articles about the future of law libraries and stumble on John Palfrey's article from volume 102 of the *Law Library Journal*, a ScholarCheck icon appears in your search history. The icon is followed by the phrase "Cited by 6." Clicking on this hyperlinked phrase will bring you to the six articles within HeinOnline that have cited to Palfrey's article, thereby broadening your overall search.

LoislawConnect

LoislawConnect's Background

LoislawConnect is owned by Wolters Kluwer, an established legal vendor known for its secondary source publications. LoislawConnect is a database of U.S. cases, statutes, rules and regulations, and court rules. For an additional cost, secondary sources such as subject-specific treatises or bar publications may be added on. Although its contents are slimmer than what is available in one of the comprehensive databases discussed above, LoislawConnect provides access to the primary sources necessary for practice or for research into cases and statutes for a comparatively low price.

Searching in LoislawConnect

LoislawConnect requires you to select a source within which to search. The resources are broken down in the following ways:

- *Case law:* federal and state, and within state, including lower courts and special courts where available
- *Statutes and acts:* federal and state codes, acts, and statutes, as well as the United States Constitution and Public Laws
- *Administrative rules and regulations:* state administrative codes and the Code of Federal Regulations and Federal Register
- *Court rules:* district and circuit court rules, bankruptcy court rules, and the court rules of the states

After selecting a source, click on the green button at the bottom of the page with the word "Continue." Multiple field search options will appear, offering the chance to search within the entire document, by case name or docket number, for an attorney or judge, or within the syllabus. Enter your search terms and click on the green button on the bottom of the page with the words "Run Search." You can use terms and connectors within your search; popular LoislawConnect terms and connectors are listed in table 17.11.

Table 17.11. LoislawConnect Wildcards and Connectors

<table>
<tr><th>Connector</th><th>LoislawConnect Symbol</th><th>Sample Search</th><th>Outcome</th></tr>
<tr><td>AND</td><td>&
and</td><td>warrant & arrest
warrant and arrest</td><td>Returns all documents containing warrant and arrest in the same document.</td></tr>
<tr><td>OR</td><td>|
or</td><td>attorney | counselor
attorney or counselor</td><td>Returns all documents containing either attorney or counselor.</td></tr>
<tr><td>PHRASE</td><td>""</td><td>"article 78 proceeding"</td><td>Returns all documents containing the exact phrase "article 78 proceeding."</td></tr>
<tr><td>NOT</td><td>%
not</td><td>vehicle bike % motorcycle
vehicle bike not motorcycle</td><td>Returns all documents containing the word vehicle or bike, but not motorcycle.</td></tr>
<tr><td>NEAR</td><td>/
near</td><td>"article 78" / judge
"article 78" near judge</td><td>Returns all documents containing the exact phrase "article 78" within 20 words of judge.</td></tr>
<tr><td>NEARX</td><td>/x
nearx</td><td>"article 78" /5 judge
"article 78" near5 judge</td><td>Returns all documents containing the exact phrase "article 78" within five words of judge.</td></tr>
<tr><td>Root expansion</td><td>*</td><td>archit *</td><td>Returns all documents containing words with a root of "archit-", e.g., architect, architects, architecture.</td></tr>
<tr><td>Word variation</td><td>?</td><td>organi?ation</td><td>Returns all documents containing possible spelling variations, e.g., organization or organisation.</td></tr>
<tr><td>THESAURUS</td><td>~
thesaurus</td><td>~automobile
thesaurus automobile</td><td>Returns all documents containing the word automobile and any synonym of automobile.</td></tr>
</table>

Evaluating and Updating Search Results

After running a search, the results will appear in the right column. Navigate through your pages of results using the bottom toolbar, or sort using the drop-down menu in the right corner. On the left, you can refine your results by searching within the original results list.

To help you evaluate results, click on the green GlobalCite button in the lower toolbar to open up all documents within LoislawConnect that cite to or are related to the document you are viewing. Within GlobalCite results, you can sort by type of document, relevance, title, source, or when the result was published. GlobalCite also offers sorting by Treatment Term; the system will search and highlight results in blue font that contain any of the terms found in table 17.12.

Table 17.12. GlobalCite's Treatment Terms

Positive Term	Neutral Term	Negative Term
Affirm	Distinguish	Deny
Agree	Limit	Disagree
Concur	Modify	Dismiss
Reaffirm	Remand	Dissent
Sustain	Remit	Overrule
Uphold		Reject
Grant		Remove
		Reverse
		Set Aside
		Vacate

LLMC Digital

LLMC Digital's Specialized Collections

The Law Library Microform Consortium (LLMC) has published microfilm for over three decades, and launched LLMC Digital in 2003 as a complement to its microfilm services, as a way to "achieve its goal of maximum access with maximum preservation."[1]

LLMC Digital contains seven collections. The title of and series within these collections are included in table 17.13. Also included are generic subseries suffixes.

Table 17.13. LLMC Collections and Series

Collection	Series within Collection	Series Label/Sublabel
U.S. Federal Government	• Constitutional • Executive • General • Judicial • Legislative	• US—Con. • US—Exec. • US. Gen. • US—Jud. • US—Leg.
U.S. States	Where available for all states, Alabama–Wyoming: • Constitution • Executive branch documents • Case reporters • Legislative codes	For example: • AL—Con. • AL—Exec. • AL—Jud. • AL—Leg.
U.S. Territories	Where available for territories, American Samoa–U.S. Virgin Islands: • Constitution • Executive branch documents • Case reporters • Legislative codes	For example: • US. Terr. Am.Sam., Con. • US. Terr. Guam—Exec. • US. Terr. Am.Sam., Jud. • US. Terr. Am.Sam., Leg.
Anglo-American Collections	• Anglo-American Case Reporters/ Encyclopedias • Anglo-American Legal Periodicals • Anglo-American Legal Reference • Anglo-American Legal Treatises • Yale Blackstone Collection	• Aacren. • AAper. • AAref. –dict (dictionary) –dig (digest) –dir (directory) • AAtre. • Blkst.
Foreign Jurisdictions	Countries Afghanistan–Zimbabwe, various types of documents of or related to the law. • Legislative codes • Treaties • Codes • Constitutions	For example: • Tonga—Leg. • Tonga—Treaties. • Switz.—Code. • Russia—Constitution.
International Law and Organizations	• International Law • International Organizations	• Intlaw. • IntOrg.
Multijurisdiction Subject Collections	• Ancient Legal Systems • British Empire Studies • Canon Law Collection • Comparative Law • Legal Bibliography • Legal Philosphy [sic] • Military Law Collection • Native American Collection • Roman Law	• Ancient Leg. Sys. • Br. Emp. S. [and/or] Pacific. • CanonL. • CompL. • LegBib. • LegPhil. • Mil. • NatAm. • RomanL.

- *Searching in LLMC Digital.* LLMC Digital allows you to search its holdings in four viable ways: by catalog record, by *Bluebook* citation, within a title, or using full-text searching. The default search page includes the first three options as well as links to help pages explaining how to use each option. The fourth option, searching full-text, is available through a few avenues, described below. Although it is tempting to fill the first available search box with your query, verify that you are searching using the best method based on your available information. The brief descriptions of the search options below and table 17.14 will assist you in formulating your approach to the data in LLMC Digital.
- *Searching the Catalog Record.* Opening the search page in LLMC Digital, the first search box is for searching the catalog record, using the parameters of title, author, Library of Congress subject heading, or keyword. You may choose to search using one or all of the first three parameters; choosing to search keywords disallows selection of the first three options, as you are searching all options by selecting keywords. It is important to keep in mind that this search option is only searching through the metadata linked to an item, and is not searching the text of an item. Use this search if you know the title, author, or a Library of Congress subject heading related to your topic. Remove any article at the beginning of a title. You may also use Boolean operators by spelling out the operator in your search query. A list of Boolean operators within LLMC Digital is included in table 17.14.

Table 17.14. Boolean Operators in LLMC Digital

Symbol	Example	Results
AND	accident AND damages	Returns documents that contain the word accident and the word damages.
NOT	statute NOT regulation	Returns documents that contain the word statute but not the word regulation.
OR	statute OR regulation	Returns documents that contain the word statute or the word regulation.
NEAR	actionable NEAR negligence	Returns documents that contain the word actionable within the same volume as the word negligence.
* ?	archit* archit?	Returns documents that contain the variations of words with the root of archit-, such as architect, architecture, architects, architectural.

Important to note: LLMC Digital supports operator order prioritization; NOT and AND are equal, and are sorted first, and then OR terms are located.

- *Searching by Bluebook Citation.* There are two citation systems that are well known in the legal world: the *Bluebook* and the *ALWD Citation Manual.* LLMC Digital allows you to search using *Bluebook* citations. Enter the volume and page numbers in the appropriate box, and select the correct source abbreviation from the drop-down menu. If you are unsure about proper citation format, reference the *Bluebook* index or ask a reference librarian at your local law library for assistance.
- *Searching within a Title.* The collections outlined in table 17.13 may be searched on the title level. For example, if you are interested in searching only the attorney-general opinions of New York state, you will click on the plus sign next to "U.S. States," and the plus symbol next to "New York." This will open all the titles available in LLMC Digital related to New York. Select "Search" next to either attorney-general opinion title to search within that title only.
- *Searching Using Full-Text.* You can find the full-text option in LLMC Digital in two ways. On the default search page, you can drill down into any series within a collection. Hitting the search button located after a series opens a new window with a full-text search option. Also, you can open full-text searching by mousing over the Search option in the upper toolbar, and selecting "LLMC-Digital Full-Text Search." After opening full-text searching, you can construct a query using Boolean operators found in table 17.14. Using this option, your query will be run against all text.
- *LLMC Digital Search Option Examples.* Table 17.15 outlines the various search strategies options available in LLMC Digital.

Reviewing Your Search Results

Your search results will vary in complexity and depth depending on which search option you choose. The general format of your results list will be as follows: Title, Volume, Part, Number of Pages, and then an Add to My File button. The Add to My File button will retain added documents for the length of your session but removes them upon exit. You can view your temporary list of selected documents by clicking My File in the top toolbar of any page.

If you want to review any queries you've run in your session, you can do so by clicking on the History button in the top toolbar of the page.

Table 17.15. LLMC Digital Search Option Examples

Sample Need	Type of Search	Sample Search	Results
"I want a case from a nominative reporter penned by Hun."	Catalog record	Type "Hun" into the author field of the catalog record search option.	Retrieves the title *Reports of cases heard and determined in the Supreme Court of the state of New York.*
"I have this citation to a case: 38 Hun 37."	*Bluebook* citation	Enter the correct volume number (38), page number (37) and select "Hun" from the drop-down menu.	Retrieves the case Shepard v. McCalmont Oil Company from *Reports of cases heard and determined in the Supreme Court of the state of New York.*
"I want the case Shepard v. McCalmont Oil Company."	Full text	Go to the LLMC-Digital Full-text Search option in the top toolbar. Type Shepard AND McCalmont oil.	Retrieves eight results spanning all collections. Review each result for case.
"I want New York state cases that mention Shepard v. McCalmont Oil Company."	Within title	In the "Search by Collection" option, drill down to New York state. Click New York, and select a case reporter, then click on search. Enter search terms Shepard AND McCalmont oil into full-text search box.	Depending on the case reporter chosen, may or may not retrieve results.

FURTHER READING

Barker, John. 2003. "Legal publishing in the United States: How to mitigate the risks of full-text searching—LexisNexis, Westlaw, and Findlaw." *Australian Law Librarian* 11:110–121.

Herman, Theodor. 1998. *Beyond words alone: The power of West Group's editorial analysis in the electronic era.* Eagan: West Group.

Justiss, Laura K. "A survey of electronic research alternatives to LexisNexis and Westlaw in law firms." *Law Library Journal* 103 (1): 71–90.

Mart, Susan N. 2010. "The relevance of results generated by human indexing and computer algorithms: A study of West's headnotes and key numbers and LexisNexis's headnotes and topics." *Law Library Journal* 102 (2): 221–250.

Marvin, William M. 1969. *West Publishing Co., origin, growth, leadership.* St. Paul: West.

Sellers, Christine L., and Phillip Gragg. 2012. "WestlawNext and Lexis Advance." *Law Library Journal* 104 (2): 341–348.

Svengalis, Kendall F. 2012. *Legal information buyer's guide & reference manual.* North Stonington: New England Law Press.

NOTES

1. https://www.llmc.com/AboutLLMCDigital.asp.

Part IV

INSTRUCTION

18

Library Instruction in the Information Age

Emily Janoski-Haehlen
VALPARAISO UNIVERSITY LAW SCHOOL

Does the existence of vast amounts of digital information actually change the way law libraries instruct patrons to use the resources? The answer to this question is yes and no. The answer is yes because the same principles of library instruction that apply to in-person instruction still exist, but the answer is no because digital content and the Internet allow libraries to explore ways to use innovative teaching techniques. In this "age of information," the constant advancement and growth of technology require law libraries to keep up with the ever-increasing desire for information in various formats. This requires law libraries to adapt to and create different methods of providing library instruction, keeping in mind the public demand for open, unlimited access to information. One way that libraries can do this is to create electronic instructional tools, also referred to as e-learning tools. Electronic instructional tools are educational tools that are distributed via various kinds of electronic media. In law libraries, these tools are typically Internet research guides, online webinars, interactive or static tutorials, simulations or video simulations, games, and mobile apps.

Because more information is becoming digitally available, the means of access and instruction on how to locate the information must also be. Thus, the ways in which law libraries are teaching patrons to use the available resources are becoming less in-person and more digital. In the past, a patron had to be physically present in the library in order to learn how to use its resources. There were no access points for library resources outside the physical space and patrons could only discover what resources the library had to offer by visiting the library. In this technology-driven world, and what is being dubbed the "information age," patrons can now access entire library collections and instructional information on the Internet without even setting foot near a library. This open availability of library resources means that in order to ensure that patrons can properly locate the information, trainings should be available in an online environment or at least outside the walls of the library. Thus, the existence of and demand for digital information mean that law libraries must rely on electronic instruction tools to teach patrons how to access and use digital content.

While traditional, in-person library instruction is ideal, adaption to electronic library instruction is not seen as a negative thing as long as the principles and core concepts of library instruction remain the same. The basic principles of helping a patron access the content and use it effectively can translate easily from in-library instruction to electronic instruction. The electronic instruction tool needs to be easy to access, easily understood, and should anticipate and answer any questions a patron might ask in an in-library instruction scenario. An example of how easily in-library instruction can translate to electronic instruction can be shown by looking at how libraries teach patrons how to use the library's catalog. Most law libraries now require the use of the Internet to access the catalog, which requires the librarian to instruct the patron using a computer. This will translate well to an electronic instruction tool, such as a webinar, since it could provide the same detailed training in a recording that an in-library training with a librarian would provide.

This chapter examines the changing roles of law librarians, with regard to instruction, and how advances in technology in conjunction with the vast amount of digital information that is available have prompted these changes. In addition, it examines teaching techniques, specifically e-learning tools, for accessing digital information in and out of the library physical space. It focuses on the tools necessary to achieve the best results when conveying instructional content in a digital environment as well as the technology currently in use in law libraries. It also examines the changing role that law libraries play in their organizations, including how social media and digital services are expanding that role.

CHANGING AND EXPANDING ROLES FOR LAW LIBRARIES

The increasing availability of information is constantly changing the role law libraries play in their organizations. Historically, law libraries relied on their print collections to support patrons and were the gatekeepers to those information resources.[1] The new law library role can be classified as gatekeeper of both print *and* digital information. As gatekeepers of this information, law libraries must provide training and guidance on how to access and utilize it. Advances in technology provide opportunities for law libraries to disseminate the training and guidance information in many forms. One of the ways law libraries can provide outreach outside the library physical space is through the use of blogs or wikis.[2] Blogs allow librarians to communicate with patrons in the online world the way a librarian might communicate with a patron in person. For example, a law library might publish a guide to using a resource and place print copies at the reference desk. If a patron wants to use a resource but does not have time to read the print version of the guide, a blog can be a useful tool to republish the information. The patron can then be pointed to the Internet where the blog can be accessed at the patron's convenience. Of course, there are pitfalls to any Web 2.0 technology and blogs are not immune. It is often difficult to advertise the existence of a blog or to identify whether it is useful to the patron group.[3]

Private law libraries may have an advantage when it comes to using Web 2.0 technology for library instruction. Most private law libraries have intranets where materials like training videos, guides, on-demand webinars, tutorials, and other instructional materials can be posted. Users can access the intranet either from their offices or remotely and use the instructional materials to learn how to access the library's resources. A private law library intranet can also serve as a gatekeeper or pathfinder for access to the library's resources, so it is doubly helpful as a teaching and access tool. Some private law librarians have adapted their roles within their organizations by offering an increased amount of access to electronic resources and on-demand access to the tools necessary to use the resources correctly. As electronic resources continue to be in high demand, more information is being offered remotely to patrons of private law libraries and librarians must adapt to this demand for electronic access.

As the role of law libraries in information instruction changes, the role of law librarians is also changing and expanding to include more responsibilities that involve technology. In addition to being masters of their collections, law librarians must become masters of technology, communication, and dissemination of all information types. Law libraries can no longer afford to rely on on-site collections and in-library usage only. This means that outreach must take on a new meaning for law libraries that includes the online world. Law librarians must also adapt their teaching techniques for library instruction by adopting new ways to communicate with patrons.

TYPES OF TEACHING TECHNIQUES

Traditionally, law library teaching methods consisted of one-on-one or group instructional sessions on how to use the law library collection and how to access databases or password-protected material. This meant the librarian had to spend a fair amount of time teaching patrons to use each individual resource and providing tours of the physical space. With the growing demand for digital content that can be accessed twenty-four hours a day, seven days a week from any location with an Internet connection, law libraries are creating on-demand instruction methods to help patrons locate information. There are several options for creating on-demand library instruction, but the most popular include webinars, e-learning tutorials, QR codes, and mobile device applications (or apps). Each of these methods is discussed and the means for disseminating this on-demand instructional content is also examined in following sections.

Webinars and Online Tutorials

Webinars and online tutorials are easy to create and useful if law libraries are trying to engage patrons on the Web or in Web-based instructional sessions. These instruction methods allow librarians to showcase library electronic resources in easy-to-view online formats. The webinars and tutorials can even be made into interactive presentations where the user can click through sample questions and searches. This provides hands-on training as well as the information about how to use the resources. Librarians can also reach a wider audience with these tools because of the ability to upload them to the Internet for distribution. Commercial vendors like Westlaw and Lexis have been using these types of e-learning tools to create tutorials for subscribers to learn how to use the two research systems. For example, the West E-Learning Center (http://www.westelearning.com/) provides free, on-demand training to any user wishing to learn about their services. Libraries are using platforms like LibGuides and YouTube to upload similar content on the Web to distribute to patrons. Continuing Legal Education webinars have also become popular for attorneys. The American Bar Association offers online, on-demand CLE trainings as well as CLE through webinar and teleconference that satisfy most state bar association requirements.[4] Law libraries are becoming more involved in developing CLE webinars and on-demand presentations for their patrons and attorneys. In fact, all types of law libraries are getting on board in creating these CLE opportunities—from academic law libraries to firm libraries to state and local law libraries. It is becoming standard practice for firm libraries to offer the trainings to attorneys as well as clients needing the continuing education credits. For example, county law libraries in Ohio are offering continuing education presentations to the public as well as to member attorneys.[5] The use of webinars and online tutorials is becoming more prevalent in law libraries, and these methods are being used for more than just instructing patrons on how to use the library. This means that law librarians must stay current with the advances in instructional technology in this digital information age.

Some academic law libraries are creating webinars on how to use an online public access catalog (OPAC). This allows patrons to watch and learn how to use the library's catalog in order to access the vast amount of information available on the holdings of the library. As an example, the Salmon P. Chase College of Law Library at Northern Kentucky University created a training webinar on using its online catalog, NKUIRE. The video was designed and recorded using Camtasia Studio and later uploaded to the Web using the social media site YouTube (http://www.youtube.com/watch?v=XEXv4bON0zc). The Valparaiso University School of Law Library also created a training webinar on how to use its online catalog, Galileo. The video was designed and recorded using Adobe Captivate and was originally only distributed on the course management tool, Blackboard, for student use. The library has since made the training webinar available to the public by adding it to its LibGuide on using the library.

There are also on-demand, step-by-step training tutorials made available by law libraries. While not as sophisticated as some of the interactive vendor online training centers, these libraries have made the information free and on-demand to reach a wider audience of patrons. By way of example, the Georgetown Law Library has a series of on-demand tutorials (http://www.law.georgetown.edu/library/research/tutorials/index.cfm) that provide research help on topics from administrative law research to searching Encore, the library's online catalog and all tutorials are available to anyone with an Internet connection and Adobe Flash Player. The Harnish Law Library at Pepperdine University School of Law used LibGuides to create a training tutorial for its OPAC (http://lawguides.pepperdine.edu/content.php?pid=238228&sid=1965906). As the user reads through the various tabs in the LibGuide, the tutorial is designed to replicate and link to searching in the OPAC. These are just a few examples of the tools law libraries have created to provide new ways for patrons to access instructional content.

Quick Response Codes

Quick Response codes (QR codes) are pictures with embedded information that can be scanned by a smartphone or device with the help of a camera and QR code reader.[6] Essentially QR codes are a simple way to access content using a mobile device. The Pew Research Center's Internet and American Life Project survey found that two out of five cellular phone owners, or 42 percent, owned a smartphone as of May 2011.[7] That same study found that nearly nine out of ten smartphone owners, or 87 percent, use their phones to access the Internet, with 78 percent of these users saying that they go online using their phone on a typical day.[8] It has been predicted that this number will only grow in the future, thus it is even more important for law libraries to provide access to library resources and instructional tools on mobile devices. The QR code is the simple solution.

By strategically placing QR code images throughout the library and on the library's webpage, patrons can use their mobile devices to access the information in the library stacks or on the Web. Essentially, the use of QR codes negates the need for those seeking information to type in the URL and provides easier access to content via smartphones and tablets. Some of the ways a law library could utilize QR codes include placing one at the front door that would tell patrons general library information such as hours and contact information or placing one at the reference desk that connects to a chat reference service or reference e-mail. QR codes can also be used to direct patrons to more instructional materials. For example, as a user peruses the shelves in the library, if the librarian wanted to lead the patron to a specific area in the library, the librarian could create a QR code and place it at the end of the row of stacks. The QR code could lead the user to an online research guide on topics covered in that area of the library or to the reference desk contact information if the patron needed assistance. This allows the patron to be helped throughout the library by the QR codes created by the librarian as if the librarian were walking around the library with the patron.

The Florida State University Law Library provides a good example of how to use QR codes effectively for law library instruction. QR codes are used to provide contact information for the reference librarians and in the stacks to assist library patrons trying to locate the electronic format of a print resource. The law library is also using bookmarks with QR codes to market library services and lead students to the mobile library catalog, law-related databases, research guides, and the library's blog.[9]

Mobile Device Applications (or Apps)

Another way law libraries have responded to the growing number of smartphone and mobile device users is by creating mobile device applications, or apps as they are more commonly known. Apps allow users to access library content on their smartphones or mobile devices, such as iPads or tablets. Patrons can download the library's free app from one of the application stores and begin accessing the content immediately. Of course, this means that law libraries must become experts in developing mobile device apps or outsource the creation of this technology. The West Virginia University College of Law created a virtual library app that is tailored specifically for students.[10] It provides access to library tutorials, an interactive map, and the library's online public access catalog. A quick search of the Android and iTunes app stores is all that is required to find more law libraries that have created similar apps. From mobile OPACs and basic library information to advanced tutorials and interactive maps, the apps created by law libraries provide a perfect method of library instruction that satisfies the needs of tech savvy patrons.

Library Tours

Responding to the increased demand for electronic access to materials and training content, law libraries have created virtual tours as well. Academic law libraries have been the biggest users of instructional technology to create library tours. Whether the library is creating a virtual tour, 3-D model, or a simple video tour, it has become fairly simple to bring the outside world into a library via the Internet. Using iPads, tablets, or digital video cameras and some type of editing software, like Adobe Premiere Elements (http://www.adobe.com/products/premiere-elements.html), the librarians can take the patron on a tour of the library just like they would if the patron were physically present in the library. Web-based virtual tours typically feature photographs of major service areas along with textual descriptions or voice narration in some cases. These tours can also include photographic images, 360-degree "virtual reality" views, textual descriptions, and contextual floor plans.[11] With the proper equipment, training, and an IT staff that is willing to help, libraries can take patrons on tours of resources and the physical space in the library like the tour provided by the USC Gould Law Library at http://weblaw.usc.edu/library/map/virtualtour.cfm. These tours are simple to put together and easy to upload to the Web to provide another access point for patrons to learn about the library.

If virtual tours or 360-degree tours are too sophisticated or too costly, many libraries are taking a simpler route and creating slide show tours that provide pictures with either voice narration or textual descriptions. The University of Oklahoma's Donald E. Pray Law Library offers a simple slide show with textual descriptions that showcases the library's reading room, circulation desk, and public access terminals at http://jay.law.ou.edu/virtour/index.cfm?action=sslibrary&ss=1#content. The Rutgers Law Library provides a timed slide show with description that allows the user to control the speed of the presentation at http://law-library.rutgers.edu/guide/tour/. As video and editing technologies advance, these presentations will continue to evolve, but law libraries of today can start providing this instruction service with a digital camera, slide show software, and a little imagination.

Law libraries are also adding interactive, virtual maps on the Internet to guide patrons through the library's physical space. At the Ruth Lilly Law Library's website at Indiana University, http://indylaw.indiana.edu/library/tour/floor1.htm, a patron can tour the library using an interactive map for each level of the library. In addition to images of the various locations in the library, the tour provides a floor plan and textual descriptions of each image and area. This type of interactive map is an excellent teaching tool that can be a substitute for a library tour if the patron does not have the time to spend on a tour.

Games or Simulations

It wasn't that long ago that the virtual world Second Life burst into the mainstream and users were listening to lectures at major university libraries and accessing virtual library collections. While the Second Life phenomena never really got off the ground in law libraries, games or simulations have become a popular means for providing library and information instruction. Many academic law libraries use games in library instruction courses to motivate students to learn. Using a Jeopardy-style game for library instruction is an easy, simple way to engage students and help them retain the information better. Law libraries can also develop anonymous classroom games using audience-response technologies like Turning Technologies and iClicker. Using these audience response systems for class participation in a game allows for anonymity in answering the questions and will make the student more comfortable with answering questions and participating.

Online games and simulations have also been developed to increase the user's interest in developing library skills. For example, some academic libraries have used Xtranormal (http://www.xtranormal.com/) to create online, animated movies that tell a reference instruction story or provide answers to frequently asked library questions. These animated videos are then posted on the library's site or on YouTube. Xtranormal is not free and requires a membership but it is a good way to reach patrons in the online world who might otherwise be wary of approaching a law librarian in person with a question.

Online Course Development

Using course management systems like Blackboard and TWEN to disseminate information for classes and to reach patrons is nothing new to academic law libraries. However, these tools are now frequently used to provide access to library instruction guides and resources. For example, if the library presents a program on how to find a paper topic or how to do tax research, then a "course" can be created where the audience can find handouts, videos, links, and sometimes even transcripts of the training. These platforms provide an easy way to disseminate the information relayed at these types of trainings, and the sites can be made available indefinitely and to a select group of users or the entire academic community.

Similarly, private law libraries can use intranets and programs like Microsoft SharePoint Workspace to disseminate training materials and provide access to useful library links and resources. Intranets and SharePoint can serve as a jumping-off point for training and information access. Private law libraries can upload content to in-house intranets or a SharePoint workspace and then make the content accessible to a select few or to the entire organization. The best selling point for law libraries in deciding whether to use course management systems, intranets, and tools like SharePoint is that the materials can be accessed 24/7 by the users. And any electronic resources the library might own or license can usually be linked in the online tool to provide more access to the information. Because of the ease of access and use, after providing a small amount of training on the systems, a user can download and upload content immediately. This immediate, open access to the material makes online training course development a popular platform for library instruction.

Electronic Research Guides

Most law libraries create electronic research guides or pathfinders that lead the user to electronic resources, print resources, Web resources, or even other research guides. The majority of law libraries use research guides created in-house or Springshare's LibGuides product (http://springshare.com/libguides/).

E-LEARNING TOOLS

Librarian Training

Law libraries that hire librarians who are experienced in instructional design and technology have a head start in creating e-learning tools. And many of the job postings in the American Association of Law Libraries career center during 2012 included technology aspects or were titles like "instructional technology librarian," "digital services librarian," and "educational technology librarian." Even most law library director positions include a technology component or require some knowledge of management of information technology. Library professionals who have graduated in the past decade or so should be familiar with using webcams, digital cameras, computers, microphones, and basic Microsoft PowerPoint software. Creating some e-learning tools like QR codes will require little training, and others like mobile device apps, or using software like Camtasia Studio or Adobe Captivate will require more in-depth training. Professional associations have been offering programs on creating e-learning tools, and most software companies offer training demonstrations on the Web as well. It might take some time before the law library professionals are ready to begin using the hardware and software required, but once the training is complete, following the steps to create an e-learning tool will be simple.

Create a Library Webinar

The first step you will want to do is to purchase the necessary hardware, much of which you probably already have. Creating this type of e-learning video file does not require much, just a computer, microphone, speakers, and a webcam or digital camera (if you want to create a virtual library tour or use images from the library). Law libraries do not have to invest in sophisticated hardware to create digital instructional content.

The second step in creating library webinars or tutorials is to choose the appropriate software. It is important to choose the software product that makes the most sense for your library. Presentation software currently used in law libraries varies from basic programs like Microsoft PowerPoint to more advanced programs like Adobe Captivate and Camtasia Studio. Camtasia Studio is the most powerful but also one of the most expensive. It allows you to edit your screen recordings and camera video by cutting, splicing, and combining clips with the video editor. It also allows you to import camera video, music, and photos to enhance your screen recordings and customize your screen recordings and videos with ready-to-use media themes, animated backgrounds, graphics, and callouts. It also functions to create interactive videos with clickable links, a table of contents, and search features, and provides the recording in a format that allows for easy sharing of videos that patrons can watch anywhere, on nearly any device. If recording a voice-over PowerPoint presentation is all a library desires, then Microsoft provides this service in most Office packages. According to Microsoft, this software can be used for a Web-based presentations, for archiving a meeting so that presenters can review it later and hear comments made during the presentation, or for self-running slide show presentations.[12] This is an inexpensive way to create a recorded presentation that can later be uploaded on the Internet or used in an instruction session. (For more information on Microsoft PowerPoint 2010 Voice Narration, see http://office.microsoft.com/en-us/powerpoint-help/turn-your-presentation-into-a-video-HA010336763.aspx?CTT=1.)

The third step involves creating a script and preparing to record the instruction. It is best to practice before trying to record the content, but most of the software used to create this content does allow for some editing if necessary. The equipment should also be tested, including the microphone and webcam. The microphone should be tested for sound quality and background noise. The library should also be certain that the software is capable of creating the type of file it would like to upload, especially if the file needs to be uploaded to iTunesU or some other platform that uses different file types.

The fourth step is to upload the content to the Internet. It is important to choose a platform that patrons will be able to easily locate and use. Social networking sites can provide law libraries with easy-to-use formats for uploading and disseminating content on the Web. YouTube channels are free and the files can be uploaded and then shared across multiple social media sites. For example, a law library could create a video file, upload it to YouTube, and then post the YouTube URL to Facebook, Twitter, Google+, or even on Blackboard. If the library subscribes to LibGuides, then the video can be embedded on the LibGuides site as well.

Social media sites are great tools to relay instructional content. These sites are free, easy to use, and do not require a computer programming degree or knowledge of HTML to use. For example, the Minnesota State Law Library uses Flickr to showcase its library tour, http://www.flickr.com/photos/9201331@N04/sets/72157600431218715/. Using pictures it uploaded and hosted on Flickr, the library created a slideshow with textual narration. Other libraries use Facebook or Twitter to connect to patrons and relay instructional content. The type of social media a law library chooses to relay the content will vary depending on user preference, but it is a good idea to choose more than one social media platform to allow for wider dissemination.

This fourth and final step is probably the most important step in the process because if the content cannot be accessed easily, then it is useless. Creating the shareable instructional content and making the content accessible via social media or the Web ensures that a library is reaching its technology-savvy patrons as well as traditional patrons.

CONCLUSION

Creating library instruction in the information age involves staying in tune with patron needs. In today's technology-driven world, law library patrons require that libraries use technology effectively to provide the best service possible. Providing a mechanism to learn in on-demand and online environments gives library patrons the option to learn how to use the library's resources without ever stepping foot into the library's physical space.

FURTHER READING

Clark, Ruth C., and Richard E. Mayer. 2011. *E-learning and the science of instruction: Proven guidelines for consumers and designers of multimedia learning.* New York: John Wiley & Sons.

Jackson, Darla W. 2011. "Standard bar codes beware—Smartphone users may prefer QR codes." *Law Library Journal* 103 (1): 153–157.

Kern, Alex, and Janice Milliken. 2012. "QR codes: Helping patrons access content." *American Association of Law Libraries Spectrum Online.* http://www.aallnet.org/main-menu/Publications/spectrum/Spectrum-Online/qr-codes.html.

Steele, Jordon, and Ed Greenlee. 2011. "Thinking, writing, sharing, blogging: Lessons learned from implementing a law library blog." *Law Library Journal* 103 (1): 113–123.

NOTES

1. See Beatrice A. Tice, "The academic law library in the 21st century: Still the heart of the law school," *UC Irvine Law Review* 1, 157 (2011): 168–172.

2. Jordon Steele and Ed Greenlee, "Thinking, writing, sharing, blogging: Lessons learned from implementing a law library blog," *Law Library Journal* 103, 1 (2011): 113–123.

3. Steele and Greenlee, "Thinking, writing, sharing, blogging," 115.

4. American Bar Association CLE, 2013, http://www.americanbar.org/publications_cle/cle_courses_by_format.html.

5. Hamilton County Law Library Association: Services, 2013, http://www.hamilton-co.org/cinlawlib/services.html.

6. Alex Kern and Janice Milliken, "QR codes: Helping patrons access content," *AALL Spectrum Online*, 2012, http://www.aallnet.org/main-menu/Publications/spectrum/Spectrum-Online/qr-codes.html.

7. Aaron Smith, "Smartphone adoption and usage," July 11, 2011, http://pewinternet.org/Reports/2011/Smartphones/Section-1/Overview-of-smartphone-adoption.aspx.

8. Aaron Smith, "Smartphones as an Internet appliance," July 11, 2011, http://pewinternet.org/Reports/2011/Smartphones/Section-2/Smartphones-as-an-internet-appliance.aspx.

9. Darla W. Jackson, "Standard bar codes beware—Smartphone users may prefer QR codes," *Law Library Journal* 103, 1 (2011): 157.

10. WVU College of Law Library, 2012, http://udm4.com/Android/WVU_Law_Library-758498.

11. For a detailed description of how to create a virtual tour, see James Duncan, "Production of the next-generation library virtual tour," *Bulletin of the Medical Library Association* 89, 4 (October 2001): 331–338.

12. Microsoft, "About recoding a voice narration for a presentation, Microsoft Office Help," 2013, http://office.microsoft.com/en-us/powerpoint-help/about-recording-a-voice-narration-for-a-presentation-HP005195030.aspx.

19

Educational Technologies

Kim Clarke
BENNETT JONES LAW LIBRARY, UNIVERSITY OF CALGARY

Nadine R. Hoffman
BENNETT JONES LAW LIBRARY, UNIVERSITY OF CALGARY

Law libraries have redesigned their roles over the past few decades, moving from emphasizing their print collections to emphasizing their expertise and services. In university libraries, this evolution is now encompassing the integration of educational technologies and the development of instructional programming for those technologies.[1] Law libraries, especially academic law libraries, are just awakening to their potential role in the implementation of learning-based technologies but are well placed to lead this discussion. "Libraries are on the front line. Working with people who are young and who are growing up with technology, we are in a wonderful place to advise on what is 'future cool.'"[2] This call to action is not a surprise given how law libraries have embraced technology over the past few decades.

Public law libraries provide computers for patron use, whether located in a computer lab or in the reference room. Academic law libraries often have IT staff to assist both library staff and patrons; in fact, they often provide all the technology-related assistance law students receive. A significant number of academic law library directors also oversee the law school's information technology staff, a recognition of the close relationship between libraries and technology. In recent years, emerging librarian positions have been created in academic and courthouse libraries to ensure someone is watching technological advances. One would be hard-pressed to find a recent issue of law library current awareness publications, such as *Spectrum*, that does not contain at least one technology-related article due to its impact on how libraries function. Educational or learning-based technologies are a natural extension of the current services offered by law libraries.

Academic law libraries may soon be required to explore the need for educational technologies at their institution. In late 2012, the American Bar Association distributed proposed amendments to the standards governing the law school accreditation process. The proposed standard 601(a)(4) states, "A law school shall maintain a law library that: . . . (4) remains informed on and implements, as appropriate, technological and other developments affecting the library's support for the law school's program of legal education."[3] As the law school's program of legal education includes both research-related and instructional-based activities, this standard may impose a duty on the library to stay informed and implement educational technologies.

In this chapter, several types of technologies that support learning are examined. Possible law-related uses are explored, many of which are future-looking as the technologies are not yet being used in the legal realm. The supporting roles that law libraries can or should play with respect to these technologies is also discussed.

COLLABORATIVE TECHNOLOGY

Collaborative learning experiences have been shown "to promote creativity, increase critical thinking competencies, build stronger interpersonal skills, enhance relationships . . . , provide exposure to diverse viewpoints, and prepare students for real-world work."[4] Is it any wonder that collaborative work is extremely common in the legal field? Lawyers frequently handle cases in teams comprising junior and senior attorneys, paralegals, and other support staff. Law students work together on competitive moot teams, for class assignments, and in self-selected study groups, among other experiences.

Traditionally, collaborative work was conducted in person, which allows for a full face-to-face discussion and debate of issues. Technology has increased the opportunity for collaboration. Cloud technology, for example, has had a fundamental impact on collaboration, allowing people to collaborate from different locations and even at different times.

There are many advantages to online rather than in-person group work. Studies have shown that online collaborators do better in their course work than students who collaborate face to face.[5] Furthermore, online technology can equalize the impact between aggressive and timid group members. If a discussion list is being used, everyone can have a voice without having someone speaking over them. If the discussion list is anonymous, no one knows who is presenting the ideas so a person's reputation or strong personality is neutralized.

Anyone who has worked on a team to complete a class assignment has experienced a situation where there was an uneven level of participation among the team members. Combine that with feelings of frustration and anger caused by many professors simply assigning the same grade for all team members regardless of output. Technology allows professors to become a member of each class group and monitor individual team members' efforts to determine participation grades.

Some online collaborative technologies support asynchronous learning, allowing team members to participate as their schedules allow. An example of an asynchronous collaborative instrument is a discussion list where group members ask questions and provide comments at their convenience. A threaded discussion list is especially effective as participants can respond to a specific comment rather than just responding chronologically. Synchronous online learning technologies require participants to be connected at the same time but from separate locations. Video communication products like Skype or FaceTime are examples of synchronous technologies.

Technology can be utilized in a variety of group activities, including brainstorming, resource sharing, and the content development process. Brainstorming can be the most contentious phase of group work. A dominant personality can overshadow the other members, disagreements can break out, and hard feelings can occur. Technology can be an equalizer in this endeavor, allowing everyone to share ideas, ask questions, or comment on earlier suggestions without the emotions. E-mail is the most common electronic communication tool; however, it is not the most effective communication method when multiple issues are being discussed by a group of people. If people are reading and replying to e-mails at different times, they may be responding to e-mails out of order and either repeating or contradicting what others have already said. This could be extremely confusing, with little resolution of matters.

Discussion forums and chat features allow team members to share their ideas and ask their questions in writing more effectively. Threaded or reference discussion forums allow individuals to insert comments at specific points in the discussion rather than chronologically with the newest comments added to the end. This increases the clarity of the discussion as participants understand the connection between the comments. Discussion forums and chat can be used either synchronously or asynchronously. That is, participants can log on at the same time to correspond or one person can begin the discussion and others add their thoughts at a later time.

These communication options are integrated in other technologies utilized by law offices and law schools, namely intranet and course management systems (CMS). The current versions of CMS, such as Blackboard or Moodle, allow the instructor to create restricted discussion forums or chat rooms for individual groups. Blogs and wikis, which may also be available through the Intranet or CMS, are other potential vehicles for group brainstorming activities, after determining the level of privacy provided by the products.

Law libraries can facilitate and promote the use of collaborative technology by learners and instructors. Library staff members often assist faculty members in developing their course page in the CMS[6] and could take those opportunities to demonstrate how the system could support group work. They can also raise the awareness of both learning and instructors of free or licensed products that support collaborative learning through the development of Web-based tutorials or guides, and informal information sessions.

Technology can also be integrated with face-to-face brainstorming sessions. Video communication products can also be utilized; of course, someone would need to take notes so there is a written record of the discussion. Interactive whiteboards allow group members to write suggestions down, save them, and share them with all group members.

The ability to access shared resources, whether the class assignment, statement of facts, interview notes, or research materials, is beneficial to team members. An electronic file-sharing platform allows materials to be accessible and updateable by all team members.

When selecting a file-sharing system, lawyers must ensure their shared systems are secure as their client's confidential information cannot be breached. In fact, many law offices have policies regarding off-site access to client files. They frequently have intranets allowing them to store materials securely. Law students have many more options for file sharing as security is less of a concern. While they do not want other groups to see their work efforts, it is highly unlikely that the other groups would attempt to hack into their files.

Most CMS systems allow professors to create separate work spaces for the groups; access is restricted to group members and the professor. One of the benefits of using the CMS is the ease in accessing the full text of relevant electronic materials. If a case, for example, is located in an electronic database, its Persistent URL (PURL) can be put on the CMS and the case can be obtained just by clicking on the link.[7]

The team can also use free cloud-based Internet resources such as Dropbox to share resources. One person can create a folder in which the team members can load relevant documents. Dropbox allows materials to be accessed offline, allowing documents to be reviewed, added, or amended even without Internet access. Any changes will be uploaded to Dropbox when the team member next connects to the Internet.

Social bookmarking products, like Diigo, allow team members to bookmark and tag interesting webpages to share. They can even highlight specific sections on the webpage and create annotations in which they can indicate the impact of the information on the webpage. Similarly, Westlaw allows individuals to create tabs containing relevant databases and share the tab with Westlaw users.

The final step in team projects is generally some work product, generally a written document. In practice, this may be a legal brief or an offer of settlement. For law students, the work product may be a memorandum or essay. Word processing software, such as Microsoft Word, allows multiple people to add sections to or amend portions of a document. Word's track changes feature records the changes made by different individuals, with each person's comments being shown in a different color. Team members can easily share their Word documents electronically, via e-mail. The difficulty in sharing drafts in this manner is keeping track of which version of the document is the latest. Teammates will undoubtedly read their e-mails and provide comments on the document at different times. One of the team members would then have to sift through all the amended versions into one document, which would be a considerable amount of work.

Restricting the work to one version of the document would allow all efforts to be consolidated. Cloud-based office programs such as Google Drive and Zoho or the cloud storage service Microsoft SkyDrive allow the work to be restricted to one version of the document. They also allow true collaboration as multiple people can be working on the document at the same time. While the choice of product is a personal choice, team members who are used to using Microsoft products may feel more comfortable with SkyDrive or Zoho, which have similar-looking toolbars.

Technology like SMART Technologies' Sync and Tidebreak's TeamSpot allow groups to collaborate more effectively in person. These software products are loaded onto a computer system installed in a boardroom or group workroom and connected to a wall-mounted monitor. Team members connect their individual laptops to the main computer. The software system allows each member to send information directly to another team member's laptop or to the wall-mounted monitor for shared viewing. They can upload individual portions of the project, then review and amend the document as a group. The document is saved on all of the connected laptops so team members have the same version.

Installing the collaborative hardware discussed above, including Sync, TeamSpot, and interactive whiteboards, requires a significant investment of money. Law library administrators may need to build a case for their parent institution to support the acquisition of this equipment. In the future, law students may begin to expect or even demand this type of equipment because they will have used it while obtaining their undergraduate degree.

SIMULATION-BASED LEARNING

Simulation or role-playing exercises have long been part of legal education. Simulation exercises began to be utilized in skills-based courses such as trial advocacy and negotiation in the early 1970s.[8] Around the same time, law schools began to develop clinical programs wherein students worked for real clients under the supervision of a practicing attorney. In simulations, students are presented with fact scenarios and are tasked with representing a party to a legal conflict. The situation may require them to negotiate a settlement on behalf of their client or represent them in court, for example. First-year students often assume the role of clients in the upper-year simulations, which brings an enhanced sense of realism to the simulations, allowing for interrogatories and direct and cross examinations.

Simulations are less common in doctrinal law courses where the focus is on teaching legal principles. Individual doctrinal instructors may utilize simulations but the exercises are a small component of the course, not the primary teaching method. Even today, "In no Common Law jurisdiction is simulation regarded as part of law's 'signature pedagogy.'"[9] One reason why law professors are hesitant to increase the use of simulations in doctrinal courses is the fear of being viewed as a "trade school"; the focus in law schools is on teaching students how to think like a lawyer. Another concern is the considerable amount of time, effort, and technical assistance needed to develop the simulation exercises. Finally, required doctrinal courses are generally quite large in size, making it very difficult to conduct and grade the simulation exercise, as the instructor must interact with all the student groups.

The 2007 Carnegie Report[10] has increased pressure on law schools to expand their experiential learning opportunities; many schools have embraced simulations as a means of meeting this challenge. Stanford is a leader in simulation education by developing a spectrum of new simulation-based courses, including some that strongly integrate doctrinal law, such as environmental law. A unique feature in the Stanford courses is having students from other disciplines playing roles in the simulations, such as graduate science students playing characters in an intellectual property fact scenario.[11] While the array of simulations has increased recently, the bulk of the simulations continue to be conducted in person. Technology does, however, allow the development of online and virtual simulations.

The most basic online simulations simply integrate some technological aspects into a traditional simulation course. The instructor can use websites, blogs, social media pages, and e-mails to share the relevant facts and evidence to the students.[12] All communication and interpersonal interactions between the student lawyers and students with instructors occur online. The key advantage to this option is the comfort level the participants would have in using these common technological resources. The disadvantage is, while these courses help students develop their written interviewing, negotiating, and communication skills, it ignores the development of in-person skills.

Legal educators have developed a software program to assist in developing law-related simulations. The Glasgow Graduate School of Law in Scotland developed "an open-source, open-standards transactional learning environment"[13] with a suite of tools instructors can use to create simulations called SIMPLE: Simulated Professional Learning. SIMPLE "provides a framework for students to engage in transactions typical of real-life legal practice, providing the kind of contextualized knowledge and skill that the Carnegie Study and others have demanded."[14] The developers created a town in SIMPLE, with businesses, citizens, and government structure, in which the simulated legal actions take place. Initially, the simulations, based on tort, divorce, and criminal law, were created by universities in the United Kingdom, but instructors in other countries have since added new simulations to the collection. The Daniel Webster Scholar Honors Program at the University of New Hampshire is the first law school in North America to experiment with SIMPLE.[15]

Virtual reality games provide a more realistic role-playing experience for the students. A multiplayer educational gaming application (MEGA) is "an immersive, 3D virtual learning environment in which avatar, or graphic representation of the user, interacts in real time with the other avatars, and virtual contexts in a visually rich simulated world."[16] They allow multiple people to participate simultaneously through an avatar, which provides a more fulsome learning environment than the other simulation models. Participants can interact in real time, which allows for a more realistic simulation. In a trial simulation, opposing counsel can debate the rules of evidence, such as hearsay information. Settlement negotiations can be conducted between student lawyers in a virtual face-to-face manner.

The integration of game-based learning with higher education is still relatively new but is gaining a solid toehold. A growing number of academic libraries in North America, including Stanford, Michigan, University of Illinois Urbana–Champagne, and Simon Fraser University, have established gaming collections to support academic inquiry and learning in various disciplines. In some courses, it is the video games themselves that are the subject of inquiry, often used to explore societal developments. Feminist courses, for example, may compare how female characters are portrayed in different video games.

In law schools, "gamification" can be used to help students enhance their skills or develop legal solutions to real-world issues, such as creating a new environmental review process. Gamification does not require using video games; it is simply the process of utilizing "game-thinking and game mechanics to engage users and solve problems"[17] while encouraging learning. Competition is one game-thinking element that can be exploited in law school to increase student engagement. Law students are, for the most part, very competitive. Instructors can encourage students to accomplish specific tasks or acquire a skill through the use of a public reward system. For example, the instructor could create a class blog and reward students for every entry they post in which they discuss an article or case they read or a legal principle mentioned in class.

Law librarians can also embrace gamification by implementing the techniques in their training. A "loyalty reward" system could be created whereby students get points for attending different training sessions. They could even use game-making software to create a legal research game to encourage skill development.

While law-related games exist, they are primarily aimed at middle and high school students. These online games are not video games; most are single player, multiple choice quizzes, while some have simulation videos the students observe before answering the question. Law students can play these games as a means of testing their knowledge of constitutional amendments or rules of evidence, for example. Two reliable sources of these games are the State Bar of Texas's Law-Related Education Department and iCivics, an organization founded by former Supreme Court justice Sandra Day O'Connor.

There is a dearth of law-related virtual reality games. One of the few is "The Game," created by a Dutch law firm and used during their hiring process.[18] The game assists the firm partners to determine how the applicants perform in fast-paced, stressful situations. Due to the scarcity of available games, law instructors are faced with creating their own games. Thankfully, game-creating software, such as GameMaker, is now readily available and quite user friendly, so programming skills are not required.

Besides implementing gamification in library-related activities, the potential roles for the library in the simulation activities of the law school are varied. The technology-focused library personnel, especially the emerging technologies librarian, can assist faculty members in identifying and utilizing software that will allow them to develop role-playing exercises, whether online simulations or actual video games. Law librarians must also consider following their nonlaw colleagues and develop gaming collections, purchasing video games, gaming consoles, and peripheral devices such as headphones, microphones, and joysticks.

TOUCH TECHNOLOGIES

Touch technology's infiltration in our daily lives cannot be denied as these "surfaces are becoming more commonplace in a range of devices such as phones, tablets and computer displays."[19] Touch devices are intuitive to use, relying on natural movements, which allows everyone from the very young to the very old to feel able to use them. As their comfort with, and reliance on, touch devices increase, individuals have begun seeking ways to utilize them in their professional and educational lives.

An explosion in legal application (app) development allows lawyers to "practice on the go." They can conduct research with FastCase, HeinOnline, LexisNexis, or Westlaw, track time with CLE Tracker, and use practice management apps like MyCase or Trial Works to stay current on their client's case. Recent graduates can access materials for their bar admission preparation courses, such as BARBRI and Kaplan, from their touch device. Several academic law libraries, including UCLA, Indiana University, and Boston University, have created information guides referencing available law-related apps for their students.

Touch technologies like the iPad, "a tool for the consumption, production, and discussion of knowledge"[20] have a natural affiliation with educational activities. They are, in fact, becoming more popular with students and educators than laptops "because users can seamlessly load sets of apps and content of their choosing, making the tablet itself a portable learning environment."[21] Their portability increases with a data plan as the tablets are not reliant on computer or wireless networks for an Internet connection. E-reader apps and services like iBooks Author increase student access to textbooks and other reading materials from any location. A few North American universities and departments have begun providing iPads to incoming students and are studying their use by students and instructors.[22]

Tablets can facilitate learning by allowing users to develop apps as a demonstration of learned legal concepts. Instead of writing papers, the students in Georgetown Law's new Technology, Innovation and Law Practice course "work in teams to produce prototypes of apps to solve legal problems. In place of a final exam, students participate in a head-to-head competition."[23] According to Professor Rostain, "These projects have significant pedagogic value,"[24] as the students must do more than merely understand the underlying legal concepts. They must "anticipate a variety of real-world possibilities, weigh choices and communicate the law effectively to the user"[25]—all skills they will use in practice.

A significant number of apps are already available to help students and nonlawyers find and understand the law. One caveat is that few of the legal apps are free and several of them are quite expensive when compared to other apps. There are apps for *Black's Law Dictionary*, some national constitutions, the U.S. Code, and court rules. Study aids can also be distilled to an app, such as the Law in a Flash app with mobile flash cards for many required law courses, and Law School Dojo, which contains a variety of law school quizzes. Future law-related app development could be the creation of concept maps or flowcharts for FILAC (Facts, Issues, Law, Analysis, Conclusion), which would assist those working through complex legal issues in a case.

Some touch devices are specifically created for educational activities and learning. Touch tables are designed for collaborative group work, as they allow multiple individuals to work together on large multitouch surfaces. Touch tables can also be connected to an interactive whiteboard to allow a greater number of people to view what is happening on the table itself. Other educational disciplines such as medicine and biology have begun to integrate these visual formats into their educational curriculum more easily than law. The *Lyndsay the Virtual Human* app, for example, is a "three-dimensional, multi-scale, interactive computer model of male and female anatomy and physiology,"[26] which is a more effective method to learn the human anatomy than with a one-dimensional diagram and more efficient than using cadavers. To date, there are no known examples of touch table use in legal education or practice. However, as touch technology is expected to gain widespread adoption in higher education for "teaching, learning, and creative inquiry"[27] activities in the near term, its integration in legal education should follow shortly thereafter. Arguably, "it would be irresponsible for [law] librarians, faculty, and administrators to ignore the ways in which the mobile environment is changing teaching, learning, and professional expectations of students. Faculty and librarians are charged with finding ways to use these devices to contribute to learning."[28]

All law libraries can support their patrons' use of touch technology. Librarians can create Web-based information guides about law-related apps their patrons may want to download. Collection development librarians can purchase electronic books that can be viewed with a variety of e-reader apps so patrons can access it from their tablet. Library technology staff could provide technical support to faculty members who want to create apps or to add an app-building assignment to their course.

Academic law libraries have begun experimenting with other uses of touch tablets. Several libraries, such as Yale, Washington & Lee, and University of Oklahoma, have developed iPad lending programs through which law students and faculty members can borrow iPads loaded with law-related apps for several days. Librarians can explore integrating iPads in their instructions, as the law librarians at Boston College did in a pilot project in January 2012.[29]

PRACTICE-FOCUSED SKILLS

Law schools have increased the number and variety of skills-based courses they offer in recent years. The focus of most of these programs remains on litigation-related activities, but transactional and practice management skills are becoming incorporated into the curriculum. Several law schools have developed institutes, centers, and clinics that explore the intersection of technology with the practice of law. The Institute of Law Practice Technology and Innovation at Suffolk University Law School, for example, is designed "to educate students, the legal profession, and the public about technology's transformation of the practice of law and the delivery of legal services,"[30] and Columbia Law School's Lawyering in the Digital Age Clinic[31] provides students with the opportunity to experience using digital technology in managing their client's case.

Skills development does not stop upon graduation; young attorneys frequently take Continuing Legal Education (CLE) courses such as advanced trial techniques to enhance their skills. To become truly competent, or even expert, in any skill, one must practice. The centuries-old adage "Practice makes perfect" should be the mantra of law students and young lawyers. Technology can assist students and lawyers develop their practical skills.

Not surprisingly, many of the skills taught in law school are verbally based. Students take advocacy and negotiation courses, join competitive moot teams, and participate in debates and clinics—all in preparation for legal practice. To become good public speakers, students and lawyers must identify and correct nervous behaviors, poor eye contact, and unwanted speaking habits. After class presentations and moot trials, the student receives constructive feedback from the instructor and perhaps other participants. Technology can also aid in this endeavor, allowing students and lawyers to record their rehearsals for playback and review. Libraries and law firms have long invested in video-recording equipment that students and lawyers can use to record themselves. Originally, the equipment might have been on portable carts and wheeled into the classroom or boardroom or, more recently, be permanently installed in those spaces.

Whereas learners previously saved their performance on a VHS tape or DVD, technology now allows them to save it as a video file on their laptop or flash drive so it can be easily reviewed later. They can also share the file with colleagues, family, and friends for a broader range of comments. Novices also benefit from watching the taped performances of their colleagues and even YouTube oral argument videos as they can identify effective mannerisms and speech patterns to incorporate use in future presentations. Law libraries can support these activities by providing workrooms to record and play back oral presentations.

Good verbal skills are only one component of effective presentations today as audience members, whether students or jurors, have come to expect a visual component. In fact, CLEs provide digital media training to lawyers as "clients and juries increasingly expect verbal presentations to be accompanied by computer graphics and visual aids."[32] PowerPoint was first-generation presentation software that informed but did not engage the audience. Prezi, second-generation software, incorporates some movement to the presentation that engages the observers more than PowerPoint. Today's digital media production technologies, however, allow even novices to create professional-looking video and audio productions. Digital media software can be used on PC and Mac computers, providing a level of technical comfort to the novice production developers. They can begin by using basic software such as iMovie, iPhoto, and Garage Band (standard Mac software), Animoto, and Adobe Photoshop, leading to utilizing more advanced products like Adobe Premiere Pro, FinalCutPro, and Neota Logic. Some of the more ubiquitous software could be installed on the computers in the library, with library personnel offering workshops ("tech talks") and creating usage guides for them.

Planning and creating these productions can actually be viewed as instructional activities as they "improve the students' ability to learn the doctrinal material in a manner that does not involve an inordinate amount of class time because the students are actively involved in preparing their own technological presentation."[33] The video developer must understand the legal concepts or process before he or she can select the best text, video and audio tracks, and images to use in the production. Video productions are useful learning tools for the viewers. They instill a deeper understanding of the subject in audience members by incorporating visual cues that bring the topic to life. A picture truly can speak a thousand words. Mash-ups, or digital collages, can be even more effective as they show relationships through a variety of images, videos, and audio clips. While to date law students have utilized these skills for entertainment purposes, such as creating parodies of their law school experiences for year-end and graduation celebrations, they could be encouraged to use these skills for educational purposes in the future such as in Georgetown's Technology, Innovation and Law Practice or Suffolk's Lawyering in the Age of Smart Machines classes.

Practice management courses are still somewhat uncommon at North American law schools but provide students with a theoretical understanding of what it takes to manage a law practice. Practice management software, such as Amicus Attorney, PCLaw, and Summation, can bring that to life for law students. A simulated law firm used for learning practice-related software with a practical assignment for tracking one's time would be beneficial in a skills-based course. Students would learn digitally drafting documents, court filings, and security of files through using practice management, e-discovery, and case management software systems to better prepare themselves for the day-to-day administrative aspects of practicing law.

Law libraries can support the development of practice-related skills without expending a significant amount of money. Practice-focused software could be purchased and loaded on terminals in the library's computer lab. Librarians could also develop a training program for this software, which would be particularly helpful for students who are not taking the practice management or clinical courses. Knowledgeable staff can offer media creation workshops and assist with creating digital media presentations. Law libraries who continue to house recording equipment on wheeled carts should consider following the examples of Georgetown and the University of Calgary and permanently install recording equipment in a few of their group workrooms so students can tape their presentations or oral advocacy exercises whenever the library is open.

THE LIBRARY'S ROLES

Potential roles the law library and its staff members can play with respect to educational technology are, for the most part, familiar roles to them. Many of today's law libraries already have significant involvement with technology at their parent institution, so educational technologies would be a natural expansion of their existing activities.

The library may need to designate physical space to house the technologies. Libraries could outfit some of their group workrooms with educational technologies for collaborative learning, such as interactive whiteboards, converted into presentation practice rooms, or even create a gaming environment.

A substantial financial investment is needed to acquire and install educational technology. It is doubtful the library's budget would allow the larger hardware and expensive software products on their own; their parent institution will need to pay for a significant portion of these costs. The library can, however, invest in software and less expensive hardware and accessories such as headphones, microphones, joysticks, and touch tablets, which would assist in equalizing access to technology among students of different socioeconomic backgrounds.

Technology-focused personnel, such as emerging technologies librarians, can be invaluable advisors to administrators and instructors in how educational technologies can be incorporated into the curriculum. These librarians stay current on research and educational technologies and would be able to advise on products that would accomplish the institution's and individual instructor's goals. Their input may also be sought regarding terms in software licensing agreements. The library's information technology staff members can also provide instructors with technical assistance in using the unfamiliar technology, such as GameMaker and SIMPLE. Law libraries can expand their training and instruction programs to include sessions on instructional technologies. Product trainers can be brought in or in-house sessions can be developed to meet the needs of the instructors or learners. Librarians can create online tutorials or user guides to provide patrons with 24/7 access to assistance.

Collection development librarians will also be impacted as they may need to begin searching for and purchasing unfamiliar new media materials, such as video games and audio files. The library may need to escalate its educational role on copyright compliance if lawyers, instructors, and students begin using digital media technology to create productions for classroom or courtroom use.

And finally, librarian instructors should consider leading by example through the implementation of these technologies in their courses. An online simulation exam for an advanced legal research course could be very effective.

CONCLUSION

Instructors, including law faculty and CLE lecturers, are starting to experiment with educational technologies;[34] however, this experimentation must be conducted in a thoughtful manner. The technology must have a clear connection with the learning objectives of the course. It must support, not replace, classroom instruction and personal interaction between the instructor and students or between the students themselves.[35] Law librarians, with their technological knowledge and their collaborative skills, are perfectly situated to participate in—or even lead—discussions to ensure a successful integration of educational technologies as learning tools at their home institutions. They simply need to wholeheartedly embrace the opportunity as the students will the technologies.

FURTHER READING

American Bar Association. http://www.americanbar.org.

Annetta, Leonard A., et al. 2006. "Serious games: Incorporating video games in the classroom." *EDUCAUSE Quarterly* (3): 18.

Barcelona, Robert J., and Donald L. Rockey Jr. 2010. "Using collaborative learning technologies to facilitate effective group work." *Journal of Physical Education, Recreation and Dance* 81 (4):12–14.

Brantley, Peter. March/April 2008. "Roles and expectations for digital libraries." *Educause Review*: 31–38.

Brown, Malcolm. July/August 2009. "Learning and technology—'in that order.'" *Educause Review*: 62–63.

Eltis, Karen. 2012. *Courts, litigants and the digital age: Law, ethics and practice.* Toronto: Irwin Law.

Fogden, Fiona. 2010. "Mashing content: A case study." *Legal Information Management* 10 (3): 222–228.

Garvey, John, and Paul Maharg. April 15, 2011. "Standardized clients and SIMPLE (SIMulated Professional Learning Environment): Learning professionalism through simulated practice." Paper presented at Future Ed 3, New York Law School, New York. http://www.nyls.edu/user_files/1/3/4/30/58/1053/Garvey&Maharg.pdf.

Goodenough, Oliver, and Marc Lauritsen, editors. 2012. *Educating the digital lawyer*. New Providence, N.J.: LexisNexis. http://www.law.harvard.edu/programs/plp/pdf/008-03358_03358-ch0007.pdf.

Guyer, Cindy. 2012. "Teaching SMARTly: A law library's experiences using SMART technologies." *AALL Spectrum* 16 (4): 11–12.

Hahn, Jim, and Hilary Bussell. 2012. "Curricular use of the iPad 2 by a first-year undergraduate learning community." *Rethinking Reference and Instruction with Tablets* in *Library Technology Reports* 48 (8): 42–47.

Higgins, Steven E., et al. 2011. "Multi-touch tables and the relationship with collaborative classroom pedagogies: A synthetic review." *Computer-Supported Collaborative Learning* 6 (4): 515–538.

Hollander, Patricia A. 1977. "The uses of simulation in teaching law and lawyering skills." *Simulation Gaming* 8:319–340.

Jaber, Laala Al, and Sarah Randag. February 17, 2011. "Dutch law firm uses video game to evaluate law graduates' talent." *ABA Journal*. http://www.abajournal.com/news/article/dutch_law_firm_uses_video_game_to_evaluate_law_graduates_talent/.

Johnson, L., et al. 2013. *NMC Horizon Report: 2013 Higher Education Edition*. Austin, TX: New Media Consortium. http://www.nmc.org/publications/2013-horizon-report-higher-ed.

Korn, Jane. 2004. "Teaching talking: Oral communication skills in a law course." *Journal of Legal Education* 54 (4): 588–596.

Law School Innovation. Blog, http://lsi.typepad.com/lsi/2007/01/simulations_par.html.

Lyndsay Virtual Human. http://www.lindsayvirtualhuman.org/.

Lysaght, Pamela, and Danielle Istl. 2004. "Integrating technology: Teaching students to communicate in another medium." *Legal Writing: Journal of the Legal Writing Institute* 10:163–180.

Maharg, Paul, and Martin Owen. 2007. "Simulations, learning and the metaverse: Changing cultures in legal education." *Journal of Information, Law and Technology* (1). http://www2.warwick.ac.uk/fac/soc/law/elj/jilt/2007_1/maharg_owen/#a4.

Miller, Willie. 2012. "iTeaching and learning: Collegiate instruction incorporating mobile tablets." Rethinking Reference and Instruction with Tablets in *Library Technology Reports* 48 (8): 54–59.

Munneke, Gary A. 2009–2010. "Managing a law practice: What you need to learn in law school." *Pace Law Review* 30 (4): 1207–1245.

Nathenson, Ira S. June 1, 2011. "Teaching law with online role-playing simulations." http://papers.ssrn.com/sol3/papers.cfm?abstract_id=1865880.

Parks, Ann W. 2012."Teaching innovations: Simulations, apps—and other new ways to learn the law." *Georgetown Law Magazine*: 29–33. http://www.law

.georgetown.edu/news/georgetown-law-magazine/upload/simulations-apps.pdf.

Ramos, Miguel. 2007. "Mash it up! Remixing the information superhighway." *Searcher* 15 (6): 17–22.

RIPS Law Librarian. Blog. http://ripslawlibrarian.wordpress.com/.

Siegle, Del. 2010. "Cloud computing: A free technology option to promote collaborative learning." *Gifted Child Today* 33 (4): 41–45.

Skalbeck, Roger V. 2012. "Tech innovation in the academy." *AALL/ITLA White Paper: The New Librarian*: 74–78.

Sullivan, W. M., et al. 2007. *Educating lawyers: Preparation for the profession of law.* San Francisco: Jossey-Bass.

Tilsley, Alexandra. January 15, 2013. "Lynn University to require all new students to buy iPads." *Inside Higher Ed.*

NOTES

1. North Carolina State University's James B. Hunt Jr. Library and the University of Calgary's Taylor Family Digital Library are leaders in this realm. Both libraries have incorporated a significant amount of educational technologies, including touch tables, technological enhanced group workroom, gaming, and digital media technologies into their programming activities. See http://www.lib.ncsu.edu/huntlibrary/explorespaces and http://tfdl.ucalgary.ca/tech.

2. Peter Brantley, "Roles and expectations for digital libraries," *Educause Review* (March–April 2008): 38.

3. American Bar Association, http://www.americanbar.org.

4. Robert J. Barcelona and Donald L. Rockey Jr., "Using collaborative learning technologies to facilitate effective group work," *Journal of Physical Education, Recreation and Dance* 81 (2010): 12.

5. Ibid.

6. Academic law librarians began assisting faculty members with their course pages when Westlaw introduced The West Education Network (TWEN). As TWEN was related to Westlaw, the librarians were viewed as the on-site experts by faculty members. This relationship has continued at many law schools, notwithstanding the transition to other CMSs.

7. Depending on the database and the law school's licensing agreement, the students may have to sign in before accessing the full text of the case.

8. Patricia A. Hollander, "The uses of simulation in teaching law and lawyering skills," *Simulation Gaming* 8 (1977): 320.

9. Paul Maharg, "Simulation: A pedagogy emerging from the shadows," in *Educating the digital lawyer*, edited by Oliver Goodenough and Marc Lauritsen (New Providence, N.J.: LexisNexis, 2012), 9-1. A signature pedagogy is one that "is recognized as being

distinctive in a profession, . . . become pervasive in a curriculum, and . . . [is] pervasive across educational institutions."

10. W. M. Sullivan, *Educating lawyers: Preparation for the profession of law* (San Francisco: Jossey-Bass, 2007).

11. Gene Koo, "Simulations, part 1: Thinking (and being) a lawyer," *Law School Innovation*, January 23, 2007, http://lsi.typepad.com/lsi/2007/01/simulations_par.html.

12. Ira S. Nathenson, "Teaching law with online role-playing simulations," June 1, 2001, http://papers.ssrn.com/sol3/papers.cfm?abstract_id=1865880.

13. Paul Maharg and Martin Owen, "Simulations, learning and the metaverse: Changing cultures in legal education," *Journal of Information, Law and Technology*, no. 1 (2007): 40, http://www2.warwick.ac.uk/fac/soc/law/elj/jilt/2007_1/maharg_owen/#a4.

14. Gene Koo, "Dungeons & lawyers: Legal simulations and the art of gamemastering," *Law School Innovation*, January 7, 2009, http://lsi.typepad.com/lsi/2009/01/dungeons-lawyers-legal-simulations-and-the-art-of-gamemastering.html.

15. John Garvey and Paul Maharg, "Standardized clients and SIMPLE (SIMulated Professional Learning Environment): Learning professionalism through simulated practice," paper presented at Future Ed 3, New York Law School, New York, April 15, 2011, http://www.nyls.edu/user_files/1/3/4/30/58/1053/Garvey&Maharg.pdf.

16. Leonard A. Annetta et al., "Serious games: Incorporating video games in the classroom," *EDUCAUSE Quarterly*, no. 3 (2006): 18.

17. Gabe Zichermann and Christopher Cunningham, *Gamification by design: Implementing game mechanics in Web and mobile apps* (Sebastopol, Calif.: O'Reilly Media, 2011), xiv.

18. Laala Al Jaber and Sarah Randag, "Dutch law firm uses video game to evaluate law graduates' talent," February 17, 2011, http://www.abajournal.com/news/article/dutch_law_firm_uses_video_game_to_evaluate_law_graduates_talent/.

19. Steven E. Higgins, "Multi-touch tables and the relationship with collaborative classroom pedagogies: A synthetic review," *Computer-Supported Collaborative Learning* 6, 4 (2011): 517.

20. Willie Miller, "iTeaching and learning: Collegiate instruction incorporating mobile tablets," *Rethinking Reference and Instruction with Tablets* in *Library Technology Reports* 48, 8 (2012): 56.

21. L. Johnson et al, *NMC Horizon report: 2013 higher education edition* (Austin, Tex.: New Media Consortium, 2013), 15.

22. Ibid., 18, describing Stanford University School of Medicine's iPad implemention; Alexandra Tilsley, "Lynn University to require all new students to buy iPads," *Inside Higher Ed*, January 15, 2013.

23. Roger V. Skalbeck, "Tech innovation in the academy," *AALL/ITLA White Paper: The New Librarian* (2012): 75. Those created thus far include *Copyright Navigator*, *Citizen Adviser*, *Automobile Search Warrant Advisor*, and *Same-Sex Marriage Advisor*.

24. Ann W. Parks, "Teaching innovations: Simulations, apps—and other new ways to learn the law," *Georgetown Law Magazine* (2012): 30.

25. Ibid.

26. *Lyndsay Virtual Human*, http://www.lindsayvirtualhuman.org/.

27. Johnson, "*NMC Horizon report*," 4.

28. Miller, "iTeaching and learning," 58.

29. Alex Berrio Matamoros and Mary Ann Neary, "Librarians, legal research, and classroom iPads—A winning combination: Integrating iPads into legal research instruction at Boston College Law School," *AALL Sectrum* 17, 1 (September–October): 27–29.

30. Institute on Law Practice Technology & Innovation, Suffolk University Law School. website. http://lawpracticetechnology.blogs.law.suffolk.edu/about-the-institute/.

31. Lawyering in the digital age clinic, Columbia Law School, http://www.law.columbia.edu/focusareas/clinics/digital.

32. Pamela Lysaght and Danielle Istl, "Integrating technology: Teaching students to communicate in another medium," *Legal Writing: Journal of the Legal Writing Institute* 10 (2004): 165.

33. Ibid, 180.

34. One of the first types of instructional technologies instructors began experimenting with several years ago was audience-response systems, such as clickers, and Web-based systems that work in concert with laptops, smartphones, or touch tablets, which allow students to respond to the instructor's questions. Jennifer Prilliman, "Live class tech—Web-Based audience response system," *RIPS Law Librarian*, November 8, 2011, http://ripslawlibrarian.wordpress.com/2011/11/08/live-class-tech-web-based-audience-response-system/.

35. Malcolm Brown, "New horizons: Learning and technology—'in that order,'" *EDUCAUSE Review* (July–August 2009): 62.

Part V

TECHNICAL SERVICES

20

Technical Services 2.0

Edward T. Hart
LAWTON CHILES LEGAL INFORMATION CENTER, UNIVERSITY OF FLORIDA

The technical services department is the heart of the law library. Although known by many different names, such as "collection services," "bibliographic services," "acquisitions," "serials," or "cataloging," it is the slang title "tech services" that best reflects the cutting-edge nature of the services this department provides in the digital age. The "coolness" of the department is evident by the accepted slang title, "tech services." This department represents the backbone of the library and is responsible for such mission-critical aspects as acquiring information resources, organizing those resources, and making them available to the library's users. It is the consistent flow of these resources to users that is integral to the success of tech services. Once these information resources are in place, the other departments of the library pick up the baton and work to connect the users with the information resources through circulation, reference, and instruction.

Advances in information technology support the work technical services librarians perform. It allows these librarians to provide greater, deeper, and faster access to information resources. These advances encourage greater interaction directly with users and a proactive approach to users' requests. This change of perception gives birth to Technical Services 2.0, which overcomes the disconnect between the department and users and the historical reactive nature of the relationship.

In the chapter that follows, we explore how advances in information have played out in tech services. We look at the traditional roles of tech services—cataloging, acquisitions, and collection management—as well as the growing role of electronic resources management. One example is the trend of overlaying the traditional online public access catalogs (OPAC) that are part of integrated library systems (ILS) with a discovery tool that provides Amazon-like features, which allows for the manipulation of results by clickable facets. Users can now enter the simplest of searches, keyword, title, author, or Library of Congress call number, and then pair the results with facets such as location, format, subjects, or time period. Being able to explore the library's holdings in such a way opens possible new connections that transcend the old-fashioned method of shelf browsing.

Like the rest of the library, successful tech services departments must embrace these advances in information technology to ensure the continued quality of our services to our users and parent institutions.

CATALOGING

Since before the rise of integrated library systems (ILS), the governing rules of cataloging are the Anglo-American Cataloguing Rules, known as AACR, specifically the second edition, 1988 revision, known as AACR2.[1] Divided into two parts, AACR2 first sets out the rules for the bibliographic description of library materials, giving guidance on how to represent the bibliographic and physical characteristics of the material. The second part covers access points, including headings and cross-references.[2] But as the growing amount of electronic resources in an ever-expanding variety of formats challenged the AACR2 parameters, there were calls for a new system of cataloging rules. What arose is Resource Description & Access (RDA).

RDA is built upon the foundation of AACR2 as well as other internationally accepted cataloging standards, such as the International Standard of Bibliographic Description (ISBD). RDA sets out how to catalog both traditional resources and digital resources. The metadata, as set by RDA, should be more easily shared across communities that had previously followed different cataloging rules and across different ILSs. RDA standards are palpable to such different metadata schemes as Metadata Object Description Standard, MARC 21, and Dublin Core, and possibly schemes outside of libraries.

For catalogers, RDA opens up efficiencies through the use of machine-captured metadata, such as metadata provided by a vendor, and allows greater integration with vendor systems as well as the electronic products they are selling libraries. When confronting an unfamiliar resource, catalogers will be able to customize their cataloging while also having online tools to access the rules covering the resource.

For users of library catalogs, implementation of RDA should provide a better display. For example, for the same title, the different editions and formats will display together. This includes different translations, abridgments, print format, or electronic formats that would have otherwise led to widely separate records.

The implementation of RDA is, at the time of writing, still in the very early stages. Besides the considerable debate regarding the value of these changes, questions are being asked about the cost of implementing RDA when libraries are already under pressure to be more cost effective.

Supporting the cataloging work discussed above is a cooperative network of catalogers connected by OCLC, a fee-based bibliographic utilities network, which supports the sharing and maintenance of original catalog records. The interdependence of cataloging is aided by the metadata provided by publishers and the work of leading catalogers such as those at the Library of Congress. Catalogers at many law libraries find themselves performing more copy cataloging than original cataloging. At the Lawton Chiles Legal Information Center (LIC), University of Florida, the cataloger usually finds she only needs to create a new bibliographic record when she receives new titles published by the Florida Bar. Over 90 percent of monographs received by the department are copy cataloged by downloading records from OCLC. These records are modified for local purposes, such as adjusting the Cutter number on the call number for the best shelf location or to avoid duplication of a call number. Maintenance of bibliographic records of continuing resources are an ongoing problem and are usually addressed in OCLC.

Classifications Systems

One function of cataloging is assigning a shelf location for a title that places it in context. In other words, titles are grouped together by subject or topic, and when one considers the possible number of topics that may arise under the broad area of law, how to assign such shelf locations is intimidating.

Although there are many classifications systems (e.g., Dewey Decimal Classification, to name just one other), the Library of Congress Classification (LCC) is the leading classification system used in law libraries. Started in 1898 by the Library of Congress, LCC broadly divides the known universe of information into twenty-one classes. For law libraries, the relative class is "K" which is broken out as shown in table 20.1.

Even though the Library of Congress (LC) started this project over a century ago, there is still a slow progression of implementation and change within LCC. For example, within the K class, LC only completed its draft of KZ–KZD subclasses in May 1997, in responses to calls for recognition of the changes in global order going back to the 1970s.[3] The significance of this latest addition to K was the transposition of international legal materials from JX, international law and relations, separating out what had become a large collection of materials during the expansion of international law through the course of the twentieth century.

Table 20.1. Library of Congress Classification (LCC) K Class

K	Law (general), jurisprudence, comparative law, conflict of laws
KD–KDK	United Kingdom and Ireland
KDZ	America, North America
KE–KEZ	Canada
KF–KFZ	United States (including the individual states)
KG–KGZ	Mexico, Central America, Caribbean
KH–KHW	South America
KJ–KKZ	Europe
KL	Ancient Orient
KLA–KLW	Eurasia
KM–KPZ	Asia
KU–KWX	Pacific Area
KWX	Antarctica
KZ	Law of Nations (International Law)
KZA	Law of the Sea
KZD	Law of Outer Space

Subject Headings

Another task during cataloging is the assignment of subject headings, which may become a lost art as original cataloging is dropped in favor of users' contributions, such as descriptive tags. Subject headings are the one word description or a phrase that indicates the theme or topic of the work being cataloged. Subject headings are important when the title of the work does not state the theme or topic contained therein. Like the title and the author's name, subject headings are an access point in the bibliographic metadata that allow a user to locate the work.

Much like the LLC, the Library of Congress promulgates the Library of Congress Subject Headings (LCSH)[4] to provide a structure of uniform headings that connects all the workings a library may have on a particular topic. As of December 2009, there are approximately 317,000 headings established by LC. Usually each title is assigned one to three headings. Many headings are a string of terms that provide a structure for both wider and narrower searches of the topic. These headings are retrieved when conducting a keyword search or a subject search of the OPAC, and once a relative title is found, clicking on the hyperlinked headings in that bibliographic record will bring up all the other holdings in a collection with the same heading, thus on the same topic.

Thanks to the expansion of metadata that can now be captured or created by catalogers for the bibliographic records in ILS, the title, author, and subject headings are not the only access points for keyword searches anymore. Many, if not most, bibliographic records now contain additional entries such as tables of contents and summaries taken from publishers' catalogs that add new access points for finding titles. With newer ILS and discovery tools, libraries may enable their users to contribute their own tags to records, providing additional metadata. The advantage of these tags is that they could build relative connections for a community of users that escape the attention of catalogers. But the addition of subject headings is a real contribution by catalogers that makes research so much easier!

A recent addition to the metadata added by catalogers is genre/form terms. These terms describe the different formats the resource may take. These descriptions can be very helpful during keyword searches when looking for a particular form or source of legal information. Three examples of the eighty genre/form terms now available to catalogers include Administrative Decisions, Hornbooks, and Writs.[5]

INTEGRATED LIBRARY SYSTEM

Probably the greatest hurdle to Technical Services 2.0 is the legacy integrated library system (ILS), also known as library management (LMS), still in use in most libraries today. Marshall Breeding describes them as "disintegrated library system,"[6] as many libraries have had to cobble together a number of separate services that do not necessarily work together to provide the infrastructure needed to manage all the information resources held by the library. Legacy ILSs were designed to provide inventory control of print materials, accounts payables, and a catalog that more or less mimicked the card catalogs of a bygone era.

There are two broad options for an ILS. Proprietary ILS are designed, built, and managed by vendors who license their use to libraries. Examples included Ex Libris (http://www.exlibris-usa.com/), which offers two systems, Aleph and Voyager. Popular among academic law libraries is the systems offered by Innovative Interfaces Inc. (III) (www.iii.com), such as its Millennium ILS.[7] Among firm and corporate law libraries, popular choices include EOS. Web (http://www.eosintl.com/) and SydneyPLUS (http://www.sydneyplus.com/). The vendors of these ILSs maintain careful control of the software coding behind their systems, and they are usually hosted by onsite servers.

A second option for running ILS is open source software. This provides an affordable option, even free in some instances, compared to the licensing fees of the proprietary systems discussed above. Particularly attractive for libraries with small collections, an open source ILS may provide more flexibility as local programmers can customize the software to meet their special needs. An early advantage of open-source software systems was they could be cloud based, so there is no need to have on-site servers, but this advantage has been overcome by newer generations of proprietary systems offering cloud networking. Examples include Koha (http://mykoha.co.nz/home) and Evergreen (http://evergreen-ils.org/about.php). Among law libraries, open source ILS is in limited use; the Indiana Supreme Court Law Library, the London law firm Pinsent Mason, and Brooklyn Law School all use Koha.[8] There are companies such as ByWater Solutions (http://bywatersolutions.com/) that provide management services for these open source systems much like proprietary software companies do for their ILSs.

But not all is well with ILSs and how they fit with libraries' needs in the electronic age. Breeding refers to ILSs as disintegrated. Beside the ILS, he points out that libraries need "electronic resource management utilities [more on this below], OpenURL link resolvers, e-resource knowledge bases, digital collection management platforms, institutional repositories, discovery services."[9] Thankfully, vendors of these systems, third-party vendors, and innovative librarians are stepping up to the challenges to adapt and eventually redesign these systems for the modern electronic age.

The goal for some time has been to build the next generation of ILS that incorporates services that have arisen to meet today's changing needs. Ex Libris is promoting as its answer a system called Alma, which they claim, "supports the entire suite of library operations—selection, acquisition, metadata management, digitization, and fulfillment—for the full spectrum of library materials, regardless of format or location."[10] Instead of an ILS, they consider this new system a unified resource management (URM). Other examples of proprietary systems include OCLC's WorldShare Management Services (http://www.oclc.org/webscale/default.htm), Intota from Serials Solutions (http://www.serialssolutions.com/en/services/intota), and III has its Sierra (http://sierra.iii.com/). An open source, next-generation ILS is offered in the product Kuali OLE (http://www.kuali.org/ole). All of these systems are attempting to bring about a cross-pollination of traditional ILS functions with new tools to manage and push out electronic resources. In tech services, these are interesting times.

The key link for all these new services is a common knowledge base of data—bibliographic, access, vendor, and licensing, just to name a few—that are in formats usable by all the possible modules, including the catalog, ERM, accounts payable, and discovery tools. But Breeding warns that dependence upon one knowledge base makes it unattractive to cherry-pick modules from different vendors to build an ILS that better meets a library's specific needs.[11] So far, adoption of these new services has been slow since most libraries have so much invested in their current ILS that they cannot or will not switch. Texas Tech Law Library has made the change, taking the leap from III's Millennium to Sierra.

DISCOVERY TOOLS

One of the latest trends in Technical Services 2.0 is the replacement of the OPAC provided by the ILS with discovery tools. As well as providing better search features of a wider range of metadata, taking their cues from Amazon, the discovery tool provides easier access to the universe of electronic resources that institutions provide to their users. Early on, discovery tools were merely an overlay of the ILS OPACs and featured facet searching using traditional ILS metadata as well as relevance ranking, spell checks, edit original search, Boolean search, and sorting. Examples of these overlay tools are AquaBrowser, Endeca, and Encore.[12]

But now discovery tools are going a step further as libraries seek to bridge the gaps between their print resources, cataloged electronic resources, and the content to which the libraries have access via licenses with third parties. Two examples are Summon by Serials Solutions (http://www.serialssolutions.com/en/services/summon) and Encore Synergy by III (http://encoreforlibraries.com/).

Libraries want to offer their communities a Google-like experience for the information universe they provide. A rough comparison can be made to the federated search engines of a few years ago. While those engines sent out the same search to the databases to which a library subscribed, the advantage the new discovery tools have is that they gather the metadata of the content of databases into their servers, creating their own metadata collection. A search is no longer slow, relatively speaking, going through every database, but is quickly searching the one discovery database containing the metadata of everything accessible to that particular researcher. Included in the discovery tool metadata are the catalog records from the institutions ILS, so the available print resources are included in the results. Researchers are presented with the results found by the discovery tool. They can use facets to limit the results and then link out to the electronic content if desired.

For the moment, at least, law libraries cannot take full advantage of these discovery tools as the metadata from the dominant players WestlawNext and Lexis Advance, which are not included in the compiled metadata gathered by these services.

ELECTRONIC RESOURCE MANAGEMENT

The predictions continue that law libraries can and will be virtual libraries. Many law firms have already moved in this direction. Because of standards for accreditation set by the American Bar Association (ABA),[13] academic law libraries have been more conservative in their movement in this direction. The ABA still offers the interpretation that dependence on one format may violate the standards.[14] However, the trend is for increasing dependence on electronic resources. The question is how to manage them. Current ILSs were not designed for electronic resources, notwithstanding many innovative attempts to do so.

Electronic resources come in many varieties from single-title serials to collections of serials, to a large database of thousands of titles. And that is before the advent of e-books! So the question is, how can you make, say, the *Federal Standard Tax Reporter* (published by CCH) findable and accessible for a library's users? This particular title is buried in CCH's electronic platform IntelliConnect. Catalogers cannot identify every title in every database, providing access information and current coverage.

To manage electronic resources, many vendors of ILSs have developed services with the generic name of electronic resource management (ERM) to fill the gaps in their ILS. ERM feature acquisitions records that include vendor information such price, licenses, and technical contacts; metadata that describe the resource; public interfaces such as an A-Z list of serials; notifications; and usage statistics. Vendors of ERM, such as Serials Solutions, which licenses 360 Resource Manager (http://www.serialssolutions.com/en/services/360-resource-manager), provide libraries a database of electronic resources that librarians can track, selecting them for inclusion in a public display. The database contains names of vendors, names of databases, and individual titles of electronic resources. Date of coverage is provided, and libraries have the option to customize this based on their particular license for the information resource. An added service Serials Solutions offers is 360 MARC, which provides a basic bibliographic record for the electronic resource that can then be imported into the library's ILS so users can locate the resource when they search the OPAC or discovery tool. Users are provided links that eventually lead to the electronic resource in the vendor's database.

A current problem with setting up an ERM to get the full use of its services is the need to populate its knowledge base, which is often an exercise of repopulating what is already in the library's ILS. This is especially true for vendor data—which brings us back to the discussions above about the new generation of ILSs or URM.

ACQUISITIONS

The acquisition of resources, except in one area, has been remarkably a more steady development of digital services as vendors improved their online systems to embrace Web 2.0. The exception is e-books, which are looked at more closely elsewhere in this book.

One of the greatest advances toward technical services 2.0 is the easier interplay of an ILS with vendor systems. An example of a vendor system is Oasis from Ingram's Coutts Information Services. The advances of the past years have made it easier to upload order records from the ILS to vendor systems and download bibliographic records and invoices to the ILS. These new connections make technical services much more efficient. Another advantage of using a vendor's system is the ability to manage approval plans on their system.

Approval plans are traditional ways for libraries to acquire monographs and are a service provided by book jobbers, such Yankee Book Peddler and Ingram's Coutts. The plans let a library establish a profile with the parameters for the automatic purchase of books, an approval book plan or notice of new titles, and an approval slip plan. The parameters of these plans can include type of publisher, subject matter, and nonsubject matter. For example, the University of Florida's (UF) LIC has an approval book plan with Coutts for the automatic shipment of new titles from university presses that meet the profile of being assigned Library of Congress Call Numbers K, KF, KFF, and KZ and are priced less than $250. UF also has an approval slip plan. On a weekly basis in the Coutts Oasis online system, information for new titles is provided that meets the following profile: exclude titles acquired under the approval book plan, all publishers; meet the subject matter profile of being assigned Library of Congress Call Numbers B65, most of H, most of J, all of K, and Z551-675.L2, or which are Interdisciplinary Studies: Law; and meet the nonsubject profile of having readership level of General—Popular or Graduate—Research, cloth or paper binding, and newly published works or new edition.

In years past, approval slip plans were literally that, a weekly shipment of paper slips representing new titles that librarians reviewed to make selections. Today, the approval slips can be managed on the vendors' systems. Profiles can be established for librarians at the same library, so they only see the slips of direct interest them. Librarians only interested in international law could limit their view of slips for books assigned a KZ call number. Selecting librarians then can use the vendor system to forward titles of interest to the reviewing authority, acquisitions librarian, or place the order themselves based on their libraries' procedures.

Other advances related to vendor systems deal with how they provide more information about titles and doing so in a hypertext environment that allows selectors to view the possible connections, such as other titles in a series. Vendors catering to libraries realize they must provide an experience much like Amazon.

PRINT SERIALS

Historically, most of law libraries' acquisitions budgets went toward print serials. Although this is quickly changing, and nearly eliminated in some private law libraries, the acquisitions and management of print serials, regardless of format, is still an ongoing effort in the electronic age and requires a great deal of attention by technical services, especially in academic law libraries. In contrast to other types of libraries, print serials are especially complicated in law libraries because of the wide variety of forms they take:

- Periodicals—weekly, monthly, quarterly, semiannually, and annually
- Loose-leafs
- Replacement binders
- Replacement volumes
- Pocket parts
- Supplements
- Serial monographs

And then there are the electronic serials, the format most commonly relied upon by users researching current topics. Electronic resources are further discussed in another section.

A description or definition is warranted for some of the terms listed above. It is fitting that because of their importance, some these terms are defined in that ultimate reference of legal understanding, *Black's Law Dictionary*.[15]

- *Loose-leafs:* "A type of lawbook having pages that are periodically replaced with updated pages, designed to cope with constant change and increasing bulk."[16] In other words, a binder for which the publisher periodically sends out replacement pages containing updated content that then need to be interfiled in the binder removing the pages being replaced. An example is the *Federal Standard Tax Reporter* for which CCH issues weekly releases that require filing.
- *Pocket parts:* "a supplemental pamphlet inserted usually into the back of the inside cover . . . to update the material of the main text until the publisher issues a new edition of the entire work."[17] An example is the pocket parts Lexis sends out annually to update the *United States Code Service*. Sometimes the pocket parts become too thick for the book's cover, so it comes as a supplement to stand along with the main volume. Lexis will eventually incorporate the updates into a new main volume.

Print serials are declining as law libraries cancel them for several reasons: budget cuts, the rise of librarians' comfort level with dependence upon electronic access, and the wider acceptance of electronic resources in legal practice. Historically, print serials in the form of reporters, statutes, law reviews, and multivolume treatises were core to every law library. Many firm libraries no longer retain print serials. Many academic law libraries are moving to only collect those print serials of direct impact on their jurisdiction or are used in instruction. Law libraries are depending more than ever on the content in the large legal databases.

One type of serial—student edited journals such as law reviews—is in the middle of an evolutionary change as many universities have discontinued printing and offer open access to their content in accordance with the Durham Statement.[18] The statement calls "for all law schools to stop publishing their journals in print format and rely instead on electronic publication coupled with a commitment to keep the electronic versions available in stable, open, digital formats."[19] The student editors in collaboration with librarians are embracing open access as a way to broaden the impact of their publications as articles are more discoverable on the open Internet than buried behind fee-pay walls.

For now, the major publishers of legal information continue to sell print serials. There is still demand by researchers for print, but the print format does evolve. Recently, Thomson Reuters started to change the format of many of its titles published as one- or two-volume loose-leafs to a comparable number of soft-cover volumes. The company cited feedback from firm librarians who wished to buy the print and preferred the soft-cover annual that did not require loose-leafing. William S. Hein & Co. (Hein) (https://www.wshein.com/) still offers its services as subscription and continuations agent. Hein is probably used by every academic law library as their agent to manage subscriptions for periodicals, especially for law reviews, which with their turnover of student staffs lead to problems with their customer service.

Related to print serials is the growing concern for preservation of these materials. A historical approach to preservation was filming print materials into microformats. This led to the formation of the Law Library Microfiche Consortium (LLMC) (http://www.llmc.com/), which among other products filmed and sold fiche copies of the regional reporters as they fell into public domain. Commercial vendors also contribute to this effort. Hein sells fiche of superseded state statutes, state session laws, and state bar journals. One modern take, now that volume count no longer dominates, are discussions and collaborative efforts for law libraries to work together to share the burden of maintaining print collections. The Legal Information Preservation Alliance (LIPA) (http://www.aallnet.org/lipa/) is coordinating law libraries' retention of last print copies of serials as well as other materials and archiving of all these materials in a repository supported by the widest possible number of partner libraries. A more modern approach is the digitization of content, which is discussed on the following pages.

E-BOOKS

E-books have had a significant impact on tech services. Adopting our work flows for their acquisitions and management is challenging and will continue to be so for some time. The central problem is the inconsistency and lack of uniformity of e-books. Although the market for selling to individual readers is settling enough for mass commercialization, publishers catering to law libraries have not settled on a standard model. To date, too many questions about who has actual control over the acquired content are still unanswered.

At UF, we are partners in a patron-driven acquisitions (PDA) plan for e-books with the Research Center at Florida State University in coordination with our universities' main libraries. We are working with Coutts and its MyiLibrary platform. Since we share a statewide ILS managed by a central office, Florida Virtual Campus, we are in a unique situation of having the bibliographic records for the PDA loaded in our catalogs. These records are discoverable by our users in our OPAC and discovery tools. The users can click through to the e-books contents on the MyiLibrary servers, triggering a purchase when the third user views the e-book's content, not just the table of contents. For the e-books that have been purchased, Coutts provides new records that note the purchase and that overlay the existing records. To these records can be attached order records to help track the purchase.

For a more traditional acquisition of an e-book based on a firm order, the complications arise when deciding where the e-book will reside. Since UF has the account with Coutts and uses their MyiLibrary service, any e-book we wish to purchase through them will reside on that platform. Our cataloger can download suitable bibliographic records or carry out original cataloging that include URLs linking to the content. The acquisitions coordinator can attach order records to process payment. How broken URLs will be fixed in the future is a problem to be addressed.

Some e-books are acquired on a subscription and are accessible via a large database. An example is the CCH Expert Tax Treatise series which is on the IntelliConnect platform. These are often better managed using an ERM, so that licenses, usage, subscription periods, and other data can be recorded, and the ERM can provide bibliographic records that are routinely updated and URLs checked.

COLLECTION MANAGEMENT

The significant advancement in this area that supports progress toward Technical Services 2.0 is the adoption by libraries from commercial enterprises the use of radio-frequency identification (RFID) for means of inventory control of physical collections. RFID in libraries uses nonpowered tags with magnetic fields containing relative data such as title and call number. An advantage over barcodes is that the RFID tag does not have to be in the line of sight. For circulation, merely passing the book containing the RFID tag over a scanner captures the data encoded in the frequency and locates the current record in the ILS and checks out the book, making for a much faster check-out process than when using barcodes. Another, and more interesting from a tech services point of view, use of RFID technology is shelf reading. A scanner can be waved along a shelf, capturing the book's data. Using the ILS, the captured data can be reconciled to identify books shelved out of order as well as books that are missing, since they are neither checked out nor on the shelf. This makes for a much quicker check that materials are in order and are not missing.

CONCLUSION

To paraphrase the somewhat corny phrase, "It's interesting times in tech services!" applies very well. With the multiple generations of both librarians and library management systems overlapping, the development of Technical Services 2.0 comes with fits and starts that depend greatly upon the experiences of librarians, support of the administration, and funding, and is often not a steady evolution across the spectrum of law libraries. But with the institutional mission of providing quality service to our libraries' users, striving toward the creation of Technical Services 2.0 is a goal embraced as a fundamental tenet of our profession.

FURTHER READING

Anglo-American cataloging rules, 2nd ed. 2002. Ottawa, Ont.: Canadian Library Association.

Chan, Lois Mai. 2007. *Cataloging and classification: An introduction*, 3rd ed. Lanham, Md.: Scarecrow Press.

Library technology guides: Key resources in the field of library automation. http://www.librarytechnology.org/.

Welsch, Anne, and Sue Batley. 2012. *Practical cataloguing: AACR, RDA, and MARC21.* New York: Neal-Schuman.

NOTES

1. *Anglo-American cataloging rules*, 2nd ed. (Ottawa, Ont.: Canadian Library Association, 2002).

2. See Lois Mai Chan, *Cataloging and classification: An introduction*, 3rd ed. (Lanham, Md.: Scarecrow Press, 2007). For an outline of the history of AACR, visit the Joint Steering Committee for Development of RDA, *A brief history of AACR*, http://www.rda-jsc.org/history.html.

3. *JZ-KZ: Historical notes & introduction to application* (Washington, D.C.: Library of Congress, 2011), 4.

4. See Library of Congress, Cataloging and Acquisitions, Subject Headings, at http://www.loc.gov/aba/cataloging/subject/

5. *Genre/form terms for law materials: Incorporated into the Library of Congress genre/form terms for library and archival materials*, prepared by the American Association of Law Libraries, Classification and Subject Headings Cataloging Policy Advisory Working Group, 2010, http://www.aallnet.org/sis/tssis/committees/cataloging/classification/genreterms/lawgenreformterms2010final.pdf.

6. Marshall Breeding, "Technical infrastructure in libraries: To tweak or transform?" *Computers in Libraries* (January–February 2013): 21ff.

7. III's Millennium has been a popular option for law libraries because of the diversity of its check in records to meet the different publication formats (pocket parts, loose-leaf, and supplements) common in law libraries.

8. http://wiki.koha-community.org/wiki/KohaUsers/NorthAmerica and http://catalog.brooklaw.bywatersolutions.com/cgi-bin/koha/opac-main.pl.

9. Breeding, "Technical infrastructure in libraries," 21.

10. Alma Overview, http://www.exlibrisgroup.com/category/AlmaOverview.

11. Breeding, "Technical infrastructure in libraries," 21.

12. See "Discovery layer interfaces," http://www.librarytechnology.org/discovery.pl.

13. "Chapter 6: Library and information resources," *ABA standards and rules of procedure for approval of law schools* (Chicago: American Bar Association, 2012), http://www.americanbar.org/groups/legal_education/resources/standards.html.

14. Ibid, Standard 606, Collection and Interpretation 606-2.

15. See Bryan Carson, ed., *Black's law dictionary*, 9th ed. (St. Paul, Minn.: Thomson Reuters, 2009).

16. Carson, "Looseleaf Service," *Black's law dictionary*, 1028.

17. Carson, "Pocket Part," *Black's law dictionary*, 1274.

18. *Durham statement open access to legal scholarship*, http://cyber.law.harvard.edu/publications/durhamstatement.

19. Ibid.

21

Collection Development

Molly (Mary) E. Brownfield
FINNEGAN, HENDERSON, FARABOW, GARRETT & DUNNER, LLP

REDEFINING COLLECTION DEVELOPMENT

What does collection development mean to today's law libraries? One of the first variations of the definition of "collect" in Webster's dictionary reads: "to bring together into one body or place."[1] But collection development encompasses far more than the act of centralized accumulation. It requires the ability to balance budgetary constraints with patrons' information needs. It calls for careful scrutiny of contracts and licensing agreements and the ability to effectively negotiate the same with publishers and vendors. It demands an understanding of one's patrons and their information-seeking habits and pain points.

Law librarians must grapple with a myriad of questions ranging from which titles to cut in order to fulfill bright-line budget constraints, to more nuanced challenges such as assessing whether e-books present a viable long-term solution for their institution. Such questions drive collection development considerations and processes, and while the challenges abound, so do opportunities to meet them. A subsequent and perhaps more apropos variation of the definition in Webster's describes "collect" as "to gain or regain control of." This goal of this chapter is to present the different facets of collection development and to offer solutions—both classic and innovative—for gaining control over this central and dynamic arena of today's law library.

CREATING (AND THEN RE-CREATING) A COLLECTION DEVELOPMENT POLICY

The underlying rule of any successful collection development policy is recognizing that it is a living document and will evolve over time. This is true not only because information is constantly changing—whether in terms of the medium, the provider, and so forth—but also because key factors such as patron base, physical library space, budget, and staff support are far from static.

Knowing one's patrons is critical to shaping a collection that will be well utilized, but how to do this? Depending on the institution—academic, private, government, and so on—it would seem a relatively straightforward process of identification of materials for the collection, but there are inevitably more nuances and dynamics involved than may initially come to mind. In the law school setting, for example, different faculty members have varying subject areas of expertise and scholarship interests. The presence of a visiting professor who has a teaching and publishing focus on law and popular culture one year may be followed by another who specializes in empirical research. Additions and phase-outs to a law school's clinical programs—or from a parallel viewpoint, to a law firm's practice groups—also create shifting collection needs. Such changes place a certain amount of pressure upon the law library to add new materials to the collection, while still maintaining the core collection—and to do it all within the confines of space and budget!

Getting Patron Input—Directly and Indirectly

Luckily, today's law librarians have technology on their side. If a new faculty member is coming on board, consider consulting an online faculty profile from his or her current institution to anticipate subject area specialties. It may be that the faculty member prefers electronic over print, or will be bringing his or her pre-acquired collection of resources—thereby negating the need to purchase resources through the law library, at least immediately. Consider sending an e-mail inquiry to the law librarians at the faculty member's current institution to learn which resources the faculty member uses on a regular basis, and then use that information to plan collection adjustments accordingly.

For law firm libraries dealing with the addition of new practice groups, consider posting a query to your local professional association LISTSERV, or on a corresponding blog to find out the core treatises and databases maintained by peer firms to support the information needs for that practice area. Furthermore, law libraries of all institutional types can seek information both directly and indirectly from their patrons as to the information resources that they would like to see in the library collection.

Advances in online public access catalog (OPAC) interfaces and integrated library system (ILS) statistical reporting tools make it easier than ever to gather patron feedback on the resources that they find useful (or not), and that the law library should therefore maintain. OPAC providers such as EOS International, as well as OPAC enhancement tools such as LibraryThing for Libraries,[2] offer the option to have patrons provide ratings and reviews for titles in the catalog. If the idea of such an option gives you cause for pause, not to worry! These options typically come with librarian control in the form of suppressing a review until approved.

If your patrons prove too disinclined to rate or review titles, consider taking advantage of the circulation usage/tracking reports available through many integrated library system tools.[3] Patrons may not think to tell you how useful they find a particular title, but circulation statistics will keep track of just how many times they check it out—providing invaluable information albeit through a more passive means. Similarly, law libraries can track which books are taken off the shelf and left for reshelving. These books can give a sense of what titles in the noncirculation collection are being used.

Location, Location

Physical space—that is, the "where" factor—presents another key component of consideration for collection development. The issue of shrinking space for the law library has been a topic of much discussion and concern in the American Association of Law Libraries and beyond.[4] Some law libraries have successfully remarketed themselves as social gathering spaces for learning and sharing, in addition to housing information resources. But between the cost of commercial office space in the private sector and the demand for more study carrels and faculty offices in academia, most law libraries are finding themselves in a losing battle over real estate.

The silver lining of limited space is that it presents an opportunity to implement new online solutions such as "ebraries." This textbook's chapter 8 by Ellyssa Kroski, titled "E-books in Law Libraries," sheds further light on how this technology offers a way to save on shelf space. In the context of collection development, physical space considerations influence the shifting balance between print and electronic resources and related collection management decisions, which is discussed in more detail in a subsequent section of this chapter.

Staff Support

Staff considerations also factor prominently into realistically determining collection scope and format. Collection materials must be unpacked, cataloged, labeled, and shelved in a timely and organized manner, all of which requires staff support. Especially in today's digital information environment, patrons expect to find out what resources are available to them with a few clicks of a button and are easily deterred from utilizing the collection if these expectations are not met. It is therefore crucial to have sufficient staff to oversee and carry out the many tasks that translate into the collection's visibility and accessibility.

In today's digital law libraries, technology like SharePoint Team Sites and internal library wikis empower law library staff by providing an online forum to discuss and record processes for collection development and management, from the formal collection development policy to title-specific retention practices.[5] One of the pros of SharePoint is that it allows any authorized member of the staff to make edits and additions to the Wiki pages and Web Parts, while retaining an archive of past versions. This facilitates a team-oriented approach to contributing to collection decisions and tracking changes to policies over time.

Budget—The Bottom Line

Patron considerations help determine the potential universe of the law library collection, while space considerations present physical limitations; but no collection development decision can be made without deference to the budget. No matter the patron demand or the plethora of open space, the dollars must be present to proceed. Budgetary considerations accompany every acquisition—whether it's a $35.00 guide on legal writing or a new database that costs thousands. The current climate of economic uncertainty has only placed added pressure on law libraries of all shapes and sizes, across all institution types, to design and implement a law library budget that reflects the utmost efficiency and accuracy.

Planning and tracking a law library budget always presents challenges; one potential example is how to control expenses for purchases made by patrons and nonlibrary staff that are allocated against a library fund account. Technology has not yet evolved to the point where a buzzer can alert law librarians of such unanticipated activities before they show up on the monthly expense report. New online tools do, however, offer organizational assistance, and organization is vital to staying on top of your budget.

In addition to classic tools such as Excel spreadsheets, newer resources such as electronic resource management (ERM) systems—sometimes offered as part of an ILS—can help law librarians to keep track of stakeholders and requestors involved in purchase decisions for new library resources, as well as renewals. ERM systems offer powerful utility beyond collection development, as is discussed in more detail in the ERM chapter (chapter 22). For the purposes of the topic at hand, law librarians can use them to add purchase stakeholders to approval phases for library resources such that patrons who play a role in purchasing decisions (e.g., practice group leaders or managing partners in the law firm setting or senior faculty members in the law school setting) can approve, reject, or request additional information via a link that submits their responses directly into the ERM, where it is recorded. Not only does this offer a means of streamlining input regarding library purchases, it provides substantive documentation that law librarians can use to support expenditures and address budget concerns.[6]

SELECTION/DESELECTION PROCESS

It's always hard to say good-bye, especially when that good-bye comes in the form of a recycling bin full of books that once purposefully resided on the law library shelves. Weeding, however, is a necessary process and must be done with consistency and regularity to avoid ending up with a cluttered and outdated collection.

A collection that is neat, well maintained, and that follows a logical organization invites patrons just as shelves overflowing with musty, out-of-date titles will likely prove a deterrent. Fear of getting rid of materials, shortage of staff and time, and even good old-fashioned inertia often result in titles residing in the stacks past their "shelf life." This makes it more difficult for patrons to identify current, relevant titles. Certain titles, especially those covering primary law such as court rules, statutes, and evidentiary rules, can even pose a liability if they are out-of-date and left on the shelf without being clearly labeled as "Superseded" or "Past Edition." Do yourself and your patrons a favor—keep your collection in good order. A bulleted list of questions to help guide your decision-making process on what to keep and what to toss follows.[7]

- Does your existing subscription/contract allow you to cancel the title(s) without penalty (e.g., affecting a bulk discount)?
- Does the title route to multiple patrons in your institution? If so, have you contacted those patrons to see if they are still finding the title useful?
- If considering switching to an electronic version, does the print format of the title present unique features or content that is of value to your patrons?
- Is the title available online through a free or a subscription-based resource available to your institution?
- Is your institution a federal depository with obligations to retain certain government documents?
- Does your institution have to deal with gifts, grants, endowments, and so on that mandate certain space and budgetary consideration?
- How much space is available to retain back issues of your periodicals?
- Is there any value for your patrons to keeping superseded editions of the title?
- Do circulation statistics or other indicators (e.g., regular requests for a title, constantly needing to reshelve certain titles) suggest that the title is being used?

As you consider the above questions, bear in mind that technology again presents options—both for efficiently documenting deselection and retention decisions for later reference, and for managing subscriptions in the aftermath of those decisions. Consider the option of using an internal SharePoint site or other wiki tool to record and track of decisions and policies, from the weeding cycle (e.g., staff will run a shelf list every six months and review unused materials for deselection) to specific deselection criteria (e.g., materials that have not circulated or demonstrated other evidence of use, such as need to be reshelved, will be moved to off-site storage or discarded) so that there is a central, accessible archive of that institutional knowledge. ILS tools also allow for inputting title-specific notes on the staff side to help keep track of retention decisions and policy.

Finally, weigh the pros and cons of available publisher/vendor arrangements such as standing orders and library maintenance agreements (LMAs). The law library community has mixed feelings about such arrangements, and not without reason.[8] Many agreements lack transparency and compromise your ability to cut costs and reduce materials when necessary. But if you are able to negotiate with your publishers for transparent agreements that provide true savings and flexibility (for example, semi-annual Swap/Cancellation periods), they can be a beneficial tool for managing your subscriptions and controlling costs. Additionally, consider available selection tools from vendors such as Hein Greenslips, which provide bibliographic information about new legal and law-related titles.[9]

CAPITALIZING ON PEER KNOWLEDGE

Maintaining a relevant collection and informed collection development policy logically benefits from knowledge of what your peers are doing with their collections. The law librarian community has long embraced and promoted the values of sharing knowledge. Today's law librarians possess the advantage of online forums, blogs, and professional association sites that facilitate sharing of collection development knowledge and insight as never before. Resources—including professional association LISTSERVS, online forums, and social media tools—abound for keeping current with peer institutions and their collection development practices. The following bulleted list highlights several such resources:

- American Association of Law Libraries Technical Services Special Interest Section Collection Development Policies Bibliography: http://www.aallnet.org/sis/tssis/committees/acquisitions/collectiondevelopmentpolicies/bibliography/
- American Association of Law Libraries Academic Law Libraries Special Interest Section Collection Development Committee Site: http://www.aallnet.org/sis/allsis/committees/colldev/index.asp

- American Association of Law Libraries Academic Law Libraries Special Interest Section Collection Development Resources on the Web: http://www.aallnet.org/sis/allsis/committees/colldev/cdweb.asp
- Law Librarian Blog: Collection Development Topical Archive: http://lawprofessors.typepad.com/law_librarian_blog/collection_development/

Knowledge of peer practices and collection holdings not only can help inform your own decisions on selection criterion, but also factors into your retention policy. Consider peer resources and open access initiatives when contemplating decisions on canceling, discarding, or converting to online. You may find yourself in a position where space and budgetary concerns require you to cancel subscriptions that you still see patron use. Knowing that those resources remain available through a law library consortium or via interlibrary loan can help you manage your collection with efficiency, recognizing that there are alternative means of meeting your patrons' needs.

Initiatives such as the Durham Statement on Open Access to Legal Scholarship, "which calls for all law schools to stop publishing their journals in print format and to rely instead on electronic publication coupled with a commitment to keep the electronic versions available in stable, open, digital formats" have contributed greatly to digital accessibility to valuable content.[10]

Similarly, knowing what peer institutions are discarding can both inform and support your decisions on which materials your collection may be ripe for weeding. Sometimes seeing that every other law firm library in town has discarded a particular print set can persuade your own patron holdouts to agree that accessing the content through an online database is a viable alternative. Some examples of law library consortiums and networks include:

- New England Law Libraries Consortium (NELLCO): http://www.bu.edu/lawlibrary/collections/consortia/nellco.html
- Mid-America Law Library Consortium (MALLCO): http://www.aallnet.org/chapter/maall/mallco/index.html
- Center for Research Libraries and the Law Library Microform Consortium (LLMC): http://www.crl.edu/collaborative-digitization/law-library-microform-consortium
- Triangle Research Libraries Network (TRLN): http://www.trln.org/

COLLECTION EVOLUTION

The question of maintaining information in a variety of formats looms ever larger as we progress further into the digital age of information. A more in-depth discussion of digitization in today's law libraries can be found in chapter 7 of this textbook, titled "Digitization" by Michelle M. Wu. Digitation of information provides many advantages in terms of accessibility and searchability, as well as saving on staff time (no more loose-leaf filing, anyone?!). However, it also presents many questions and challenges for law libraries as they endeavor to manage the collection. Cost, access, changes in content, mergers and acquisition between publishers and providers, licensing agreements, and much more come into play.[11]

From a collection development standpoint, law librarians must consider these and a number of other factors when deciding how to develop a collection in which the combination of print and electronic resources best supports the information needs of the patrons. In the law firm setting, certain areas of legal practice create a continued need for certain print titles. For example, patent attorneys frequently need to consult older scientific and technical dictionaries and other references. Such materials are typically not available online, and therefore must be retained in print. In the law school setting, faculty members often produce scholarship that requires reference to historical legal materials that may only be available in print through a special or rare book collection.

Furthermore, every electronic resource comes with its own set of rules concerning access, permissible uses, and so forth. Law librarians must take care to read the fine print when purchasing electronic resources for the collection to ensure that they are deployed appropriately. Factors to consider when contemplating the additional of electronic resources for your law library collection include:

- Cost
- Ease of use
- Customer support and training
- Licensing restrictions
- Access options (e.g., are individual logins required for each user? Is IP authentication an option? How does remote/off-site access work?)
- Date range of content
- Delivery and format of content (e.g., can you e-mail or download materials from the database as Word and PDF documents?)
- Administrative control over out-of-contract content

IF YOU COLLECT IT, WILL THEY COME?

The best of collections can still go unused and unappreciated if not make known and accessible to patrons. With that in mind, take steps to ensure that your time, energy, and other resources spent on collection development are paralleled by efforts to construct and maintain a user-friendly collection experience for your patrons. Use your OPAC or library website—or better yet, both—to enable your patrons to easily find and access print and electronic resources.

Always put yourself in the mind-set of the patrons and consider their workflow and search habits. As information professionals, we eat, sleep, and breathe the library's resources. Take stock of the fact that a law student or a new associate may not be fluent in "library-speak" and therefore won't relate to library terms such as "call numbers," "periodicals," and "superseded materials." Present your collection in a way that resonates with your patrons. If they refer to a title by a popular nickname, make sure the catalog record includes that moniker as an alternative title to make the record easier to find.

Work with vendors to deploy available technology to make the collection as accessible and user-friendly as possible. Options include use of IP authentication to circumvent log-in requirements when behind the institution's firewall, compatibility of online resources with mobile apps, and ensuring ready access to remote access log-in instructions.

Capitalize on library orientations and training sessions to highlight points of access to the collection. Include a tour of the physical library space for new patrons so that they possess a baseline level of comfort when navigating the shelves, and take this to the next level by constructing a virtual tour that patrons can access from the library's website. Depending on your institution and resources, this tour need not be on par with the next 3-D blockbuster sensation. Technologies such as Captivate, WebEx, or even a running PowerPoint slide show present feasible and effective options for bringing your collection to patrons, regardless of their location. Also consider taking advantage of free social media tools, such as automatically tweeting new acquisitions on Twitter with the help of Twitterfeed, to raise awareness of the collection.

Such virtual tour options are especially relevant for academic law libraries that offer distance learning or study abroad programs, for law firms with multiple offices sharing a physical library collection that resides in the main office, and for government law libraries supporting patrons who are regularly traveling out of the country for work.

Don't remain wedded to traditional methods of library organization if they don't make sense for your organization. For a private law library supporting a particular practice, such as tax or environmental law, consider whether there is any utility to arranging by Library of Congress call number, or whether there is a superior organizational method for guiding patrons to titles on specific subtopics within the collection.

CONCLUSION

Law libraries need only look back a few short years from the present to see just how much and how quickly change occurs. Though hard to imagine, it was not so long ago that collection development focused almost entirely on print resources. The Internet, wireless technology, digitization of resources, new online platforms, and mobile devices and apps have forever changed the look of collection development, but the mission remains much the same.

As today's and tomorrow's law librarians, we must plan and maintain a collection that meets our patrons' information needs, fits in our available space, can be supported by existing staff, and stays within the budget. We must remain attuned to shifts and trends in the legal information world and let those inform—but never dictate—collection development decisions. Finally, we must take advantage of available technology, ranging from Web-based tools to social media sites, to facilitate these tasks and to assert and maintain control over the law library collection.

FURTHER READING

American Association of Law Libraries Technical Services Special Interest Section. "Law library collection development policies: Policy Documents and Resources Acquisitions Committee." http://www.aallnet.org/sis/tssis/committees/acquisitions/collectiondevelopmentpolicies.

Gregory, Vicki. 2011. *Collection development and management for 21st century librarians.* New York: Neal-Schuman.

Hodnicki, Joe, and Mark Giangrande, eds. "Collection development topical archive." *Law Librarian Blog*, http://lawprofessors.typepad.com/law_librarian_blog/collection_development/.

Lastres, Steven. June 2011. "Collection development in the age of the virtual law firm library: How to expand your footprint beyond your 'walls.'" *AALL Spectrum* 15 (8): 20–23. http://www.aallnet.org/main-menu/Publications/spectrum/Archives/Vol-15/No-8/pub-sp1106-Collection.pdf .

Lenz, Connie, and Helen Wohl. 2008. "Does form follow function? Academic law libraries' organizational structures for collection development." *Law Library Journal* 100 (1): 59–116. http://www.aallnet.org/main-menu/Publications/llj/LLJ-Archives/Vol-100/pub_llj_v100n01/2008-04.pdf.

Russell, Gordon, and Michael Chiorazzi. 2002. *Law library collection development in the digital age.* New York: Haworth Information Press.

NOTES

1. *Merriam-Webster's Collegiate Dictionary*, 11th ed., s.v. collect.

2. See FAQs: General on Library Thing for Libraries at https://www.librarything.com/forlibraries/about.

3. See, for example, Jennifer E. Knievel, Heather Wicht, and Lynn Silipigni, "Use of circulation statistics and interlibrary loan data in collection management," *College and Research Libraries* (January 2006): 28–49. http://crl.acrl.org/content/67/1/35.full.pdf. The article focuses on circulation in a university library setting and provides a useful case study on utilizing circulation statistics to support collection development decisions.

4. See Lucy Rieger, "The new normal library space," *AALL Spectrum* 16, 3 (December 2011), http://www.aallnet.org/main-menu/Publications/spectrum/Vol-16/No-3/new-normal.pdf; State, Court, and County Law Libraries Special Interest Section Toolkit Site on Space Issues, http://www.aallnet.org/sis/sccll/toolkit/spacing.htm .

5. See, for example, a 2012 AALL annual conference program titled "AALL/ILTA program: SharePoint as a showcase for library resources and knowledge repositories," which mentions use of SharePoint in "creating team sites to increase efficiency for collection development" in the program description, http://aall12.sched.org/event/4dec3992d961d501e5f33af931c6b6f0#.UQWvmvKBqSo.

6. For a practical guide to implementing an ERM system, see Nancy Scibelli Bouthilet, "Implementing an ERM system: A practical guide to implementing an electronic resource management system in your law firm," *AALL Spectrum* 17, 7 (May 2012), http://www.aallnet.org/main-menu/Publications/spectrum/Archives/Vol-16/No-7/erm.pdf.

7. Another useful resource to consult for retention policy and deselection decisions is the PowerPoint from an AALL 2009 ALL-SIS alternative program titled "Weed, shelve, or store: Making the hard decisions," by Michelle Pearse, http://www.aallnet.org/sis/allsis/annualmeeting/2009/weeding_periodicals.pdf.

8. See, for example, John P. Joergensen, "Deal or no deal? Are LMAs really in the best interest of law librarians?" *AALL Spectrum* 15, 5 (March 2011), http://www.aallnet.org/main-menu/Publications/spectrum/Archives/Vol-15/No-5/pub-sp1103-LMA.pdf.

9. See http://www.heingreenslips.com/faq.php for more details about Greenslips.

10. The Durham Statement on Open Access to Legal Scholarship: http://cyber.law.harvard.edu/publications/durhamstatement.

11. For a thoughtful discussion of this topic, see Michelle M. Wu, "Why print and electronic resources are essential to the academic law library," *Law Library Journal* 97, 2 (2005): 233–256, http://www.aallnet.org/main-menu/Publications/llj/LLJ-Archives/Vol-97/pub_llj_v97n02/2005-14.pdf.

22

Electronic Resources Management and User Authentication

Catherine M. Monte
FOX ROTHSCHILD LLP

Electronic resource management (ERM) "is a term that refers to the practice of managing data and metadata about electronic resources acquired by libraries."[1] Electronic resources are materials made available electronically in a digital form, such as an e-book, database, or journal article, and include text as well as images.[2] These e-resources are accessible on computers or other electronic devices, including e-readers and tablets.[3] University and other academic libraries have hundreds if not thousands of these electronic subscriptions, and all titles need to be tracked and managed in a systematic way. Private law firm libraries have similar needs on a smaller scale and also require detailed usage statistics for client billing and training. Ideally ERM is a seamless integrated solution encompassing all stages in an e-resource lifecycle—whether renewing digital journals and books, analyzing new electronic titles, organizing other Web-based tools (aggregators, subject webpages, etc.), or gathering usage statistics on these resources.[4] With the ever-growing shift toward online subscriptions, budgets focused on new digital acquisitions, and the increase in staff time allocated to related projects, law libraries need to better manage their digital inventory.

Although managing the acquisition of print resources involves a standard methodology and protocol, managing the acquisition and workflow of e-resources is quite complex and undefined. In addition to content analysis and budgetary approvals, there are also the accompanying technological challenges of working in an electronic world. These challenges involve gaining a better understanding of users' demands, needs, and expectations. In this digital era, it is essential for librarians to understand the differences in print vs. electronic subscription options, and they must also be better prepared to navigate/negotiate through the world of copyright and digital rights management.

This chapter provides a basic overview of the ERM life cycle, describes a variety of current ERM systems, and discusses what the future of ERM will mean for legal libraries. Tracking workflow across all of these systems is complex and challenging. Current ERM technologies endeavor to solve some of the basic problems in workflow, but there is no one system that is a cure-all. The discipline of ERM is currently in a state of flux and it will continue to mature and develop.[5] This chapter strives to elucidate a better understanding of the tool options so you will be able to choose the best solution that will mesh with your library needs today and into the future.

BACKGROUND OF CURRENT INDUSTRY

Since 1992 there has been a rapid evolution of electronic titles, especially in the realm of current awareness news.[6] In addition, law librarians have experienced an increase in demand for e-books and other electronic media with a result in increased budgets and expenditures.

With these changes comes an even greater need to better manage the collection development process and better understand the use of these services—hence the need for an ERM system. In fact, the American Lawyer 2012 survey debuted questions about the adoption of integrated library systems (ILSs), news aggregation, and online monitoring products. Approximately 80 percent of law firm respondents are currently using an online monitoring system. Support for e-books and e-readers also increased.[7]

Law libraries currently operate with both print and electronic collections. This dual environment creates a host of complexities on many workflow fronts: physical space, budget, user interfaces, staffing skills, and access control for individual and group passwords and licenses. With the shift to purchasing more electronic media, librarians need to adjust their approach toward electronic resource management and support this growth and change.[8]

COMPONENTS OF A FULL SERVICE ERM SYSTEM

Fully functioning or all-encompassing ERM systems offer solutions to support the complexity of managing digital items. The following are primary functional components ideally found in such a system:

1. Workflow management (also known as project management)
2. Selection of e-resources:
   a. Acquisitions and collection development of e-content
   b. Budget and financial management

3. Licensing and copyright
4. Organization of e-resources
5. Access to e-resources
   a. User authentication
   b. Passwords
6. Management of e-resources: Statistics and reporting
7. Integration across systems[9]

ERM = PROJECT MANAGEMENT

Organizing e-resources throughout their overall life cycle involves detailed project management skills. There is technical preparation in terms of hardware, software, and staffing needs. There are phases of configuration, testing, rollout, and training plans. And with all projects of any size, there is a need for proper messaging, communications, and change management. Feedback is also essential to ascertain if users are satisfied with content, design, and interface functionality.

SELECTION OF E-RESOURCES: ACQUISITIONS AND COLLECTION DEVELOPMENT OF ELECTRONIC CONTENT

The process of acquiring a new resource can be initiated in several ways. A user may read an advertisement for a specific product; a user may attempt to click on content he or she does not have access to and receive a marketing e-mail solicitation from a vendor; an upgrade may need to be made to an existing resource (potentially involving an increase in cost) and this prompts the librarian to explore other options. Or an entirely new product has been launched that seems to overlap with existing content. Each of these instances requires organization, standardized protocols, and an action (also known as project) plan to obtain an objective assessment of content scope, user needs, and so forth.

Once a decision has been made to explore a potential new resource, the analysis phase begins. Often this involves comparing the new option with existing resources that are part of the current collection. Librarians need to research reviews of the product and often need to conduct beta tests to ascertain functionality and fit. It can be helpful to have a standard template for comparison as a starting point as well as a ranking system for comparing multiple tools. Budgetary implications such as overall costs and per user/license costs are important to consider here as well. ILSs such as Innovative Interfaces Decision Center (http://www.iii.com/products/decisioncenter.shtml) assist with collection management workflow by making recommendations based on current titles and topics.

As previously discussed, a standard approach to acquisitions is key for any title or subscription, but particularly an electronic title. Detailed functionality is required to capture key fields such as subscription effective dates, expiration dates, and licensing terms. Equally important to capture are costs associated with the initialization of the e-resource, price increases over a multiyear subscription contract, and any potential hardware or software expenditure. Budgeting and financial analysis now take on even greater significance. Digital versions of items often come with higher price tags, and libraries must continue to justify expenditures and control costs. Regular, ongoing analysis of existing collections and cancellation of underused services is paramount.

There are a variety of ways ERMs can minimize costs and control budgets, including:

- Chargebacks: requiring users to include a reference for incurred charges (i.e., client/matter)
- Validation of charge codes, client numbers, and project descriptions
- Charge code lists that autofeed and update a pick list of current and valid codes

LICENSING AND COPYRIGHT

Copyright

One area that differs greatly from print and requires detailed review and consideration is copyright. Compliance with granted digital rights and restrictions will determine the types of tools a library can offer to users—whether it is on an individual, group or institution wide level. The management of IDs and passwords and the sharing of content both internally and externally can put the librarian and organization at risk. Therefore it is critical that librarians share the restrictions and compliance language with end users and articulate these messages wherever possible to reduce risk for their respective organizations. Several ILSs include modules that capture this information and have the ability to "push" terms to users. The Copyright Clearance Center also has a tool to support this function: RightSphere[10] (http://www.copyright.com/content/cc3/en/toolbar/productsAndSolutions/rightsphere.html). Other info on copyright in law libraries can be found in chapter 3.

Licensing

Similarly, knowing what your organization requires in terms of digital use is essential when negotiating license agreements. When feasible, librarians should ask vendors for specific standardized language to support these needs. The library staff needs to be familiar with how e-content may be restricted, and often there is an obligation to inform each end user about limitations for reuse or distribution. Electronic distribution rights will vary widely—especially for sharing content beyond the organization itself—whether to colleagues, clients, or on intranets/ extranets, and websites.

Keeping track of the growing number of electronic licenses is another challenge. Integrated library systems (ILSs) such as Innovative Interfaces ERM, EOS International ERM, and others, have modules to support this area of management. These tools are able to centrally store license terms for each digital subscription in a standardized way and can automatically report on upcoming renewals/expiration.

For more information about licensing specifics, see the American Association of Law Libraries website: Principles for Licensing Electronic Resources (http://www.aallnet.org/main-menu/Advocacy/recommendedguidelines/licensing-electronic-resources.html) for guidelines and links to further reading.

ORGANIZATION OF E-RESOURCES—ACCESSIBILITY

The details on how users access and navigate through a particular resource are paramount to acceptance and success. Sign-on protocols should be straightforward with clear tech "Help." Likewise, research assistance areas and user interfaces (UIs) need to be intuitive. Training is often needed to ensure thorough awareness of content and advanced functionality. Usage statistics can be very informative about how to pinpoint individual training issues or trends among user groups such as students, practice areas, or office locations.

There are several ways librarians can organize the plethora of resources for end users. A–Z Systems are Web-based tools that automatically organize links into an alphabetical list. A few examples include:

- EBSCO A-to-Z (A–Z): http://www2.ebsco.com/enus/ProductsServices/atoz/Pages/index.aspx
- California Western School of Law: http://libraryguides.cwsl.edu/a-z_resources
- Serials Solutions' 360 Core (A–Z): http://www.serialssolutions.com/en/services/360-core

Link resolvers provide immediate access to the full text of titles from a citation, abstract, link, or reference source using seamless authentication.

- EBSCO's LinkSource: http://www2.ebsco.com/en-us/ProductsServices/linksource/Pages/index.aspx

- Ex Libris's SFX: http://www.exlibrisgroup.com/category/SFXOverview
- Innovative Interface's WebBridge: http://www.iii.com/products/webbridge_lr.shtml
- Serials Solutions' 360 Link: http://www.serialssolutions.com/en/services/360-link

Subject indexes and webpages are topical indices organized by major subject categories. Resource materials and guides are organized by topic often to benefit a specific subgroup.

- Judicial Branch of Arizona, Maricopa County Superior Court Library http://lawlibrary.maricopa.gov/search/y
- ProQuest Research Library: http://www.proquest.com/en-US/catalogs/databases/detail/pq_research_library.shtml
- PubMED: http://www.ncbi.nlm.nih.gov/pubmed
- LibGuides by SpringShare LLC: http://springshare.com/libguides/

Browsing lists are directories or catalogs of links arranged to be read "at a glance":

- Seattle University School of Law Library: http://lawlibguides.seattleu.edu/
- Library of Congress Databases and E-resources: http://www.loc.gov/law/find/databases.php

USER AUTHENTICATION AND ADMINISTRATION OF DATA

Being responsible for e-resources also involves a certain degree of database administration skills. Academic libraries typically use a software programs called EZproxy to validate students and faculty and "authenticate" their login and password credentials into the online network. EZproxy provides these verified users with remote access to Web-based licensed content, allowing users at home or elsewhere to log in and gain access to library resources.

It also captures detailed usage statistics using standard Web log analytics and tracks which databases are being used remotely to justify budget expenses. See the following websites for overview of features, benefits, and additional technical details.

- University of Virginia School of Law: http://libanswers.law.virginia.edu/a.php?qid=214127
- Northern Illinois University College of Law: http://law.niu.edu/law/library/help_faq1/ezproxy.shtml

Law firms typically use a metering or monitoring tool such as Research Monitor and Onelog for user authentication and statistics. These tools help to organize user IDs and passwords, verify client/matter numbers, and provide usage statistics, among other features. Passwords can be individual or group based and with the increase of electronic titles the need for an organizational system becomes essential. Often user IDs are sold in groups or sets (e.g., 0–5 users or 1–25 users) and determining open licenses can be cumbersome.

Having an automatic process to authenticate users on a granular level—office, department, practice area, job title, or similar grouping—is part of the basic functionality. Authentication can be defined as a process in which a system validates the identity of a user who wishes to gain access to a resource and hence is critical to ensuring robust security.[11] Other benefits include eliminating the risk of lost or forgotten passwords and lessening the risk of computer hackers stealing passwords.[12] See figure 22.1, which illustrates a client validation screen from Research Monitor.

Additional information about resource metering/monitoring tools:

- Innovative Interfaces ERM: http://www.iii.com/products/electronic_resource.shtml
- Lookup Precision: http://www.lookup-precision.com/cost-recovery/portal.aspx

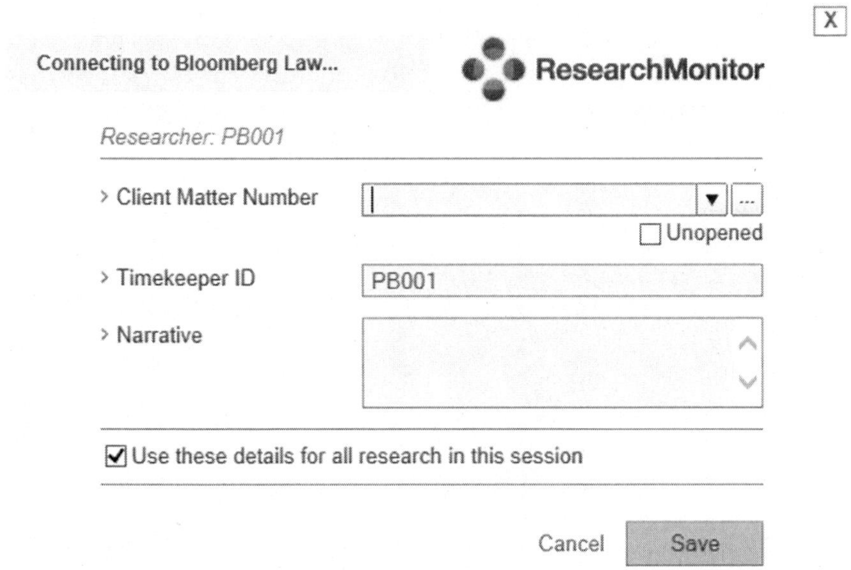

Figure 22.1 Research Monitor Client Validation Screen. *Courtesy Priory Solutions Limited*

- Onelog: http://www.onelog.com/
- Research Monitor: http://www.priorysolutions.com/ResearchMonitor
- Serial Solutions (management piece): http://www.serialssolutions.com/
- SWETS: http://www.swets.com/

MANAGEMENT OF E-RESOURCES: STATISTICS AND REPORTING

Maintenance

Ensuring proper performance of e-resources is also challenging. Librarians must be familiar with related technologies and understand how each resource works both for the end user and on the server side. They will be called upon to troubleshoot various issues such as administration setup, password and IP authentication, and perhaps desktop software that may be required for reader access. A system for cost recovery may also need to be established to capture per use fees, which will be informative when renewal needs are being considered.

Renewal

Analysis of usage and statistical trending are key components when considering renewals of any e-resource. Statistical reports and analysis can determine a return on investment (ROI) for a title and hence budget needs and cost implications. Having a system in place that captures and tracks basic data such as expiration dates of e-resources allows for better organization and management of license agreements.

Statistics and Reporting

Statistics and reporting also play a critical role in e-resource administration. Usage metrics provide evidence for maintaining and renewing an existing resource, which dovetails into budget preparation. Statistics provide references for which tools are more highly utilized by certain subsets of users and can pinpoint training needs. ROI figures can assist with promoting the library's tools to an organization and also showcase initiative in prudent cost consciousness and strategic planning. There are other advantages as well. Reports compiled and received directly from vendors are often incomplete and one dimensional with little option for customization. Using a metering tool allows unfettered access to usage on a vendor independent platform so the information is more comprehensive. This allows librarians to see a 360-degree view of their e-resources.

Librarians can also take this information one step beyond these statistics and apply more context and analysis for a practice group or department annual report. In addition, evidence of usage comes in handy when questions arise as to client charges and costs per session in law firm libraries. See the screen shot in figure 22.2 from Priory Solutions Smart Recharge for a summary depicting monthly database charges.

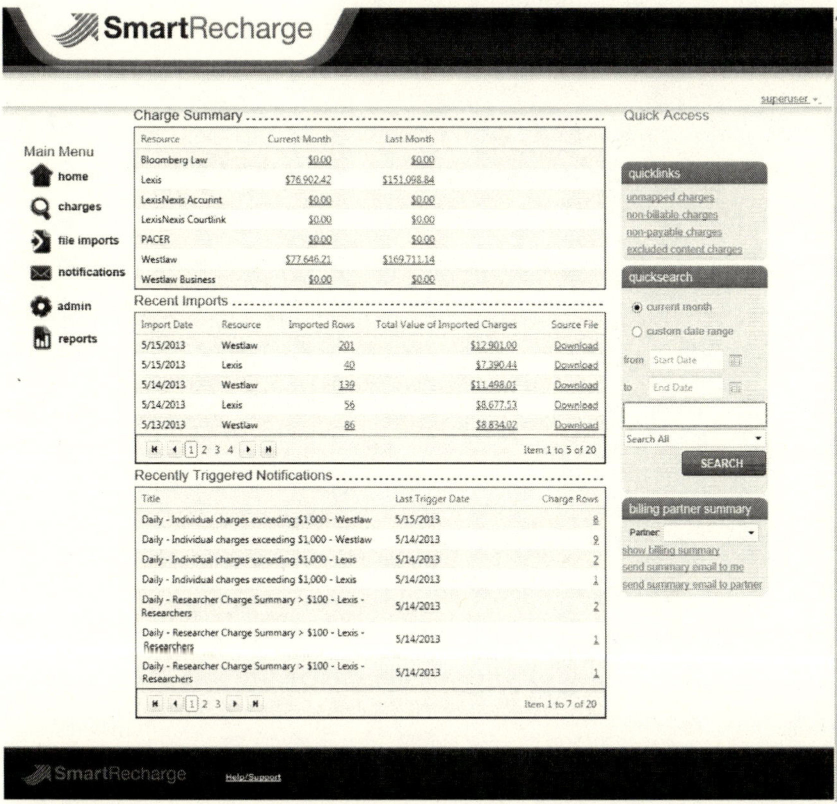

Figure 22.2 Smart Recharge Billing Summary. *Courtesy Priory Solutions Limited*

INTEGRATION ACROSS SYSTEMS: TECHNOLOGY IMPLICATIONS

Coupled with the content differences between print and electronic are the technology challenges associated with digital titles and how end users adapt to these varied platforms. ERM systems can help in several ways. They can control how users perform research, for example, by preventing access to out-of-contract content. They also have the ability to display customized messages and notifications, for example, to inform users about alternative resources. ERM systems can also help to build more awareness about features of a specific e-subscription.

Then there is the added component of integration across systems. Quality assurance and standardization of all related data also plays a role. ERMs usually have modules to assist with linking usage to billing systems (such as ELITE) to streamline the capture and billing of e-resource charges. These modules automate the workflow for download, processing, and reporting of data. Additional features include the option of viewing summaries of specific charges by session, matter, or office as depicted in the Onelog screenshot shown in figure 22.3 that includes session detail. Alerts can also be set up to inform users when they are nearing budget thresholds.

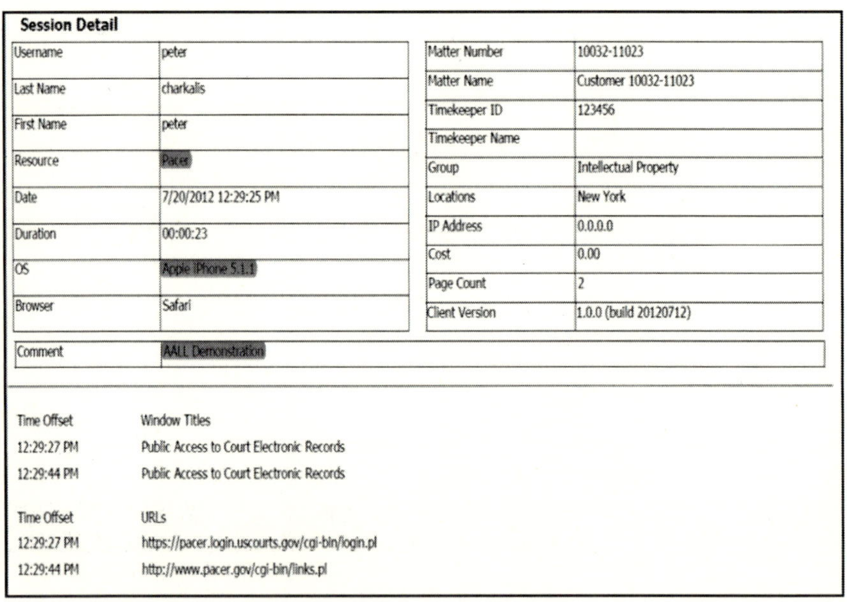

Session Detail

Username	peter
Last Name	charkalis
First Name	peter
Resource	Pacer
Date	7/20/2012 12:29:25 PM
Duration	00:00:23
OS	Apple iPhone 5.1.1
Browser	Safari

Matter Number	10032-11023
Matter Name	Customer 10032-11023
Timekeeper ID	123456
Timekeeper Name	
Group	Intellectual Property
Locations	New York
IP Address	0.0.0.0
Cost	0.00
Page Count	2
Client Version	1.0.0 (build 20120712)

Comment	AALL Demonstration

Time Offset	Window Titles
12:29:27 PM	Public Access to Court Electronic Records
12:29:44 PM	Public Access to Court Electronic Records

Time Offset	URLs
12:29:27 PM	https://pacer.login.uscourts.gov/cgi-bin/login.pl
12:29:44 PM	http://www.pacer.gov/cgi-bin/links.pl

Figure 22.3 Onelog Session Detail. *Courtesy Priory Solutions Limited*

Librarians also need to factor in how to link and integrate resources within an intranet page or information portal. Whether utilizing an A-Z list of alphabetical titles or a grouping of resources by topic or practice, users should be able to easily locate the information they need. Ideally automatic feeds of relevant data and current updates would be part of this workflow.

ADDITIONAL FUNCTIONALITY AND BENEFITS OF ERMS

Training and Awareness of Resources

Librarians must be aware of upgrades to existing resources and often retrain users on new searching platforms. With practice-specific print resources there is often a learning curve for new users but with electronic resources, it is usually greater. Often a library will have several tools to support a group, and training becomes paramount in order to understand the pros and cons in addition to content overlap of each option. Users are hesitant to learn a new system as they consider it an inconvenience. Now more than ever, librarians are being called upon to understand the nuances of each tool, to ensure people use the right resources for content and budgetary reasons, and to be able to clearly articulate why one should be purchased over another. Targeted "just in time" training in various forms such as online guides, e-learning modules and tutorials, podcasts, and the like become part of the overall curriculum. User behavior and user preferences are also taken into account. Along with training, awareness building for these resources publicizes the library's value to the organization.

Mobility

Another benefit of ERM systems is compatibility with mobile apps. Mobile access to e-resources broadens the availability of research tools beyond a PC networked environment. Users can have easy access to the tools they utilize in a platform they prefer (e.g., iPad, iPhone) along with the secure user authentication organizations require (see, for example, the Onelog Mobile product http://www.onelog.com/modules/mobile). As an added bonus, this functionality will work even if the main tech network is down. Statistical reports also provide the librarian with a complete view of data usage across all access points: network, Citrix, and mobile devices.

FUTURE OF ERM AND LAW LIBRARIES

How will ERM systems continue to evolve? Here are a few logical predictions. The current dual world of print and electronic information will continue to morph into primarily digital library collections. The iPad and e-readers such as the Kindle and Nook will gain even more in popularity in the legal market, and sales for e-books will continue to rise and eventually surpass print books. These trends will impact the legal publishing industry and will bring more complex license negotiations for librarians. Consolidation of vendors in the legal information market will most likely occur in the next five to ten years.

Technological advancement and improvement will take place with enhanced computer interface design. Librarians will increase their utilization of messaging and alert functionality in ERM and other systems to communicate with users as they are navigating through pages and resources. This will assist users in finding what they need more quickly, as it will be built into their workflow. The library staff will have opportunities to solicit feedback more directly using this functionality—whether via a survey upon sign-off or during an active research session. Search and discovery tools (e.g., Enterprise Search) will have more direct meaning for libraries and e-resources.

Virtual reference desk staff already use e-mail, online chat, wikis, and related tools to communicate with end users. These trends will continue to increase, as will the demand for 24/7 access to all materials. They will offer more distance education, utilizing course management software. Likewise, staffing needs, titles, and skill sets will also change in response to the proliferation of e-resources.

As is evident, technology-based customer service, support, and training will be a large part of the librarian's role. Staff members will need to provide assistance with use of computers, printers, and software along with research training. Strong negotiating and licensing skills will be key, as will familiarity with metadata standards. Successful library professionals will possess a "package" of information management skills. Collaboration and integration of functions will occur between subject specialists, electronic resources staff, and information technology support.

What will this mean for the future of law librarians and ERMs? The consolidation of certain roles will result in greater efficiency in terms of time savings and streamlining workflow. For example, subject matter experts will be able to understand collection development assessment, analysis, and licensing, in addition to understanding research protocols. This combination of skills will provide a more valuable service to end users. The University of California has a good example of the roles law libraries will need to fill in the next decade that can be found at http://www.cdlib.org/services/collections/licensed/. With e-resource information management, content analysts are key, but those with digital rights acumen, negotiation experience, knowledge of metadata standards, and superior customer service will thrive and be successful in this new paradigm.

FURTHER READING

Articles, Blogs, and Websites

"2012 Law Librarian Survey." ALM Media Properties, LLC.

Accardo, Janet. 2007. "Electronic resource management." *Practice Innovations: Managing in a Changing Legal Environment* 8 (3).

Bouthilet, Nancy Scibelli. May 2012. "Implementing an ERM system." *AALL Spectrum.*

Breeding, Marshall. July–August 2008. "Helping you buy: ERMS." *InfoToday.*

———. January 2004. "The many facets of managing electronic resources." *Info Today.* http://www.infotoday.com/cilmag/jan04/breeding.shtml.

Collins, Marian, and Jill E. Grogg. March 1, 2011. "Building a better ERMS." *Library Journal.*

DYNIX White Paper. "Managing the challenges of electronic resources." June 16, 2004.

Farb, Sharon E., and Angela Riggio. 2004. "Medium or message? A new look at standards, structures, and schemata for managing electronic resources." *Library Hi Tech* 22 (2): 144–152.

Kingan, Rory. 2010. "Electronic resource management systems: Manage online library spending and increase the value of the library." *Legal Information Management* 10 (4): 271–274.

Lastres, Steven A. June 2011. "Collection development in the age of the virtual law firm library." *AALL Spectrum.*

Myers, Anne. "Innovative law users group's list of law libraries with innovative systems." http://libftp.nyls.edu/iiilawlibs/iiilawlibs.html. *Innovative Law Users Group.* Includes many academic institutions with details of which modules each owns.

O'Grady, Jean July 10, 2012. "ALM 2012 law library survey: Budgets, billables, insourcing and outsourcing up. Online cost recovery down. embedding and ebooks emerge." *Dewey B Strategic.* http://deweybstrategic.blogspot.com/2012/07/alm-2012-law-library-survey-budgets.html.

———. November 6, 2012. "Digital resource management in a mobile world: Onelog Mobile—There's an app for that!" *Dewey B Strategic.* http://deweybstrategic.blogspot.com/2012/11/digital-resource-management-in-mobile.html.

Pearson, Sarah. 2011–2012. "Electronic resources management (ERM) systems showcase." *HeinOnline* 37 (1): 36–38.

Books

Conger, Joan E. 2004. *Collaborative electronic resource management: From acquisitions to assessment.* Westport, Conn.: Libraries Unlimited.

Elguindi, Anne C., and Kari Schmidt. 2012. *Electronic resource management: Practical perspectives in a new technical service model.* Oxford: Chandos.

Schmidt, Kari. 2006. *Selecting and managing electronic resources: A how-to-do-it manual for librarians.* New York: Neal-Schuman.

Weir, Ryan O., and Library and Information Technology Association. 2012. *Managing electronic resources: A LITA guide.* Chicago: ALA TechSource.
Yu, Holly, and Scott Breivold. 2008. *Electronic resource management in libraries: Research and practice.* Los Angeles: California State University.

NOTES

Special thanks to Grace Boivin and Elisa Ferrante who assisted me in the research and production of this chapter.

1. Darlene Hawthorne, "NASIGuide: Electronic resource management," North American Serials Interest Group, March 2005, http://www.nasig.org/publications_erm .cfm.
2. www.library.fau.edu/depts/ref/vocab.htm.
3. http://en.wikipedia.org/wiki/Ebook.
4. Maria Collins and Jill E. Grogg, "Building a better ERMS," *Library Journal* (March 1, 2011). This article includes survey results on ERM priorities and also a detailed chart on tools.
5. Ryan O Weir and Library and Information Technology Association, *Managing electronic resources: A LITA guide* (Chicago: ALA TechSource, 2012).
6. Emma Gardner, "Lean back 2.0: A visual look back at digital publishing in 2012," Economist Group, December 19, 2012,. http://www.economistgroup.com/leanback/advertising/a-visual-look-back-at-digital-publishing-in-2012/
7. "ALM 2012 law library survey: Budgets, billables, insourcing and outsourcing up. Online cost recovery down. Embedding and ebooks emerge," *Dewey B Strategic*, July 10, 2012. http://www.almlegalintel.com/SurveyDescription.aspx?id=Q/K8nmK4pG4=&type=fEFglaD+grg=.
8. Anne Elguindi and Kari Schmidt, eds., *Electronic resource management: Practical perspectives in a new technical services model* (Oxford: Chandos, 2012).
9. Collins and Grogg, "Building a better ERMS." This article also includes survey results on ERM functionality.
10. http://www.copyright.com/content/cc3/en/toolbar/productsAndSolutions/right sphere.html.
11. http://hitachi-id.com/concepts/authentication.html.
12. http://www.ehow.com/facts_7482699_definition-automatic-user-authentication .html.

Part VI

KNOWLEDGE MANAGEMENT

23

Knowledge Management

Steven A. Lastres
DEBEVOISE & PLIMPTON LLP

Don MacLeod
DEBEVOISE & PLIMPTON LLP

WHAT IS KNOWLEDGE MANAGEMENT?

Rapid changes in technology and the way organizations conduct business are propelling librarians and informational professionals forward to a world where they must adapt to new ways of thinking about and managing the information over which they are stewards. This changing world presents new opportunities for librarians. To take full advantage of these opportunities, however, we must align ourselves with the language and values of the community we serve. It is not enough to understand the needs of our clients—we must also understand the organizations we serve.

Today's successful organizations employ knowledge workers, individuals who are valued for their ability to use and share knowledge within a specific subject area. Knowledge workers benefit the organization by making change happen. As they perform better, change happens faster. To anticipate these changes, librarians must transition from managing books, magazines, and databases to managing knowledge and supporting the information needs of the knowledge workers in the organizations we serve. Since law librarians outside of the academic setting are a part of a law firm's ecosphere, they need to adapt to the new business realities that lawyers face. By adopting a business perspective, librarians can better serve the organizations where they work.

In short, law librarians find they need to adapt traditional skills to a changed marketplace. And for many librarians, the most promising professional next step is to become knowledge managers.

One approach that enables librarians to achieve these goals is knowledge management. The precise definition of knowledge management (KM) is an elusive one, but one pillar of knowledge management holds that it is "the leveraging of the organization's collective wisdom (know-how) by creating systems and processes to support and facilitate the identification, capture, dissemination and use of the organization's knowledge to meet its business objectives."

TYPES OF KNOWLEDGE

Knowledge is distributed and dispersed throughout the organization as either explicit or tacit knowledge (see figure 23.1).

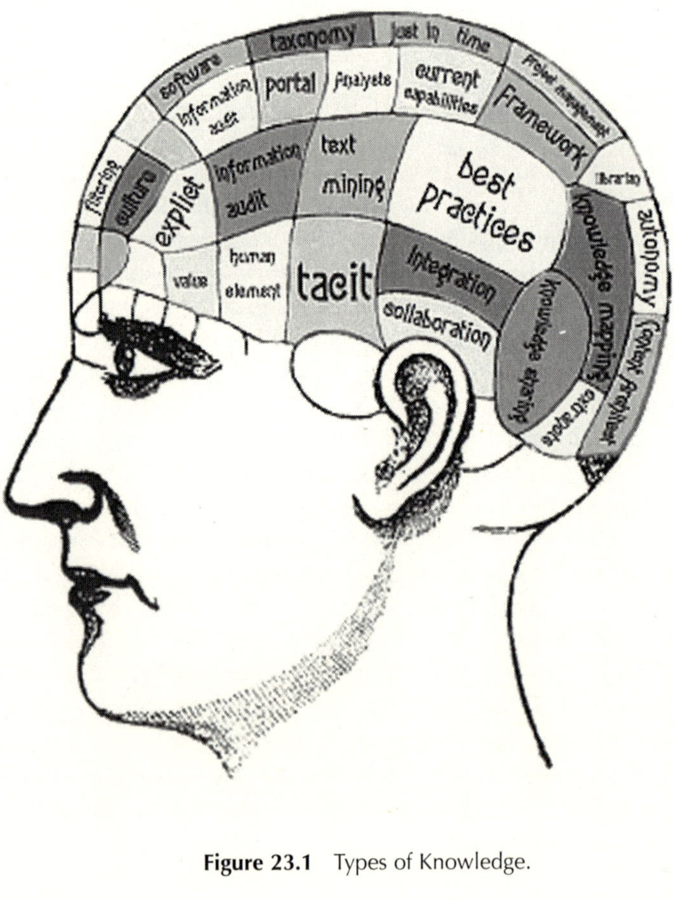

Figure 23.1 Types of Knowledge.

Explicit knowledge is knowledge that is written and stored. It can be readily discovered and passed onto others. Examples of explicit knowledge in law firms include legal work products, models and forms, business records, financial information, marketing research, client lists, practice group blogs, and wikis.

Tacit knowledge is the kind of knowledge that is difficult to transfer to another person by means of writing it down or verbalizing it. This kind of knowledge generally gets revealed through legal practice in a particular context and transmitted through social networks. To some extent it is "captured" when the knowledge holder joins a network or a community of practice. Examples of tacit knowledge include the experiences people possess and share via instant messaging, e-mails, in-house training, business referrals, and matter summaries.

The leveraging of explicit and tacit knowledge is a significant strategic business driver of modern law firms that seek to meet the sophisticated needs of savvy clients.

WHAT IS KM'S VALUE PROPOSITION?

Over the past several years, with clients increasing pressure on law firms to provide efficient and proactive legal services at a competitive lower cost, knowledge management has been embraced in the legal industry as a serious discipline. The same pressures that law firms are experiencing are being felt in law schools. With declining enrollments and reduced staffing, law schools too need to do more with less, just like private firms. Although the academic world hasn't yet embraced KM as enthusiastically as private firms, the same need for easy access to critical information is there. Law firms are finally catching up to what large corporations have known for years: Knowing what you know and being able to discover it is a competitive advantage. No one summed it up better than Hewlett-Packard CEO Lew Platt. "If only HP knew what HP knows, we would be three times more productive." By creating a culture of knowledge sharing, a law firm can realize a number of benefits. A program designed for knowledge sharing can:

- Provide a competitive advantage by making know-how easily accessible
- Increase productivity: knowledge workers don't waste time searching for information
- Improve organizational effectiveness by sharing best practices
- Improve collaboration by fostering knowledge sharing across the organization
- Speed response time to clients
- Enable organizations to learn by leveraging past experiences, successes, and failures

- Retain tacit knowledge within the organization
- Provide a practical solution based on the essence of the business that will be adopted
- Address cultural changes to create a knowledge-sharing environment (users are engaged)
- Lead to tangible outcomes (these should include increased productivity and the development of client loyalty)

KM BUILDING BLOCKS

Even small legal organizations are becoming more complex, with silos of knowledge dispersed throughout the organization and a myriad of communications platforms used to share this know-how. For those new to knowledge management, it will be helpful to understand some of the basic KM building blocks and the various repositories of knowledge that may exist in the typical legal environment.

Collaborative Communications

In the age of instantaneous communications and with the recent adoption of smartphones and tablets in legal practice, lawyers have been forced to find quicker and more efficient ways to deliver legal services. KM processes and systems have provided a collaborative technology foundation that enables lawyers to work more efficiently and to meet perceived client demands. Most law firms today have a number of collaborative tools at their disposal. They include:

- *Intranets*—an intranet is a computer network to share information, operational systems, or computing services within an organization. Much of the tacit knowledge in legal organizations is shared via the intranet.
- *Extranets* extend a private network onto the Internet with special provisions for authentication, authorization, and accounting. Many law firms provide clients with access to extranets that distribute relevant content and information about a matter or transaction between clients and the firm.
- *Online discussion forums* provide value by allowing lawyers from a designated practice to have internal conversations in the form of posted messages. These posted messages are archived and can be tagged or indexed for later retrieval. Within a forum's topic, each new discussion started is called a thread and can be replied to by multiple people. Online discussion forums have proven successful in capturing and sharing tacit knowledge and reducing e-mail overload.

- *Social media* is one of the newest forms of interactive communication among lawyers and legal staff in which they create, share, and exchange information and ideas in virtual communities and networks. These networks tend to be external facing and depend on mobile and Web-based technologies to create highly interactive platforms via which participants share, create, discuss, and modify user-generated content. Social media introduces substantial and pervasive changes to communication between organizations, communities, and individuals.

Intellectual Capital

The legal industry is a service industry. Law firms sell the intellectual capital of their lawyers, who may represent clients in litigation or transactions. Most law firms today have firm-wide precedent or form document libraries. These precedents, also referred to as "best practices" documents, enable lawyers to save time in repurposing and creating a first draft of a new document. Additionally, it supports a firm's risk management initiatives to ensure quality is applied to the drafting process by using an appropriate model or precedent by younger and less-experienced lawyers.

These databases of legal opinions and best practice documents are examples of how law firms have been successful in adopting knowledge management strategies related to explicit know-how. Law firms recognize that work performed for a specific client may be repurposed for another client with a similar engagement. Because of continued specialization in the law, best practice document collections are further refined by KM staff into practice collections to meet the specific needs of particular practice groups or even specific matters where the value extends beyond the specific client engagement to future similar matters.

Expertise

The scope of knowledge management has expanded beyond document databases. Law firms are taking a much broader view of KM that now includes the knowledge lawyers need to not only practice but also run their business. As a result, law firms typically now have myriad databases to track expertise (lawyer experience and skills), tap into client information, provide financial data and analysis, and retrieve helpful professional development presentations that teach lawyers on an on-demand basis.

Conclusion

As knowledge management initiatives continue to successfully support the business of law, the industry is seeing that leveraging the knowledge of your law firm differentiates you from your competitor and leads to reduced costs, increased revenue, increased profitability, better trained lawyers, and more satisfied clients.

NEW BUSINESS IMPERATIVES CHANGING THE ROLE OF LAW LIBRARIANS

There is no mistaking the major realignment that law firms have undergone since 2008. Declining revenues, reduced employment opportunities for new law school graduates, and downward pressure from clients to reduce fees all have forced law firms to adjust the way they have traditionally conducted their practice. Even the billable hour, long the dominant way that lawyers earned their fees, is giving way to a new pricing model; alternative fee arrangements (AFAs) are becoming commonplace at even the most elite firms, who had long been immune to price pressure.

In the early days of KM, the large scale Web development and database interactions that most KM projects required made it seem that the technology team was the right group to lead these efforts. In some firms, KM initiatives were driven by practice support lawyers or experienced counsel. While there is nothing inherently wrong with IT or senior lawyers driving a KM project within a firm, an experienced law librarian brings skills that are ideally suited for leading KM efforts.

Skills Librarians Possess

- Librarians are skilled in the development of taxonomies and classifying information.
- Most librarians are fluent in Web design and database development, and are usually technologically savvy.
- Librarians are long accustomed to anticipating lawyers' information needs better than IT staff or counsel (or lawyers themselves). Most practitioners are focused on a single practice subject. As generalists, librarians usually have a wider grasp of resources and practice needs.

Mapping of Librarian Skills to KM Skills

Traditional librarian skills include the ability to discern a real need from a perceived need—think reference interview—and so provide an ideal base for leading KM initiatives. Good KM leaders are also good project managers. They can define the requirements for a KM effort, get a handle on the scope of the project, and then understand the resources that can be brought to bear on creating a successful program.

Librarians tend to be organized by nature and can communicate well with lawyers. They know the standard sources for legal information from outside the organization, but more importantly, also bring to the table an institutional understanding of how the firm's lawyers communicate with each other, how they mine the document management system (DMS) for precedents and models, and how they use other "business of law" resources such as contact relationship software to stay in touch with clients, colleagues, and co-counsel. This "mental map" of how the firm works is a key skill that librarians possess (whether they know it or not) and prepares them for the process of modeling attorney workflows as part of a KM program.

LEARNING TO CONDUCT A KNOWLEDGE AUDIT

Sometimes referred to as an "information audit" or a "management review," the knowledge audit looks at three elements:

- Organizational structure
- Knowledge services in the organization and delivery
- Marketing, awareness-building, and identifying advocates and champions

The information audit is the process by which knowledge managers become expert in how information is created, distributed, and consumed throughout the organization. The effective information audit convenes an audit team to define the scope of the information flow in the enterprise. As part of this same exercise, knowledge managers will also gain insight into how knowledge changes.

Most information audits will follow a standard six-step model:

1. *Planning.* At the planning stage, KM managers should be able to formulate a mission statement that describes what the audit hopes to achieve and what the scope of the audit will embrace. This is the point at which the team should decide how much of the firm's business processes will be examined and analyzed to see where KM strategies can streamline the process and deliver a better product to business users.
2. *Data Collection.* Collect data by talking to the stakeholders. See how information is collected, stored, and disseminated by the business owners. Take a single business process as an example. Say your office manager routinely sends out printed updates to the firm's telephone directory. As part of the data collection review, the information audit will reveal how the office manager collects the data (runs a report from the HR database), compiles it (puts into a Word document), and disseminates it (prints out 1,000 copies and delivers the printout to all firm employees). A good information

audit will identify each business process and its life cycle from creation to final disposition.

3. *Data Analysis.* With the data collected, the KM team can analyze where the process can be improved. In the case of the telephone directory, data analysis should see if the firm's users still rely on the printed directory. Does the update contain all the information a user would want to see? (Phone number? Office number? E-mail? Secretary's contact?) The KM team should examine the data thoroughly with the content owner to see what is working, what can be made better, and that the interests of all stakeholders have been considered. This means looking at the audience for the deliverable information as well as the process by which that information is made available. Perhaps instead of circulating a directory, a better approach might be to provide a search box and look-up function that queries data from the HR system or from a data warehouse of current employees.
4. *Data Evaluation.* Data evaluation means making sure that the information is accurate, current, and being managed in the most efficient way possible.
5. *Communicating Recommendations.* After the data have been analyzed and recommendations made for next steps, the KM manager needs to clearly communicate the recommendation for next steps to the stakeholders. Communication is the key ingredient for making sure that information audits succeed. The stakeholders who "own" information have been successfully managing their data before KM strategies came, so keeping them apprised of any changes, especially if the recommendation requires any kind of change in workflow (which it should!), is critical. Get buy-in by fully informing information owners about new, presumably automated workflows. Communications should be clear about deadlines, and expectations reasonably set for what a new KM initiative can achieve and how.
6. *Implementing Recommendations.* Once business owners are comfortable with the new workflow, it is up to the KM manager to see to it that the new system is implemented. That requires organizing functional teams across the firm to work together to launch the new system, including IT developers, business owners, and high-level users who can encourage adoption of the new way of doing things.

CREATING AN OPTIMAL KM VALUE STRATEGY

The key components of a KM strategy document include:

- An executive summary
- Definition of knowledge management
- Current state report

- Strategic vision for KM (this includes business drivers, strategic focus, key enablers, risks and benefits)
- Implementation plan

The first step in creating an optimal KM value strategy is to define the scope of knowledge management within the organization using the KM maturity model. Logically, an organization that is further along on the maturity model will be able to take a more aggressive approach to the range of KM initiatives it may wish to undertake. Therefore, it is important to clearly articulate the maturity level of the organization.

The next step is to align KM initiatives with firm business drivers. Your KM initiatives should help solve or support critical business drivers. This can be achieved by conducting lawyer and client interviews and/or surveys and understanding the firm's competitive positioning in the legal market. Relevant questions that should be included in an interview or survey include:

- Which of the firm's current KM services do you think are essential to your practice?
- What changes in the legal marketplace create risks for the firm's ability to provide client services?
- Are clients asking for KM services?
- Are peer firms offering KM services that the firm cannot currently match?
- Do you think your firm/practice group has a knowledge-sharing culture?
- Successful KM initiatives often require changes in work flow. What motivates lawyers to change the way they work?

Once the strategic KM goals have been formulated, it is important to describe what KM will look like at the firm at the end of the planning window. Document, report, and "socialize" the strategy within the organization so that you build buy-in and support for the KM strategic plan.

Next, identify the key enablers to achieve the KM strategy success. This includes the people who will need to be engaged in the KM initiatives. This may include librarians, IT staff, and select attorneys or other support staff. Then identify all processes or work flows required to carry out the initiatives as well as a new technology that will need to be acquired. A good plan will discuss the benefits and risks of each initiative.

The final step in composing a KM strategy is to document how the plan will be implemented. Each initiative should have a defined purpose or mini-project plan; projects should be prioritized with quick wins in mind to build on early successes that will lead to greater confidence and buy-in for the overall strategic plan, including important financial support for longer-term initiatives.

When prioritizing the list of initiatives, it is important to consider the value of these initiatives to the firm versus the ease and cost of implementation. Those initiatives that strongly support the firm's strategies, provide a significant return on investment (ROI), provide a positive behavioral change (e.g., creating a knowledge sharing culture), help provide a competitive advantage, or meet clients' needs should lead your list of initiatives. Factors that may lead to lowering the order of an initiative is if the ease of implementation is significant and the cost prohibitive.

HOW TO START UP A KM INITIATIVE AND ENGAGE KM SPONSORSHIP

When embarking on a new KM initiative, it is important to start small and build on your successes. This will buy credibility for the librarian or information professional. Confirm that your KM implementation is focused on solving real, pressing business issues. If so, it is likely to be closely aligned with the firm's strategic goals.

Develop a business case for each initiative that includes ROI analysis and clearly states the value to the organization. This includes establishing at the outset the metrics that will prove your success. Remember that KM implementation requires an organizational change program. What incentives will be provided to migrate lawyers and administrative staff to new workflows?

Goals should be defined in terms of organizational goals. Reports should show how the KM initiatives contribute to the organizational goals and be analyzed on a regular basis so that methodologies and outcomes can be adjusted as necessary. Remember that you can't manage what you can't measure.

From a human resources perspective, identify an excellent in-house change agent to lead the program. This can be a library or KM director who is supported by a collaborative team comprised of lawyers, librarians, IT professionals, and other knowledge workers.

Obtain sponsorship at a high level so that you can address and embed the required organizational change and insist on a steering committee to ensure necessary support. Similarly, obtain stakeholder support and segment your audience to ensure that your initiatives address their specific concerns.

Remember that KM success can only be accomplished when KM is fully embedded into company processes, accountabilities, and governance. Changing organizational culture is a long-term process and will take perseverance. Therefore, remember that KM is an evolution, not a revolution.

Since KM depends heavily on the automation of business workflows, technology to assist with the process is readily available. Most KM efforts require a technology platform that will deliver information to end users and use a number of search tools, collaboration spaces, and other software to achieve the seamless connection of information to the user, which is the hallmark of a successful KM program.

SharePoint

Microsoft SharePoint is rapidly becoming the default platform for developing KM initiatives at most large firms. It's widely available, most IT departments either know how to work with it (or can figure it out), and all the information vendors who intend to provide information inside the firewall are developing widgets that can be easily installed on a SharePoint page. SharePoint brings to the table granular security on Web parts to make it much easier to produce a customized user experience and integration with other Microsoft application such as Outlook, Word, PowerPoint, and Excel. It can be used as a document management system, wiki and blog host, and collaboration site for team workspaces. That's the good news.

But like many singer-dancer-actors, SharePoint doesn't do any one of these things particularly well. Out of the box, SharePoint is little more than a framework that can be used, but as for getting librarians or KM staff a place to work, SharePoint itself doesn't do all that much. An effective SharePoint deployment with interactive pages, collaborative workspaces, or document management capability is an IT-heavy project. Developers will be required to configure sites, to connect to the underlying stores that provision the various widgets (called "Web parts" in SharePoint parlance) that will populate the pages, and to work with Microsoft's typically bloated design templates.

Even with its shortcomings, a well-planned intranet deployment can deliver much better information delivery to the desktop thanks to SharePoint's ability to mash-up content easily on a page. Think of a single webpage containing news, links to internal documents, blogs, and financial data and you get an idea of what SharePoint can do, albeit with plenty of developer help.

Work Product Retrieval

Recommind, West KM, and Lexis Search Advantage are the major players in the space. They are powerful search agents designed to work effectively with a large body of documents. Work product retrieval is distinct from enterprise search because these tools are designed specifically to examine legal documents and surface features of those documents that are unique to the legal profession. All of these applications can extract legal-specific metadata from documents, such as judge, jurisdiction, or specific entities like company name; on the corporate side, finding definitions and clauses in deal documents with precision is the value of work product retrieval tools.

Lexis Search Advantage and WestKM both have a powerful search capability that equips lawyers to find useful materials from the firm's collection of know-how without having to depend on the often unreliable (or nonexistent) document-level tagging that previous document search tools depended on for filtering.

The benefits of using work product retrieval tools are many. With a reliable search tool, it's no longer necessary to surface possible candidates for precedents to cover every predictable need. Instead, users can confidently search the firm's own unstructured collection of material with a high degree of confidence of finding a responsive document. Many early KM initiatives relied on teams to first identify firm work products that would provide guidance to attorneys as they drafted new documents, in the form of precedents, models, or samples. And in fact, that strategy made sense because it overcame the shortcomings of the search engines installed in most document management systems. But that manual process is both labor intensive and costly. Better to have software that can locate that needle-in-a-haystack document when needed and then show the user enough metadata about the document—jurisdiction, clause language, legal concept—to make an informed decision about the value of the document as a precedent.

Search

Enterprise search software, especially such heavyweight, industry-leading applications as Microsoft's FAST, Recommind's Decisiv Search, and HP's IUS (IDOL Universal Search) offers the key to unlocking the information socked away in disparate silos throughout the firm. When you consider that useful client information can live in the document management system, the accounting system, the contact management system, and records, it's very challenging to pull related information together without some technical way to connect the dots. Point the enterprise search engines at those various repositories to leverage the content stored in them and very quickly, a multifaceted portrait of a client or a transaction becomes accessible. Setting these search systems up effectively is far from a trivial matter, but the data-mining values that can be realized using the power of enterprise search can justify the considerable cost of acquisition and manpower to make them work.

Document Management Systems

iManage and its flexible and powerful WorkSite can be a very useful tool for KM administrators. With the ability to set up folders around subjects, practices, or particular matters, KM managers have a way to offer lawyers a place to find

KM-worthy documents. Practice groups themselves can nominate a particular document to serve as a model or precedent. They can merely drag the document into the KM folder and then, by exposing that folder either via a SharePoint Web part or in the lawyers' folder view, make that document readily available.

Social Collaborative Platforms

While it's not obvious that texting has caught on in any law firm environments for actual work-related communications, the social media revolution is no longer unfolding outside the firm's firewall. SharePoint and iManage have collaborative tools baked right into their architecture. Imaginative KM managers can set up workspaces for client teams to work together by offering a secure space to share documents, e-mails, and blog/wiki posts. Both systems are flexible enough to give managers control over the scope of the workspace and security settings.

Current Awareness for the Enterprise

Getting the right news into the hands of readers who can get the most value from it is the job for current awareness tools. Sifting through the nearly overwhelming amount of information that pours forth daily is an immense undertaking. Not only traditional media outlets but also government agencies, courts, other law firms, bloggers, and websites produce prodigious amounts of information. Winnowing out the good stuff from this tidal wave of data is where content aggregators and disseminators come in.

Librarians and KM staff now have tools that can make sense of this monsoon. Using a tool like Ozmosys, Linex, or Manzama, information managers can easily set up distribution groups. They can also define packages of information to feed to those groups, tailored to the group's specific information need.

CASE STUDIES

The principles of knowledge management we've covered need to guide the development of actual working tools. Turning vague, aspirational ideas of "helping lawyers practice better" into concrete deliverables does take a fair amount of planning, design, project management skills, and communications with the ultimate users. Our experience at Debevoise & Plimpton is one approach. Not all firms need to or ought to follow the same strategy, but our experience will highlight many of the same successes and pain points that other firms will likely encounter. But KM is very much a custom process that needs to be tailored to each firm's unique set of needs. There is no one right approach; it's not a cookie-cutter process.

KM Document Collections: Organizing the Firm's Know-How

Most KM programs at law firms focus initially on creating the "electronic binder." This is the Web-based version of the binders that most partners and senior lawyers amass, which contain the firm's collected wisdom. The print version of these binders, which can still be found on the shelves in every partner's office, features examples of beautifully crafted memos, firm-approved models, go-to forms that provide starting points for drafting standard documents, and a host of other materials pulled from the obscurity of the document management system.

In many first-generation KM installations, a document collection of best practice or training documents always seems like the low-hanging fruit, and it is. After all, there generally is an existing workflow in place to develop and circulate these materials in binders; it seems like a simple matter to organize the collections into some taxonomy for easy browse-and-retrieve.

In our first-generation intranet, a great deal of time and effort was spent in constructing a detailed and robust collection of documents. Over the course of seven years, the collection of materials grew to more than 17,000 individual items, each of them recommended by the practice or by the senior KM counsel who worked with the practices to identify the materials. Every single item, be it document or Web link, needed to be coded manually by trained editors. These human editors added metadata around each item in order to (1) distribute the item correctly in one or more of the thirty-plus individual practice sites and (2) provide more accurate search results.

Over time, we learned a number of important lessons about managing the firm's know-how collection:

- Manual tagging is a not an efficient process.
- KM document collections ought to be managed at the practice level and not by KM staff. Precedent documents and models ought to be managed by an informed information steward, such as a senior associate, counsel, or practice support lawyer who knows the practice well and can make informed judgments about what ought to be collected and what can be safely ignored.
- Effective document management system search and foldering can make it much easier to maintain an up-to-date and reliable collection. Many KM systems evolved simply to overcome the shortcomings of a DMS search engine. Today's DMSs are flexible and come with more muscular search capability than their predecessors. Findability of useful materials is no longer the major driver of KM strategies.

Actively managing the collection to add new materials and weed out old content is critical to maintaining the credibility of the know-how collection. Identification of stale content or documents that no longer have relevance (think of such previously "hot documents" as that Y2K memo) is key. Users will be turned off a system that they perceive as out of touch with current practice. Documents nominated for the limelight of KM precedent can even be specially tagged within the document management system itself; materials marked as having precedential value can be identified by checking off a box in the document profile. As our firm adopts a new document management system, the ability to identify and manage KM-worthy models, precedents, and practice guides is baked right into the new DMS. At the time of profiling, a document can be tagged and then added to the appropriate folder in the DMS for display on the firm's intranet.

Current Awareness

The problem most librarians and KM managers face is not the lack of current awareness materials but its opposite problem: overabundance. One of the goals of any KM effort is "narrowcasting" information to attorneys. Lawyers hate to waste time, and sending them headlines or stories that they either won't read or have no interest in receiving is one way to aggravate, not enlighten, them. So we decided to find a way to deliver only high-value information in an easy-to-digest package. Of the different vendors working in the news distribution space—Linex, Manzama, and Ozmosys to name a few—we selected Ozmosys.

Ozmosys provisions practice group pages with current information. Like any good RSS feed, we can expose the most recent headlines and offer a link to more stories. News is updated around the clock, so lawyers can check in at their convenience. And with its integration with EOS, our ILS vendor, we can now use the library catalog to serve up feeds to attorneys and enable them to subscribe and unsubscribe to any of the 1,000+ feeds we make available.

New-generation current awareness software comes equipped with social media features like tag clouds, user preference indicators, sharing, smart filters, personalized information delivery, and better search across time-sensitive materials.

Blogging

On the spectrum of urgency for information sharing, most law firm KM efforts try to find some sweet spot on the information continuum between sending an e-mail (immediate!) and requesting the closing set from Records (when you get around to it). Some information needs to be shared—but not right now—and stored—but not forever. For providing timely information that ought to be saved for later reference, a blog is the ideal platform.

A blog is immediate without being intrusive. It's searchable and it turns out to be an ideal way of capturing institutional know-how on the fly without asking lawyers to do extra work. With the same effort that it takes to send e-mails flying back and forth, lawyers can instead have a running discussion with each other in the normal course of business. If one lawyer asks a group by e-mail, "Have you ever seen this issue?" to which others reply with answers, references, and opinions, why let that institutional insight fly out the window? Blogs also come with the happy extra features of tagging for easy filtering and retrieval and robust search by full text, index (i.e., tags), or author.

At our firm, we have set up internal blogs for a variety of purposes and configured to meet different needs using Traction Software's TeamPage. The adoption of blogs has been slow but steady as each of the practices discerns a need to share information among itself. Blogs are usually dedicated to a specific use case: announcements only in one practice (such as "Did you see this court case from Delaware Chancery?") to logging business development activity. Blogs also serve as a place to temporarily store deal information so that it can be quickly shared by the team during a matter's active life.

Restricted versus open access. One popular blog is operated by the partners in one of our practice groups. Only partners can post, but all associates and staff in the practice can read them. Typically, the partners post articles on which they comment on case decisions and news stories or share surveys and other publication within the practice. Remarkably, partners post frequently and even more remarkable, have grown accustomed to tagging each one of their posts.

Some of the blogs we established are intended to be forums in which anyone can post and anyone can answer (so long as that "anyone" is part of the group for which the blog was established). This is useful for areas of law where associates or partners don't regularly practice, such as certain pro bono activities or where new associates need to come up to speed quickly on a specific topic and need a place to ask questions.

One of our biggest challenges to deploying blogs is blogging by e-mail. Like most lawyers, our attorneys live in e-mail. Expecting them to fire up the blog software every time they want to post an article is asking a lot from busy attorneys; our "next best" solution was to install a copy of the blog home page on the practice group pages and allow attorneys to access the blog that way. A number of technical reasons and some business decisions early in the process precluded us from enabling blog-by-e-mail, but we soon may be able to enable attorneys to send their posts directly into the blog right from Outlook.

What was striking to us, though, was how quickly even some of our most technology-challenged users quickly grasped the need to tag new posts and to post messages that contained links to the outside world as well as to documents on the document management system. And while we wouldn't go so far as to say that the blogs are universally loved, we have found that, when deployed for very specific and narrowly focused purposes, they can quickly become a valuable resource for capturing the practice group know-how with little additional effort. Critical to the success of blogs is the same as for any other new technology rolled out to the lawyers: keep it simple. It requires no more than ten minutes to show users how to post, search, or subscribe to the blogs and this ease of use has been a key part of successful adoption.

Access to Administrative Support

The legal profession doesn't run on brilliant legal work product alone. Getting those dazzling documents into the hands of the client or court requires an administrative staff. The staff supports everything from proofreading those sterling documents and providing legal research to supporting those propositions of law to taxi dispatch and conference room booking so that lawyers can talk to their clients. These "business of law" responsibilities are ripe for automation themselves. Clearly, there is no need make repeated calls to book a conference room when it is much easier instead to link to the conference room booking software.

In many ways, automating, or at least streamlining, access to administrative information is one of the key benefits of knowledge management. To mash up two old aphorisms: If time is indeed money and a penny saved is a penny earned, then saving lawyers time is very literally saving them money as well. Good KM ultimately aims for efficiency.

The best way we've found to deliver services to attorneys is by asking every administrative department to conduct an information audit of the materials that they routinely send to or collect from attorneys. Those items that are routinely requested by attorneys (or routed to them) can instead be delivered by intranet, RSS feed, or answered with a query to a database (e.g., "Is conference room 41N available for me at 4:00 tomorrow?"). But the process is not a wholesale one: it's more of a retail, door-to-door effort to identify candidates for automation and then setting to work to build the better mousetrap.

We try to populate each practice group home page with links back to their primary contacts in marketing, the library (for research needs), and accounting. This is not only convenient for the lawyers but also highlights to the legal staff the many ways that the administrative staff supports the practice. Thanks to some judicious linking, lawyers can quickly put their hands on glossy marketing materials to send to current or prospective clients, biographies of lawyers in the firm, and the forms required to open a new matter. This type of "one-stop shopping" offers attorneys easy-to-access firm know-how on a given legal subject plus administrative information. This helps solve the problem of lawyers having to figure out where to look in multiple systems for the information that they need.

The Future of KM

There is still a great deal of work to be done, both technically and culturally, to achieve what we see as KM's greatest benefits, which are client and client-team webpages. We're starting work on client team pages to provide the global view of a client on a single streamlined website, mashing up external data and internal information into a one-stop shop for client teams. Next step, hopefully, will be the matter-centric pages that allow small groups of lawyers and staff to share their matter-specific work and communicate easily about a given matter. We're still working out the details. But, in every step we take and every project we manage, our KM goal is always the same: help the firm's lawyers get great work completed in less time and with much less effort expended in the search for useful information. A KM success might be something as simple as pushing out a client's phone number and e-mail to a client page to an enterprise-wide solution that provisions rich practice pages with sophisticated work product. KM is the way we find ways to make the practice of law smoother, more friction-free and, ultimately, to deliver the best legal product to the client that we can.

CONCLUSION

Although the precepts of knowledge management have been applied in different forms at law firms in the past, the state of technology today and the sophistication of the end users who need information means KM is still a fairly new idea at most firms. But in an environment where every dollar counts, the introduction of new business procedures that result in practice efficiencies is critical. KM therefore needs to be considered not as a "nice to have" feature but as an essential business strategy. It's a way of using news tools in creative ways to deliver information most efficiently to those who can make the best use of it. Techniques of knowledge management allow firms to streamline business of law workflows and make the firm's intellectual capital easier to use. For us at Debevoise, our KM initiative lets our lawyers "know what Debevoise knows."

FURTHER READING

Books

Collins, Heidi. 2003. *Enterprise knowledge portals.* New York: AMACOM.

Hawley, Tim. 2012. *Making knowledge management work for your organisation.* Peoria: Ark Group.

Rusanow, Gretta. 2003. *Knowledge management and the smarter lawyer.* New York: ALM.

Articles

Lastres, Steven A. June 2011. "Aligning through knowledge management." *Information Outlook* 15:4. http://www.questia.com/library/1G1-259079447/aligning-through-knowledge-management-by-adopting.

———. March 2009. "Knowledge management in a changing world." *Law librarians in the new millennium.* http://futureready365.sla.org/02/28/knowledge-management-in-a-changing-world/.

Lastres, Steven A., Don MacLeod, et al. October 2012. "Knowledge management: A strategic role change for legal information professionals." *AALL/ ILTA Digital,* white paper 34-89. http://read.uberflip.com/i/87421/32%.

24

The Law Library Intranet

Emily R. Florio
FISH & RICHARDSON

Michael J. Robak
LEON E. BLOCH LAW LIBRARY AND UNIVERSITY OF MISSOURI-KANSAS CITY SCHOOL OF LAW

THE BASICS

Law librarians, whether in academia or private law firms, work within their respective organizations providing, for the most part, similar services. The law librarian, in a law school or law firm, can no longer serve as passive keeper of a print material warehouse. That role is inconsistent in a world where the majority of resources are digital and available online twenty-four hours a day, seven days a week, and 365 days a year. Law librarians in both settings must transform their role to proactively and creatively deliver information resources and push out new and critical information to their constituencies. An intranet is one of the strongest tools within the librarian's information arsenal to help with that mission, because it provides a superior platform for communication and information storage and retrieval. In basic terms, an intranet is a computer network for use within an organization, often as a collaboration tool. Intranets "provide organizations and institutions with a single electronic access point to a large and diverse array of internal . . . information."[1] An intranet allows for sharing information, transmitting organizational knowledge, and supporting knowledge processes.[2] It can also be described as an internal website for which only authorized users have access.

This chapter provides an overview and rationale for creating an intranet in both the law school and private law firm library. It offers a general primer and considerations for building the intranet. Also provided is a guide for the law librarian to make the case he or she should be the one to oversee either, but preferably both, the construction of the intranet container and/or the creation of the content for the intranet.

Many people use the words *portal* and *intranet* interchangeably, though they are distinct but related terms. The *Oxford English Dictionary* defines an intranet as "a local or restricted computer network; spec. a private or corporate network that uses Internet protocols"[3] and a portal as "a server or web site that provides access to a number of sources of information and facilities, such as a directory of links to other web sites, search engines, email, online shopping, etc."[4] Therefore, an intranet may exist without a portal, but a portal requires an intranet. Within the context of the law firm or law school, the entire institution might have an intranet and then a library department might have their own portal, which serves as entry into their resources. In this current digital age of libraries, librarians utilize intranets and create portals:

> to deliver the same guidance to those users who are far beyond the walls of the reference room. . . . These gateways on the Web can be thought of as maps to the library's information space and are designed to guide the user to the correct resources. They attempt to divide resources into familiar categories and keep similar kinds of resources together.[5]

Law librarians within firms and universities have been using and supporting intranets for many years. An article from 1999 titled "Intranet and Knowledge Management Solutions: Law Librarians Lead the Way"[6] addresses many key concepts still true in 2013, including the role of librarians as online content managers. The usefulness of an intranet for law librarians has only grown since the article's publication, with the opportunities and benefits always evolving and expanding. Benefits for supporting or managing an intranet within a firm or university include: providing relevant solutions to existing information access problems, offering resources that are easily accessible and adaptable and promoting efficiency through streamlined access. Law libraries might use intranets to provide a centralized access point into their resources, a document repository, a knowledge base, online public access catalog (OPAC) integration, current awareness support, or more generally driving utilization of services.

Successful intranets or portals require a variety of contributing factors, including technical support, attention to layout, and a consistent approach throughout the organization. In law firms and universities, librarians work along with their IT departments to develop, implement, roll out, and support their organizational intranets or departmental portals. This cross-departmental teamwork remains crucial, as members of IT are usually the technical experts and librarians are the information or knowledge managers.

THE INTRANET AND THE ENTERPRISE

As discussed earlier, the word *intranet* means something very simple: a Web-based internal information resource to help organize and transmit information through an institution or organization, with the emphasis on the word "internal." At the very least, law schools and law firms will deploy a public website, use e-mail, and create folders for storing information on a shared network drive to serve as the backbone of the communication system. This infrastructure will also include the library's catalog and other digital resources, but in the law school those will most likely be presented as external facing (i.e., anyone can find and access the website but some of the materials may be behind a secured log-in). Law schools that rely on this kind of infrastructure are primary candidates for development of an intranet to improve both communication and information storage and retrieval in order to develop their business processes and workflows. Law firms benefit from an intranet for similar reasons and some of their own unique situations, including sharing content and resources across multiple offices and various departments.

If the law school or firm does not have an intranet, and the librarians would like to design one, then one of the first questions to ask about the design is whether it is a standalone intranet, (i.e., dedicated just to library information) or is it part of a larger intranet for the entire organization. The most efficient intranet developments are those developed to place as much information as possible about the institution in one place. An institutional intranet rather than a standalone departmental offering allows for consistency and fluidity along with efficiency, if there are defined teams to support both the technical and information management needs.

Regardless of whether the intranet is just for the law library or for the larger institution, law librarians are information professionals and are in the best position to advise and assist with content management. Librarians have always been trained to organize materials for access. That is even more important with digital materials. With law libraries, both in the law school and law firm, moving away from becoming simply print warehouses and working toward collecting more and more digital content, the librarian's job is more important than ever. Law librarians have always served as guides and navigators to the materials in their charge.[7] But digital materials present a special challenge since, without proper instruction, guidance, and most importantly, a well-crafted platform, they could remain invisible to the benefiting population, whether law students and faculty or lawyers and administrative staff. Law librarians need to be especially vigilant about making their content easily accessible, findable, and straightforward to use. An intranet is the perfect platform to ensure that law students, faculty, lawyers, and their support staff have the best possible information to support the instructional and scholarly missions of the institution.

Intranet administration usually falls to both the library and IT departments. IT staff hold the technical expertise and access to hardware and software required to support the intranet. However, librarians and other information professionals are the true managers of information and knowledge, which makes them a natural fit as managers of intranet content. Library staff often works directly with legal and administrative staff, which allows them to be knowledgeable about the various practice groups within their enterprise. Librarians pair their expertise and resources with the demands of their practice groups, which promotes synergy across the firm. Librarians remain the experts on legal resources, including content, cost, and accessibility; therefore, librarians remain the best prepared to work with practice and administrative groups along with vendors or custom solutions. This holds true for law schools as well—in fact, more so. In law schools, librarians interact with the student, faculty, and staff as well as the other administrative offices in supplying their information needs. They also have an understanding of resources available at the university level so that pairing resources with the demands of students, faculty, staff, and administration is best left in their capable hands.

RATIONALE AND BENEFITS FOR HAVING AN INTRANET

A librarian will likely need to provide rationale for purchasing an intranet solution or devoting staff time and resources to creating an intranet presence. In the private law firm library, customized library pages may lead to organizational efficiencies through the development of practical applications. There are fewer hurdles for a library department, if the firm or organization already utilizes an enterprise-wide product, such as Microsoft SharePoint. If that is the case, then creating a centralized library portal should be a priority and goal for most departments. Having a well-organized library portal page may lead to enhanced accessibility and visibility of library resources while also allowing for better collaboration with practice and administrative departments.

An intranet in an academic library adds another important tool for producing and retaining customized information for students, faculty, and other library customers. Because the intranet can also be designed to push information in a much more targeted manner, this gives the academic law librarian the ability to provide much more than a one-size-fits-all LibGuide or other external-facing website. Information and resources can be tailored and provided in a way the law school's website simply doesn't allow.

A centralized university or firm intranet may provide cohesion and strengthen consistency within an organization, which, on a smaller scale, can be seen in the library. By having one location for all information on the library, there is less chance that material will fall through the cracks or be underutilized. This also allows for patrons of the library to know where to go for answers to their questions, rather than not knowing where to turn. This leads to a more informed, productive, and efficient user community.

Many firms have seen a decrease in their full-time library staff since the recession, which means that providing the same level of service with fewer staff and resources has become a regular occurrence and everyday requirement. Librarians can take this opportunity to put resources in their users' hands, enabling the end users to perform their own research. By creating user guides and video tutorials on how to perform certain tasks and then embedding them right on the intranet pages, librarians aid in the instruction of their users.

Having a centralized intranet benefits librarians and other firm employees, from secretaries up through senior partners. By providing all the readily accessible information, library staff time can be freed up for more complex tasks and responsibilities, in addition to making the information available even if library staff are unavailable, such as at night or on weekends. Library resources need to be accessible 24/7/365 and from any location, including the workplace, a home office, a courthouse, or an airplane. Mobile computing remains commonplace in the legal community, thus directly affecting law librarians.

In recent years, the legal information industry has seen a proliferation of new paid database offerings from existing (Lexis and Westlaw) and newer providers (Fastcase and Bloomberg), along with free resources (Google Scholar). Every new subscription database means another potential URL and ID/password to remember. By creating a resource listing or directory on an intranet, law librarians push their users to the most authoritative and applicable resources. In the same vein, curated library pages for each practice or administration group promote efficiency by showcasing the most relevant resources. Librarians should make every effort to ease access and drive utilization of existing resources rather than attorneys wasting time using free resources. Working with practice and administrative group leaders may also lead to enhanced visibility for the library department, which could facilitate involvement with other high-level projects.

In an academic library setting, the largest subscription resources are often listed on the website and hidden behind a proxy server. But some academic libraries have moved to a model of contracting for a smaller number of seats for a resource, either because these resources support a limited set of the law school community or because a full enterprise license is too expensive. Making the intranet the access point allows for sharing, where otherwise students might have to seek out a librarian or go to a central terminal in the library for access. In addition, the intranet provides a method for sharing the research and other findings of those accessing the limited resources in a way a law school website cannot be designed to accommodate.

Benefits of Supporting an Intranet

There are numerous benefits for a law school or law firm library to support and maintain an intranet or portal. These include:

- *Providing relevant solutions to existing information access problems:* By gathering feedback from user needs assessments, a librarian may create a portal that eliminates gaps in information access that might include decentralized information on all available internal and external resources or no clear directory of services provided by the library, and that provides tools for increasing efficiency.
- *Accessibility:* With the assistance of IT, a portal can offer limitless accessibility, allowing users to utilize content whether they are in the office, in their home office, sitting on the train, or at trial.
- *Scalability:* A portal can be scalable and evolvable based on content availability and user feedback. It is adjusted based on new content and is always current.
- *Promote efficiency:* Efficiency ensures that the most authoritative and comprehensive resources get to the right people when they need it, while also providing quick and streamlined access to content.

Challenges of Supporting an Intranet

Of course, there are drawbacks and roadblocks to most situations, including the implementation and support of an intranet.

- *Communication and adoption:* Educating users that the product exists and getting users to read announcements regarding the new portal are definite hurdles. Seek feedback from users throughout development and any pilot period to ensure that the site is uncomplicated, integrated with the user's workflow, and interoperable with other firm systems. Use different media to post announcements about the new site, including e-mail and posters. If the portal is drastically different from anything the firm has seen, provide comprehensive educational training sessions for your users.
- *Soliciting relevant and sufficient feedback:* Embed feedback tools into your intranet pages so it is easy for users to provide feedback. Leverage established relationships with your users and keep open the lines of communication regarding development.
- *Staff bandwidth for creating and developing the product:* Obtain management buy in to make creating a portal a crucial and supported project. When possible, choose a project team that has the most technical skill and content knowledge that will be necessary throughout development.
- *Complexity and timeliness of development and implementation:* Utilize existing vendor relationships when looking into customizing or creating an intranet. The vendor might have something in development or might team with you to create your desired tool.

USES OF INTRANETS

There are many intranet products on the market, including offerings from Microsoft, IBM, OpenText, and Oracle. Microsoft SharePoint seems to be the most talked about in the private law firm library community, because it has been the "go-to standard" and "is a platform with an intuitive tech interface and hence provides a shorter ramp-up time in terms of learning."[8] There are several versions of SharePoint currently utilized in law firms, including 2003, 2007, 2010, and most recently, 2013. The core functionality and resources mentioned exist in all versions, with obvious enhancements in newer offerings. As with most applications, and this is particularly true with SharePoint, there can be a very large difference between the "out of the box" setup of the application versus a more specifically designed and customized version of the application. For this reason, and to help avoid oft-expressed frustration, as part of developing the plan for the intranet, a sufficient budget and resources should be allocated for this customization or, in the alternative, sufficient training on the tool to avoid user difficulty.

In the Law Firm Library

According to the 2012 International Legal Technology Association (ILTA) Technology Survey, law firms are using SharePoint for intranet usage, collaboration, custom apps, extranet usage, workflow, and document management.[9] In addition to the uses highlighted in the survey, library departments in law firms and universities see additional practical applications for intranets, including keeping statistics, providing content, centralizing resources, and the intake of new requests. Using SharePoint to track metrics and keep statistics only scratches the surface of available options. For instance, at Fox Rothschild, the Knowledge Services Department utilizes SharePoint to create pages and organize content, track projects and tasks, and also generate statistics for reporting and analysis.[10]

Law firm librarians already use SharePoint in some fantastic and unique ways. Collaborating within a library department remains a crucial component within most firms; therefore, SharePoint can be leveraged in this regard. It can be used as a workflow management tool that keeps a log of all reference/technical services requests, such as the intake of new requests, tracking of projects in process, and then used to deliver and possibly store completed requests. This is a great example of remote team communication, which is highly prevalent and common within the modern-day law firm library.

> The library department at Ropes & Gray uses a SharePoint discussion board to manage requests for research, materials and library services. They list some of the benefits as:[11]

- Requesters can use an online form or send e-mail.
- Library staff can centrally manage requests for services and materials.
- Librarians are automatically notified by e-mail of their new tasks.
- Exporting to Excel enables reporting and data mining.

Some of the disadvantages include:

- Library staff receive e-mail alerts, but no workflow solution.
- Reporting requires strict consensus and commitment from staff.

Another form of collaboration could be creating a knowledge base for exemplary research requests, processes, or procedural documentation and the general capturing of organizational knowledge. Utilize lists and fields within SharePoint to store information that will be beneficial and relevant to the library staff or firm employees. A document repository would be more inclusive than a knowledge base, though definitely offers similar benefits to the end user. At the institutional level, a repository or knowledge base on an intranet can be used to organize documentation and forms for different administrative groups, such as the health care benefits offered through human resources or the payroll schedule from accounting. More generally, an intranet could be used as a rudimentary document management system.

The intranet can be used as a training tool and aid where written instructions or educational videos are embedded into your library pages that introduce users to the resources or tools available. The library department at Fish & Richardson maintains a trainings and webinars page on their library portal that connects to external database trainings, print user guides and tutorials, and a full catalog of available library courses. This catalog links to the firm's online educational system, which includes on-demand trainings and the opportunity to request those not currently scheduled.

Self-service offerings allow law firm employees, including secretaries, paralegals, attorneys, and other administrative staff to access certain content without going through the library. This allows librarians to work on higher level, value-added projects while also not limiting access to their constituents. Some legal information providers have out-of-the-box SharePoint (often called Web parts or widgets) solutions that can be downloaded from their website and added to SharePoint. These Web parts deliver additional functionality into a SharePoint page that enhances the experience for the end user. Westlaw currently offers portal Web parts that allow the quick lookup of cases by KeyCite, Find by Citation, and a few related tools.[12] LexisNexis has similar Web parts for locating dockets in CourtLink or cases using Get & Print.[13] Wolters Kluwer and Bloomberg BNA have Web parts in development, but none were commercially available at the time of this book's publication. Many law firms have IT development departments that can create custom applications and tools based on the specific needs of their legal staff.

Many integrated library systems (ILS) have the ability to integrate with SharePoint. In fact, the top four systems used by firms (EOS International,[14] Sydney Plus,[15] InMagic,[16] and Softlink[17]), according to the ALM Law Librarian Survey,[18] all offer some form of SharePoint integration. A Web part for your ILS provides another access point into that content and provides for easier access since the end user does not need to remember a different URL. Librarians can set up RSS feeds for newly added titles or subject-specific titles, then have those RSS feeds embedded in practice or administrative group pages.

In the Academic Law Library

Just as in private law firms, the desire in the academic law library is to drive utilization of existing authoritative and paid resources. The online catalog is now only one part of the library information distribution system. Content hosted by an online provider will still have a catalog record for the individual content, but there will most likely be a companion databases page on the library's external website. So, for example, HeinOnline, which describes itself as the "[w]orld's largest image-based legal research database" and contains complete collections of a majority of academic law reviews, would be listed on the page of "subscription" databases as well as a link from the catalog record for the individual publication. Most law school libraries will have, as part of an external website, some portion of the collection advertised. This makes sense since the library's scope of services and collections are potential reputational drivers for the law school and should rightly be highlighted on the external website. And it makes particular sense if the institutions, like many across the United States, are moving at the system level to build and deploy digital libraries and institutional repositories for their system.[19] In a digital, virtual space, an institution's collections can be "linked" together in ways that allow them to be independent yet managed together.

But the intranet allows for taking what is particularly unique for the law students and faculty and finding ways to push it out in a more effective manner. While the online catalog still remains at the heart of library information organization, the intranet allows for creating wholly new ways to package and present the information and make it findable. This in turn allows academic law librarians to market themselves to the students, faculty, and administration and to continuously demonstrate they are vital to the life of the institution. The development of robust intranet pages goes beyond simply providing links to resources. It also surpasses the limitations imposed by external Web applications, since it allows for the creation of customized internal webpages as well as the opportunity to link users together around projects and resources using such features as a wiki.

No law school's or university's external website has the capacity to integrate lists of links or allow for the targeting of resources by interest and expertise level.

In the academic world there is a saying: Priorities get funded, so make yourself a priority. By developing or maintaining an intranet, academic law librarians can more successfully establish their status as the keepers of law school information. The American Association of Law Libraries, as part of its strategic plan, seeks to establish law librarians as the "recognized authority in all aspects of legal information."[20] An intranet within the law school provides academic law librarians this opportunity plus expands their reach to other areas of the law school's information needs. And an intranet platform allows academic law librarians to create new ways to collaborate with students, faculty, and administration.

For example, law school student organizations exist, often for years and years, within the institution. And in many law schools, the organizations, while continuing to exist, suffer from a classic example of institutional memory loss. That is, when the students most involved with the organization graduate, they often take away the knowledge of the organization's activities for the period of time they attended school. While there is some charm to the notion of students continually reinventing the wheel, it is not terribly efficient. Law students, like members of any organization, are best tasked if they are working on new and innovative strategies for the organization. An intranet allows for organizations to have a place for their own content. In addition, it provides an excellent opportunity for the law librarians to assist with the collection and organization of the materials on the intranet site. As an additional benefit, as information is collected and curated over time, the law school's advancement or development group has easy access to historic information about alumni activity during their time in law school. This information is particularly prized by the administration when looking for ways to connect alums to the school and school activities.

The above example demonstrates how an intranet can create an organizational efficiency that involves students and law school administrative offices. Another example of creating organizational efficiencies includes the ability to create a centralized and robust collection of library offerings tailored for and tied to a specific class offering. Instead of the one-size-fits-all approach to presenting information, there exists a greater degree of control over who, how, and when information can be distributed. The intranet should allow for a greater school-wide approach to pushing the information out to particular constituencies, because the design should focus on allowing the utmost flexibility for both ease of setting up as well as presenting content. Creating more and varied tools for self-help and instruction on use of resources will allow them to be sent to users in a just-in-time information delivery. Again, the library's external website can be a repository for information and users can be directed to these resources, but the intranet allows the librarian to more appropriately choose to pair specific resources with students or faculty and present them when they are needed. Just as the librarians no longer can serve as storekeepers of print warehouses, they can also no longer passively await students and faculty to appear in the library to seek assistance. The intranet allows for librarians to embed themselves throughout the law school in a way that serves as a reminder of their connectedness to the resources students need to succeed in law school and faculty need to thrive with their teaching and scholarship. It is not only about providing more with fewer human resources but about more crucially making an explicit connection between the work of the organization and the resources. Applications for the intranet are limited only by the imagination and the technical skills of the librarian or IT staff relative to the intranet platform chosen. The examples described below are those utilizing the SharePoint platform but they are certainly adaptable to other intranet solutions.

SharePoint allows for both blog and wiki functionality within the intranet. One immediate application is the ability to create space for collaborating with faculty, students, or administration. So, for example, if you have a law professor who is working on an interdisciplinary project with faculty from a different school on campus, you could set up a wiki for the professors to share material relative to the project and allow the librarian to monitor and provide additional materials for the project. Once the project is complete, it can be archived or become a subset of a larger wiki that tracks projects either an individual librarian is working on or a department or unit. Not only does this serve as a terrific repository for project information but it can also be expanded any number of ways.

Perhaps one of the most exciting applications that can be developed is a "ticketing system"—that is, a way of tracking requests for service. This can apply to any department or function within the law school for which service requests occur. One area very amenable to such a system would be requests for reference assistance. But rather than be confined to a physical location (i.e., people need to visit the reference desk to initiate a request), a ticketing system could be easily implemented to allow requests for research assistance to be done from any location. The ticketing system could be easily designed as a patron-initiated service in which forms are created for the information request. The ticketing system would then store the forms and allow for tracking completion of the request. The additional benefit from such a system would be the ability to track the number and types of requests. The forms would also be stored and available for query so that the librarian could build a knowledge base. The ticketing system provides a more streamlined approach to customer assistance. In addition, it would replace e-mails sent to individuals who may or may not respond timely, as well as replace a third-party reference tracker such as Gimlet.

If an academic library is utilizing one or more third-party applications to track requests, or the law school is using multiple ticketing systems for a variety of student needs, an intranet would allow the consolidation of these into one access point: a creation of a dashboard for students, faculty, and staff to follow any of their service requests. If the academic library is working with a standalone intranet, the ability to offer a place for a centralized ticketing system for the entire law school is an incredible opportunity for the law library to show it can move beyond its brick and mortar walls and serve as a conduit for a whole host of information important to the law school's efficiency as an organization.

Librarians need to pay attention to metrics. There is nothing more valuable to the librarian then the ability to quickly and easily provide evidence of the work accomplished by the librarians and staff. Academic law libraries are under constant pressure to show they are providing ROI, or return on investment. This is becoming increasingly so as budgets have been tremendously impacted in recent years. An intranet, particularly if developed with SharePoint, allows for a whole host of analytics about the information and resources presented through the intranet. Web-based analytic tools, like Google Analytics, can be deployed through the intranet. These are powerful tools for the law librarian in the digital age as they offer robust and seamless methods for, at a very granular level, understanding what resources are being utilized, who is utilizing the resources, and when are they being utilized. This is functionality that allows for truly evidence based librarianship and for gathering information of high value to any administration.

IMPLEMENTATION OF AN INTRANET

This chapter does not go into detail about designing, building, and implementing an intranet solution or portal. But it is important to remember that implementing an intranet is as much about people and the organization's culture as it is about technology. This holds true for either the academic law library or a private law firm. Adopting an intranet means changing how the people in the organization conduct and, more broadly, think about work. Any change, no matter how small, will likely cause concern. Here is a checklist of items to keep in mind throughout the process that will get you started on the right track.

1. Learn what version of SharePoint, or whatever platform that you'll be using, and attend training, if available. If working with SharePoint, an intermediate understanding of the features and specifics will be beneficial when working with vendors and solutions. The level of technical support varies by vendor; therefore, a librarian's robust knowledge can prove worthwhile.
2. Obtain permissions to edit library page(s).
3. Identify the stakeholders and those most likely to be the earliest beneficiaries of the intranet and invite them to participate in design at the earliest stages possible.

4. Design a user-friendly and intuitive interface.
5. Create a communication and training plan for rollout.
6. Request feedback from users that could lead to future enhancements.

To learn about two librarians' experiences with upgrading SharePoint and designing new portals, read "Driving Resource Usage through SharePoint Customization" by Molly Brownfield and Emily Florio.[21] The authors go through the steps they followed when upgrading to SharePoint 2007 while also providing insight and suggestions for guaranteed success.

Training Opportunities

There are many opportunities to receive training or learn about intranets and SharePoint. Having a basic understanding of how intranets work while also being familiar with some of the key terminology will show your organization's stakeholders that you have done your due diligence and are serious about the project. In addition, check for new and recorded webinars or other educational programs from library organizations, including:

- American Association of Law Libraries: http://www.aallnet.org/main-menu/Education/webinars
- Special Libraries Association: http://www.sla.org/content/learn/members/index.cfm

In addition to national organizations, your local library association might have a SharePoint or intranet user group. If one does not exist, then poll the members and start one yourself.

- The Law Librarians' Society of Washington, D.C. PLL-SIS SharePoint User Group: http://www.llsdc.org/en/users/login.asp?

If you currently work in a law firm that is a member of the International Legal Technology Association (ILTA), then it is free for you, as an employee, to join. There are knowledge management and other technology communities that might be relevant to SharePoint users: http://www.iltanet.org/. For a higher-level view, KMWorld has a free print newsletter and provides various free webinars: http://www.kmworld.com/.

POST-IMPLEMENTATION OF YOUR INTRANET

After you have successfully developed, implemented, and rolled out an intranet or portal, it is imperative that you continue to promote your new intranet and educate your users. The communication plan you created for the implementation should also include post-implementation action items, which might include ideas for targeted e-mails and trainings. You do not want there to be a lull in the adoption or interest in your newly created intranet; therefore, there are a few things you can do to keep your users engaged and interested. You might include a column in your firm's practice group newsletters or perhaps send out an electronic library newsletter all about your new intranet that highlights the resources and services available. Create quick videos that showcase how one of your new SharePoint Web parts works, so your end users will be able to make immediate use of the new tools.

Following the implementation and roll out, incorporate an overview of your intranet into your orientations for newly hired legal and administrative staff. For the academic law library, you will want to make sure to take advantage of student orientation sessions as well as faculty and staff meetings to present the intranet or get feedback on how people are using the intranet. In the law school library, you will benefit from the natural cycle of student progression through school. That is, if you can capture the attention and interest of a first-year student, it will make it easier to keep him or her interested in the second and third years. Crucially, although the law library serves both students and faculty, if the students can be sold on the benefit of the intranet, it makes an additional selling point for the faculty. Students and the student attitude toward the intranet are vital to its implementation and adoption. If students are sold on the benefits of the intranet, and in turn find it useful and convenient, it makes it much easier to bring otherwise reluctant faculty on board for adopting and using the intranet.

More generally, take every opportunity to have a teachable moment. In addition to incorporating an overview during orientations, ad hoc trainings can be done any time an appropriate question arises. If a law student visits the reference desk in need of locating the reference guide for contracts or an attorney asks the law firm library where they go to access HeinOnline, then use those moments to introduce the intranet and where the answer to their question can be found.

ADDITIONAL AND FUTURE OPPORTUNITIES WITH INTRANETS

The benefits and examples presented in this chapter only begin to scratch the surface of how intranets can be used within the law firm or law school library. Some firms are already addressing future uses through the support of extranets, the creation of dashboards, and with some Web 2.0 technologies, including wikis, blogs, and social media integration. Fish & Richardson's IT and Library Departments use SharePoint for their client extranets, which enhances the firm's commitment to case management and client relations. Extranets allow internal users to collaborate and share files with select external users in a controlled environment.

The Library and Knowledge Management Department at Debevoise & Plimpton LLP already uses their portal to support Web 2.0 technologies, including blogs, wikis, RSS feeds, and social tagging. These Web 2.0 tools leverage existing content, whether it is internally created or through paid subscriptions, to add value to their organization.[22] Other firms utilize third-party tools and custom development to create financial and human resources/benefits dashboards. Through custom development, the knowledge services team at Fox Rothschild has created dashboards for attorney profiles, experts, and outside counsel. Over the past few years, internal social media has increased in the legal community; so therefore, there is a definite opportunity for integration within an intranet.

Academic law libraries, especially those that are part of a state-funded institution, may not have the resources to keep up with all the potential new opportunities for the intranet. However, that does not suggest academic law librarians could not track and work to incorporate the newest opportunities. The key is to look for low-cost, high-benefit opportunities. A perfect example of a low-cost opportunity that could yield high benefit is developing higher value information about people as resources. One example, especially with SharePoint, is to create a robust list of adjunct faculty who teach at the school and include information on the courses taught over the years as well as their external connections. Not only can the administration utilize this tool, it is a tremendous resource for students seeking to find lawyers to discuss potential career opportunities. In addition to the previously mentioned power of metrics the intranet allows, increasing feedback on material and resources within the site through ratings—the use of "thumbs up" or "thumbs down"—can provide real-time feedback that is as useful as deploying survey instruments.

In addition, more and more, librarians are turning to video as a mechanism for both capturing information and training materials. Whether using a tool like Camtasia or YouTube, these materials can be embedded within the intranet and deployed to bring more just-in-time teaching opportunities. A caveat here: the growth of audio and video files will also need increased storage and bandwidth for sending, receiving, and storing materials. Make sure to consult with your IT department to ensure the infrastructure exists to proceed along this path. Perhaps a shorthand way to think about the changes that will be coming is to divide the intranet content between static and dynamic. Although an intranet allows material to be updated in a much easier and more efficient manner, webpages that offer information and do not change often are seen as static. Adding elements to the intranet that make it more dynamic, more responsive to collaborative and real-time information needs, are the elements to be on the lookout for future enhancements.

CONCLUSION

As noted in the introduction, law librarians, whether academic or private law firm, work within their respective organizations providing, for the most part, similar services. One of the most crucial services they can provide to either the firm or the law school is to develop or enhance their organization's intranet.

Necessary prerequisites for developing the intranet include understanding the IT infrastructure for the organization and who controls how IT projects are implemented, understanding the communications infrastructure and whether an intranet needs to be developed or if one exists and can be enhanced, and determining who is in control of adding content to the intranet. Law librarians in an academic setting may be more likely to have some control over the development and implementation of an intranet given their long-standing relationship with technology in the law school. Law librarians in a law firm may have to follow the lead of the law firm's IT group.

For both the law school and the law firm, the scope of the intranet development could be as simple as setting up the library's own page or adding library content to the organization's intranet, or the law librarians may develop and maintain the entirety of the organization's intranet. Much will depend on the communication infrastructure existing in the organization and the desire or willingness of law firm management or law school administration to ensure resources are available in the most efficient manner possible. And, although the library offers its own set of digital and other resources, the better approach is to think of those resources as part of the overall information ecosystem existing within the organization. Law librarians are information professionals whose job is to organize information for access. They are obviously in the best position to design the intranet for their materials and to help ensure the organization's legal research needs are being met. But if the intranet is utilized by the organization for providing access to other information, then the argument can be made for the law librarian to take an active role in the overall development of the intranet.

Finally, while there are several types of software that can be developed to create the intranet, the Microsoft SharePoint platform has become one of the most popular. Regardless of the platform chosen, the key to making the case for the intranet is that legal research, and other materials, will continue to move to digital format. Developing a method for 24/7/365 access will be crucial to ensuring those materials are properly leveraged within the organization. Ultimately, the law librarian, by assisting with the effective development of the intranet, can assist the overall organization to achieve the highest levels of productivity and efficiency. Regardless of whether that is a law firm or law school, this will make the law librarian a key player for the organization in the digital age.

FURTHER READING

For additional information on the law library intranet, we recommend the following resources:

Drisgill, Randy W. 2010. *Professional SharePoint 2010 branding and user interface design.* Indianapolis: Wiley.

Ennis, Lisa A., and Randy S. Tims. 2010. "Harnessing the power of SharePoint for library applications." *Computers in Libraries* 30 (5): 6–12. http://www.infotoday.com/cilmag/jun10/Ennis_Tims.shtml.

Florio, Emily, and Molly Brownfield. October 2012. "Driving resource usage through SharePoint customization." *AALL/ILTA The New Librarian.* http://read.uberflip.com/i/87421/26.

Friedmann, Ron. February 1, 2013. "New generation of law firm portals and good design principles." *Strategic Legal Technology* (blog). http://www.prismlegal.com/wordpress/index.php?p=1260&c=1.

Monte, Catherine. October 2012. SharePoint: A librarian's most versatile tool. *AALL/ILTA The New Librarian*. http://read.uberflip.com/i/87421/26.

NOTES

1. William D. Neill and James E. Richard, "Intranet portals: Marketing and managing individuals' acceptance and use," *Australasian Marketing Journal* 20 (2012): 147–148.
2. Marina Trkman and Peter Trkman, "A wiki as intranet: A critical analysis using the Delone and McLean model," *Online Information Review* 33 (2009): 1087–1088.
3. *Oxford English Dictionary*, s.v. "intranet," http://www.oed.com/view/Entry/248422.
4. *Oxford English Dictionary*, s.v. "portal," http://www.oed.com/view/Entry/148114.
5. Kathleen M. Kluegel, "Electronic resources for reference," in *Reference and information services: An introduction*, ed. Richard E. Bopp and Linda C. Smith, 150 (Santa Barbara, Calif.: Libraries Unlimited, 2011).
6. Sabrina Pacifici and Roger V. Skalbeck, Intranet and knowledge management solutions: Law librarians lead the way, *LLRX*, August 1999. http://www.llrx.com/features/lead.htm.
7. Robert C. Berring, "Deconstructing the law library: The wisdom of Meredith Willson," *Minnesota Law Review* 89 (2005): 1381–1406
8. William Caraher, "What's the point of SharePoint 2013 in a law firm?" *Law Technology News* (January 2013). http://www.law.com/jsp/lawtechnologynews/PubArticleLTN.jsp?id=1202584247390.
9. *ILTA'S 2012 Technology Survey* (Austin, Tex.: International Legal Technology Association, 2012), http://www.iltanet.org/techsurvey.

10. Catherine Monte, chief knowledge officer of Fox Rothschild, LLP, in discussion with the author, February 2013.

11. Linda Wells, electronic services supervisor at Ropes & Gray, in discussion with the author, February 2013.

12. "Westlaw portal Web parts," *Thomson Reuters*, http://info.legalsolutions.thomsonreuters.com/support/westlaw/portals/default.aspx.

13. "Download and support center," *LexisNexis*, http://www.lexisnexis.com/community/downloadandsupportcenter/media/g/us-legal-web-parts/default.aspx.

14. *EOS.Web SharePoint* (Carlsbad, Calif.: EOS, 2013).

15. "SydneyPLUS SharePoint Web part," *SydneyPlus International* (blog), August 6, 2010, http://sydneyplus.blogspot.com/2010/08/sydneyplus-sharepoint-web-part.html.

16. "Current Newsletter," *Inmagic*, http://www.inmagic.com/current-newsletter?q=sharepoint.

17. "Optional Modules," *Softlink*, http://www2.softlinkint.com/?nz/public_optional_modules.

18. *ALM Legal Intelligence 2012 Law Librarian Survey* (New York: ALM Media Properties, 2012). http://www.almlegalintel.com/SurveyDescription.aspx?id=Q/K8nmK4pG4=&type=fEFglaD+grg=.

19. See, for example, MoSpace, the University of Missouri Institutional Repository and Digital Library, at https://mospace.umsystem.edu/xmlui/.

20. "American Association of Law Libraries Strategic Plan 2013–2016," http://www.aallnet.org/main-menu/Leadership-Governance/strategic/strategicplan-2013-2016.

21. Emily Florio and Molly Brownfield, "Driving resource usage through SharePoint customization," *AALL/ILTA The New Librarian* (October 2012). http://read.uberflip.com/i/87421/26.

22. Steven A. Lastres, director of library and knowledge management, Debevoise & Plimpton LLP, in discussion with the author, March 2013.

Part VII

MARKETING

25

Digital Age Marketing

Carol Ottolenghi
OFFICE OF THE OHIO ATTORNEY GENERAL

Marketing is the third prong in the triumvirate—service, funding, marketing—that today's law libraries must keep in mind if they are going to thrive within their organizations and communities. These three prongs are tightly interrelated. Marketing cannot hide poor service. However, marketing can ensure that high-quality service is recognized and sought after. That recognition is necessary for libraries to maintain their funding because it is not enough for libraries merely to do good work. Libraries must be *seen* doing good work.

The need to "be seen doing good work" is a new imperative for law libraries. Law libraries traditionally promoted programs or materials, but, until recently, did not defend their existence. They did not have to—when it came to providing legal information, they were the only game in town. The digital age has changed that. The online standards—BNA, Hein, Lexis, LoisLaw, Westlaw—have added shiny new interactive features, cool buzzwords, and simplified searching. They and hundreds of other legal information providers are beating the drum loudly to promote their products; some have even suggested that law libraries are no longer necessary.[1]

Librarians have responded vigorously to that challenge. They are using digital age tools to demonstrate that, while law libraries may no longer be the *only* game in town, they are still the *best* game in town. In this chapter, we examine core marketing concepts and explore ways in which librarians are mashing-up their skill set with digital age tools to create inexpensive and effective library marketing.

WHAT IS MARKETING?

> Marketing's ultimate assignment is to serve customers' real needs and to communicate the substance of the company.
>
> —Regis McKenna[2]

Marketing Is Everything

In 1991, when Regis McKenna famously declared that "marketing is everything, and everything is marketing," computers were beginning to revolutionize corporate capabilities in the United States. Computer programming made product variations inexpensive, and consumers suddenly had real choices. McKenna enthusiastically hailed this new consumer power as the end of "the 'any color as long as it's black' school of marketing."[3] He predicted that marketing that did not incorporate significant consumer input would fail, and that advertising would become obsolete.

Some aspects of his theory are overstated. For example, advertising is still with us. However, McKenna's emphasis on consumer input, and on using a company's institutional knowledge and experience, signaled a seminal change in marketing thought. Marketing was no longer the sole responsibility of the marketing team but was an "all-pervasive, part of everyone's job description, from the receptionists to the board of directors."[4]

Currently, librarians are responding to a significant increase in possible user choices. Much of what was true for corporate business in McKenna's day is true for libraries today, so we can develop a definition of library marketing based on McKenna's article. For example, library marketing:

- Addresses a user's real needs
- Educates users as to what the library is and does
- Involves every staff member
- Is a conscious part of every library action

This definition is accurate, as far as it goes. It encompasses most actions that a library might take to market itself. However, it does not account for user–library relationships. Such relationships are implicit in McKenna's theory. We need an addendum that addresses those relationships *explicitly*. We find it by considering the aptly named relationship marketing theory.

Relationship Marketing Theory

McKenna's work reinvigorated research into relationship marketing (RM) theory. This school of thought had begun years earlier[5] and had fully integrated the importance of information exchange between providers and users. McKenna's article changed that. He established information exchange as a variable to be considered, even if a theory only considered it in order to dismiss it.

A review and analysis by Moller and Halinen[6] posits that RM emerged from the cauldron of several different marketing frameworks. Each of these frameworks emphasizes different aspects of information exchange. As they merged to create RM, the dynamics of provider–customer relationships became explicit.

Subsequent RM research was divided into two primary camps: consumer relationship-related marketing (CRM), and interorganizationally oriented marketing (IOM).[7] CRM research focused on the relationship between providers and individual users who were independent of each other. IOM research focused on the relationships between providers and users who were mutually interdependent. According to Moller and Halinen, CRM assumed:

- Low relationship complexity
- Short-term relationships
- A large number of users, giving both buyers and sellers multiple options
- Many alternative resources (people, products, information) that could be substituted easily
- Action was primarily taken by the provider

Conversely, IOM assumed:

- High relationship complexity
- Long-term relationships
- A limited number of users
- A limited number of providers
- Linked and specialized resources that made it costly to change providers or users
- Action could be taken by either the user or the provider

What Lessons Does Relationship Marketing Theory Have for Librarians?

The tension between the CRM and IOM schools of thought offers guidance for law libraries trying to stand out in a field crammed with legal service providers. It's fairly clear to librarians that library–user relations fall squarely in the IOM camp. We see that:

- Relationships routinely span years.
- Users are limited by school, organizational, or geographic boundaries.
- There is a limited number of providers.
- Library and user resources are linked and specialized.

It might not be so clear to our users. Users who view library–user relationships through an IOM lens, as we do, identify the law library as *their* library, and acknowledge the co-mingling of resources. Perhaps most importantly, they recognize that the law library offers something unique, something that they cannot get from other providers.

However, users who view us through a CRM lens see law libraries as one of many providers, *all of whom are equally valued.* They consider themselves independent of the library, with no sense of ownership or joint resources. They do not support the library any more than they would support a particular fast food restaurant. Perhaps the most discouraging thing is that they do not give us opportunities to support them because they are not aware that we have anything special to offer. This is where libraries must not only do good work but be *seen* doing good work.

We will use these insights from relationship marketing theory to augment the definition of library marketing developed above. For the purposes of this chapter, library marketing is the conscious, organization-wide effort to build and maintain library–user relationships that results in a measurable increase in the library's ability to:

- Address a user's real needs
- Educate users as to what the library is and does
- Be recognized by the user as a unique provider

HOW TO BE SEEN DOING GOOD WORK

No two libraries are the same. So, the first step in an effective library marketing effort is to answer two questions: "What does this library provide?" and "Who are the library's users?" The answers to these questions give you the information necessary to address a user's real needs, educate users as to what the library is and does, and be recognized as a unique provider. Without answers to these questions, you may waste time, money, and credibility on unsuccessful programs or publicity.[8]

What Does the Library Provide?

All law libraries can respond similarly to the first question. At their most fundamental, libraries provide a place, a collection, and, most importantly, the librarian skill set. One of the primary differences between libraries is the way in which they provide these three basics.

A Place

Libraries provide a place. The place can be real or virtual, large or small, old or modern, welcoming or not. It might be a gathering place or a place of solitary study. The emotional responses that users have when they enter the library's space can range from fear of appearing foolish, to relief at an answer, irritation with a dropped connection, or feeling of comfort (see the sidebar "Inspiring Positive Social Media Posts about Your Library #1"). Those feelings can encourage users to engage with a librarian or deter them from ever returning. An effective marketing effort will acknowledge—and possibly improve—the sense of place that your library provides.

Some libraries are using digital age marketing tools like Quick Response (QR) codes to expand the virtual place they provide. In California, the Contra Costa County Library System[9] teamed with area transit authorities to produce "Snap & Go," an initiative that uses QR codes to link mobile users to the library. People can snap a picture of the QR code on a bus or wall poster, and access e-books, library information, text-a-librarian services, museum passes, and more. The QR codes expanded the library's "place" to everywhere the buses go.

DIGITAL AGE WORD-OF-MOUTH ADVERTISING: INSPIRING POSITIVE SOCIAL MEDIA POSTS ABOUT YOUR LIBRARY #1

Many law libraries try to make their users comfortable. The Lillian Goldman Law Library at Yale offers an innovative example. Staff there consciously and regularly reevaluate user needs and how those are reflected by the space the library provides. In addition to legal resources, students and faculty may check out a variety of unusual items, including bicycles, phone chargers, umbrellas, and iPads. In 2011, recognizing that law students often feel highly stressed, staff brought in Monty, a certified therapy dog, for short, in-library visits.[10] A warm, fuzzy media frenzy ensued, with stories on NPR and in papers around the world.

The media response—however welcome—is incidental. What matters in terms of marketing is the effect that Monty has had on the library's users' perception of the library. That effect has been positive. Monty is "checked-out" on a regular basis. In addition, the students share pictures of Monty on social networking sites; in effect, they are telling their friends that the law library is a great place to go.[11]

A Collection

Of course, a place without a collection is not a library at all—it is just a place. Law library collections include print, digital, fiche, maps, tape, and CD/DVD materials. Many of these items do not get used because people do not know about them. Digitization, which is discussed elsewhere in this book, can dramatically increase the use of your collection. However, unless you have money to waste, your choice of items to digitize is critical. From a marketing point of view, libraries should digitize only those materials that fill their users' needs.

Many digital age tools can help you make it easier for people to use (or even know about) your library's collection. We discuss several later in the chapter, but briefly:

- Start with a mobile-friendly library website. If people cannot find you on the Web, they may not even consider you as a resource.[12]
- Incorporate do-it-yourself library aids, such as audio tours and Quick Response (QR) codes.
- Add short instructive pieces on how—*and why*—to use specific resources.
- Use mobile-friendly interfaces for your library's electronic resources and databases.

The Librarian Skill Set

The first two basics that libraries provide—a place and a collection—are what is often identified as "the library" by our users. However, it is the librarian skill set that transforms a place with a collection into a library. Librarians are often the only organizational members who have this skill set; this makes us unique.

Our users often do not use, or even recognize, the depth and breadth of our skill set—so I make a point to define it when I present the library's portion of the Ohio Attorney General's Office's "new hire" orientation. Everyone who joins the Ohio Attorney General's Office in any capacity hears about the collection, our home-grown online libraries, the online services, and the fact that library staff welcomes all research questions, "The weirder, the better." These are all valuable library products and services. However, the most important information that I convey is an explanation of the librarian skill set. The set includes the ability to:

- Define the issue and needed information. The reference interview is an excellent template for clarifying what information is needed about the who, what, when, where, why, and how of any issue. This skill is an invaluable tool for answering information requests and targeting current awareness services. However, its use goes far beyond that. You can use it to plan classes, programming, collection, and space changes. It can also help you determine what emerging issues like eDiscovery or changes in intellectual property law mean for your users.
- Locate responsive information. This includes deep, complex research, as well as simple retrieval. Librarians are trained to scour resources in many different formats to find needed materials.[13]

- Evaluate information resources. New evaluation challenges have risen with the digital age, prompting the occasional, "You probably don't want to cite Wikipedia in a brief." We value accuracy, authenticity, and authoritativeness. Our users do, also, but they may not recognize it by those terms.
- Organize information in a searchable manner. The digital age makes terabytes of information readily available to all of us. But if users cannot isolate the information they need, it does not matter how much information they find. Librarians are trained to organize materials so people can find and use them. As Nunn and Ruane phrase it, it's not that "librarians have all the answers, but that . . . librarians [are] a guide and a mediator through the labyrinth of information that is now available at the click of a mouse."[14]
- Disseminate information. Once librarians find and evaluate information, they share it appropriately. The digital age has made it easier to share information of all types, from all times (see the sidebar, "Inspiring Positive Social Media Posts about Your Library #2").

This skill set can be applied ubiquitously. The topic does not matter. It does not matter what technology is involved. It does not matter when or where something occurred or is occurring. These skill sets can be applied to anything, *provided users remember that they exist.*[15] However, busy people often forget things—like libraries—that are not immediately in front of them. To counteract that, I augment my orientation spiel with regular e-mails, tweets, and other digital marketing. Anecdotal evidence indicates that these reminders encourage people to ask for the specialized skill set that we bring to the table.[16]

DIGITAL AGE WORD-OF-MOUTH ADVERTISING: INSPIRING POSITIVE SOCIAL MEDIA POSTS ABOUT YOUR LIBRARY #2

Poland's Weilkopolska Digital Library collection includes daily newspapers and other documents from multiple Polish libraries. Many of these documents date back almost 200 years. The delicate physical state of the documents meant that this part of Polish library collections had been available only on a *very* limited, on-site basis. When the collections were digitized, "a situation which in a cultural sense is entirely new [was created]. The general public . . . obtained access to practically unknown collections."[17]

In 2009, researchers Gorny and Mazurek evaluated the Weilkopolska Digital Library collection's usage. They began by defining the user base. Users would be those:

- Interested in the very specialized collection.
- Who choose to read. In 2008, 62 percent of Poles did not read for pleasure.
- Had the technology to access the collection. In 2008, less than 57 percent of Poles used the Internet for anything. In rural areas, where over one-third of all Poles lived, only 32 percent had access to the Internet.[18]

Gorny and Mazurek used an online survey to analyze the digital collection's actual users. They found that in 2008 there were between 25,000 and 30,000 regular users. These users made an estimated 2,212,000 visits. Most interesting from a marketing viewpoint is the significant amount of user-generated social media activity that researchers noted. In 2008, approximately 8,000 library users, almost one-third of the regular readership, had established reader accounts. These accounts allow users to interact with the library and other users by evaluating resources and adding tags. Researchers found that very active social networks had evolved, linking academics, genealogists, students, local historians, and others using the library. It will be interesting to see what effect this digital age "word-of-mouth" has on collection usage.

A place, a collection, the library skill set: These are the three basics that every library provides. These are the "products" that every library must market. The next question is, "To whom are we providing these products?"

Who Are the Library's Users?

Public law library users are different from firm library users, who are different from law school library users, who are different from government agency law library users. Our legal collections consciously reflect those differences. Effective marketing efforts must consciously consider user makeup, as well. For the purposes of this chapter, users are divided into three primary categories: current library users, potential library users, and interested third parties.

Current and Potential Library Users

When depicting current library users for marketing purposes, dig deeply into their demographics. Try to determine their:

- Information needs
- Ages
- Education
- Abilities and willingness to use new technologies (see the sidebar "Knowing Your Audience = Knowing Their Technology")
- Locations, including online
- Cultural needs, including expanded hours to offset religious observances

KNOWING YOUR AUDIENCE = KNOWING THEIR TECHNOLOGY

A key measurement of a marketing effort's success is: How many members of the target audience got the message? Obviously, you want to maximize that number. In order to do so, you must deliver the message in a way that your audience can receive it. When marketing with Web 2.0 applications, you need to know what technologies your audience can and will access.

National surveys on the subject can only point to trends. For example, in February 2012, 46 percent of American adults owned smartphones.[19] What would that mean for your library's marketing efforts? Would mobile applications and Quick Response (QR) codes be worth the effort to develop and implement? The answer depends on your specific audience. As of this writing, if you are marketing to:

- Law school students, then you might already be using QR codes and mobile apps. Many colleges and universities have implemented them because large numbers of students love discovering convenient new uses for their smartphones.
- Middle-aged (or older) partners who can affect your budget, include executive training sessions in your marketing plan. They have to know how—and why—to use the technology *before* you implement it.
- Public and county law library users, know that some of your audience can access mobile marketing, and some cannot. Among those surveyed, blacks, Hispanics, and urban dwellers had higher smartphone ownership than did whites. People with lower incomes had lower smartphone ownership.[20] Your audience is likely to be less technologically homogenous than a school or law firm; effective marketing efforts will recognize that disparity.

- Government employees, check for agency policies that limit the effectiveness of certain marketing tools. For example, the cameras in many government-issued smartphones are purposely disabled. That makes certain mobile apps superfluous.

Superfluous for now. But perhaps not forever. Technology and organizational attitudes evolve. Librarians must be conscious of this evolution. We must employ marketing tools that match a target audience's existing technological capabilities, and plan for tools that match developing capabilities. When we do that, our target audience will get the library's message.

When defining the characteristics of potential users, use the demographic list above and add:

- Perception of the library
- Reason(s) for not using the library now

Not all of these aspects will matter to every library. However, the more you know about your current and potential users, the better you will be able to address their real needs.[21]

Interested Third Parties

Interested third parties are users who have a financial or organizational influence on the library. An interested third party could be a university administrator, a legislator voting on a budget, a firm's managing partner, or a chief information officer who has suddenly taken over the library.[22] Interested third parties may or may not be current users.[23] When defining the characteristics of these influential users, use the lists above and add:

- Perception of the library's role in the organization or community
- Perception of the importance of that role to the organization or community
- Perception of the unique value that the library brings to that role

Let us return to the Contra Costa County Library System's QR code initiative for a moment. It provides an excellent example of how knowing your users can pay off.[24] Before launching Snap & Go, library staff identified their potential users as all county residents. They learned that:

- Forty percent of the county's residents commuted outside county borders to work.
- County residents were above the national average for owning smartphones.
- Sixty-seven percent of the population was between the ages of fifteen and sixty-four, an age-range heavy with mobile service users.

Library staff used this and other information to determine (1) what services were needed by their potential users, and (2) how best to deliver and publicize those services. This background research resulted in services that were welcomed by county residents. Usage of library's mobile site rose by 16 percent.

These results brought the library numerous regional and national awards, including several from ALA for technology innovation and marketing excellence. These awards and the accompanying media hoopla caught the attention of the library's interested third-party users. Thus, even if the third-party users never used the library, they saw it filling a unique role that was needed by the community. They saw the library doing good work.

DIGITAL AGE MARKETING TRENDS AND TOOLS

In the remaining pages of this chapter, we look briefly at some of the marketing trends and tools available to us in the digital age. Many of these are discussed in depth in other chapters of this book. It is important to note that law librarians should be aware of digital tools and trends *even if they cannot or will not apply them in their libraries.*[25] Furthermore, it can be beneficial to inform our interested third parties and other users of why we make certain technology choices. Why is it beneficial? Because it subtly underlines the fact that law librarians look forward as well as backward. It emphasizes our evaluation skills. It establishes librarians as an up-to-date, go-to resource on trending technology issues.

It also—and this may be the most important benefit—gives librarians a way to convert interested third parties and potential library users into active users. Think back, for example, to McKenna's assertion that marketing the organization is everyone's responsibility. Law librarians are part of a larger organization. With a little research, librarians can determine ways in which our skill set can complement the marketing department's skill set. In fact, many of us are already using our research and analytical skills in this manner; digital age tools help us do that faster and more effectively. As your read the section below, consider how you can apply these trends and tools not only to your library, but within your larger organization.

The Trend—Listening

Effective marketing is research driven. Part of that research is "listening" to what is being said about your library, your larger organization, your users, your competitors, and specific topics. Beth Kanter[26] identifies four skills needed to listen effectively online. Listeners must be able to:

- Refine keywords
- Analyze patterns
- Synthesize or digest findings
- Interact effectively with the "speakers"

The first three skills Kanter identifies—the ability to refine keywords, analyze patterns, and digest findings—fall squarely within the librarian skill set. The fourth skill—the ability to interact effectively—is more subjective. If you are listening on behalf of your library, the person who eventually interacts with the "speakers" should have a thorough understanding of the library, its users, and tact. If you are listening for your larger organization, let the marketing department or practice area specialists handle the interaction phase.

The Tools

Monitoring, tracking, and analytics tools evolve quickly. Many, including Google's suite, are free. You will need some sort of:

- *RSS reader.* RSS stands for Real Simple Syndication. This lets you know when the blogs, journals, microblogs, and other sources that you follow are updated.
- *"Alert" function.* This scans the Internet for relevant blog posts, articles, Web posts, Tweets and other microblog posts, columns, and breaking news. Then it pops the results in your e-mail or RSS reader. Current tools include Technorati, Google Alerts, Board Reader, and BackType.
- *Social media search tool.* These are similar to alerts, but they focus on social media outlets. Current tools include Twitter Search, Social Mention, and Social Searcher. Social Searcher can be adjusted to search multiple social media platforms, including Facebook, Twitter, and Google.[27]

The Trend—Joining the Online Conversations

Establishing a library Facebook page or Twitter/microblog account is very easy. Encouraging users to follow it may not be as simple. In 2010, Harvard Law School librarians asked students which library message stream—e-mail, Twitter, library blog, or Facebook—they paid attention to. E-mail emerged as the big winner.[28] Other school libraries have also learned that while students like the library, they don't want to "Like" it on Facebook.[29]

However, Twitter and blogging offer an innovative opportunity for law librarians to strut their librarian stuff. For example, many law firms have Twitter accounts. And many of those Twitter accounts are so general and self-referential as to be completely unimpressive. The consensus in the blogosphere is that these microblogs fail because they do not address specific user needs.[30] Librarians could help their firms identify what information their users want. This would result in more effective firm social media marketing and highlight the librarian skill set and the value of the library.

The Tools

There are myriad tools and sites to help you blog, including weblog tools WordPress and Movable Type. These tools evolve rapidly. Fortunately, many librarians are already testing them, and, being librarians, they like to share what they discover. So, when you start researching which tools will be best for your library users, check out some of the library wikis. *Library Success: A Best Practices Wiki*[31] is an excellent resource. It has clear explanations of what many of the tools are, how they work, how to apply them, and links to libraries that are using them.

Some of the current "online conversation" tools include:

- *Pinterest.* Blogs and tweets are primarily verbal. Other tools, like Pinterest, are visual. Pinterest lets you "pin" photos to a virtual bulletin board, add descriptive text, and then share it. Librarians at Central Methodist University (CMU) in Missouri use Pinterest to highlight new acquisitions. They began posting on Pinterest when they realized that few users were reading the library's "new resources" e-mails. Now they take photos of new books, add annotations, and "pin" it online. Faculty and student response has been measurably positive.[32]
- *Linkbacks.* Linkbacks, also called pingbacks, alert you when someone links to your blog, article, or other message. They provide some of the metrics that you need in order to determine if your users are "hearing" you.
- *Twitter.* Developed in 2006, Twitter has been both a boon and a boondoggle. As discussed above, tweeting that ignores user needs wastes time, money, and goodwill. But tweeting that meets or exceeds user information expectations can be a marketing boon. For example, the Defiance Public Library System (DPLS) in Ohio uses Twitter to promote new resources, encourage tweeted book commentaries and microdiscussions, and alert people to canceled programs.

Several years ago, Sarah Milstein offered "Twittiquette" tips for institutions that are still excellent guidelines for librarians using Twitter and similar microblogging tools.[33] These Twitter etiquette tips include:

- Let people know who is doing the tweeting. Fill in the bio and URL fields. Librarians know that sources matter; make sure our users do, too.
- Treat your microblog as a "conversational" opportunity, not as an electronic bulletin board. Many libraries just post lists of new acquisitions. That's nice, but it does not allow for much user response. Libraries like DPLS (see above) that ask open-ended questions encourage user response. This reinforces a sense of user ownership in the library.
- Post daily, but do not overwhelm people with too many posts. Be part of your users' daily lives, but not an irritating part.

The Trend—Taking the Library to the Users, Anytime, Anywhere, and without Wheels

"Taking the library to the users" used to mean loading up the bookmobile. No longer. Increasing library outreach has never been this easy.

The Tools

- *Quick Response (QR) codes.* Quick Response (QR) codes are two-dimensional bar codes. Smartphones and mobile phones with cameras can read them, if the phone has the application (app) to do so. The bar code holds information—anything from a link to an e-book collection, to the nutritional content in a fast food hamburger. Many law schools libraries are using QR codes to guide students through the stacks, and to link them to additional practice area resources.
- *Mobile websites.* Traditional library websites can be very information heavy. The amount of information on them can overwhelm a smartphone or tablet. This frustrates our users. To alleviate that, libraries such as Mendik Library at the New York Law School are creating trimmed-down mobile websites that reflect the technology their users are using. In 2011, Mendik developed a mobile app that is available for several smartphone platforms. The free app lets students search the catalog, find library contact information, and access law-related websites.[34] Other libraries are testing responsive websites, which automatically adjust to the size of the user's screen.
- *Podcasts, webcasts, and vodcasts (video).* Law libraries post a wide variety of audiovisual materials on YouTube and similar sites. Some are strictly instructional, like Temple Law Library's YouTube video series of very short

introductions to different legal resources.[35] Others, such as those posted by Cornell Law Library, explore specific legal issues based on books authored by Cornell faculty.[36] It does take an initial investment of time and money to produce instructional podcasts, webcasts, and video. However, there are three distinct marketing payoffs. (1) Students and other users can access YouTube anytime, anywhere, so there is an increase in outreach; (2) Staff will not spend as much time on general instruction; and (3) When the library honors faculty and other interested third parties, those interested parties see the library doing good.

- *Robust, multiuse platforms like SharePoint.* These platforms provide collaborative space, intranet and Internet portals, and social networking capabilities. Many school libraries use the platforms to embed articles, books, and multimedia in online classes pages. Other law libraries use platform document management functions to create and organize online collections. These can greatly improve outreach. For example, the Ohio Attorney General's Office (AGO) library has two SharePoint collections: "eDiscovery Resources at the AGO" and "Law Enforcement Resources." These are used by more than 300 staff members each month—almost 50 percent of the legal and investigative staff—with an average of over 4,000 downloads per month.

The Trend—Aggregating Data

Librarians and attorneys are using digital age tools to wrangle unrelated datasets from multiple sources into useful new resources. Many of the resulting resources provide cross-discipline insights. Others just make the legal life a little bit easier.

The Tools

- *"Big Data" technologies.* Many legal issues, such as eDiscovery, require the collection, organization, storage, search, analysis, dissemination and often the visualization of datasets so enormous that they overwhelm traditional processing methods. Big Data technologies address that very issue.[37] Many librarians are working with attorneys and litigation support teams to create eDiscovery search terms and organizational schemes.
- *Mash-ups.* In music, a mash-up combines parts from several different songs to produce a new song. A digital mash-up combines the data, presentation, and function from several different resources to create a new resource. Some original resources, like Google Maps, provide the application programming interfaces (API) that you need to use their data.[38] Librarians are using mash-ups to market themselves in a variety of ways. For example, the Portland State University Library of Oregon created a current awareness service that

mashed-up RSS feeds with different datasets to produce a single RSS feed. The mash-up both simplifies and enriches their users' experience. Users who subscribe to that single RSS feed receive librarian-selected resources without having to manage multiple RSS feeds.[39]

A Word about Collecting Social Media Statistics

Effective marketing is research driven. Our definition of library marketing specifically includes the word "measurable" because hard numbers tell us if our marketing efforts are effective, and help us justify them to management and interested third parties. Currently, social media applications have varying abilities to capture meaningful metrics. As of this writing, there is no single aggregator of statistics; if you post on a blog, website, Twitter account, and YouTube account, you have to use multiple programs such as Technorati, Google Analytics, TweetStats, and YouTube to get all of the metrics, none of which is qualitatively equivalent to another.

As with everything else digital, the capability for metrics collection will evolve rapidly. Libraries using social media will have to evaluate which metrics are truly important. For example, is the number of blog's RSS subscribers more or less important than the number of Twitter followers? How much weight should retweets get? How much weight should be given to Facebook-launched online public access catalog (OPAC) searches? The answers will vary with your message and users.[40]

CONCLUSION

Properly applied digital age technology can make necessary marketing easier and more productive. The new technologies are quick, relatively simple, and have huge outreach potential. The downside of that quick simplicity is the huge number of digital "pokes" that everyone gets from family, friends, school, coworkers, organizations that we belong to, and companies that we might have visited once.

Your library's message can easily get lost in this glut of information. To avoid that, give your users what they need in ways that they can and will access it. Let them know that you will not waste their time. Answer the questions, "What does the library provide?" and "Who are the library's users?" Use the answers to inform all aspects of your marketing. Every blog entry, tweet, e-mail, podcast, QR code, promotional piece, publicity release, program, public relations campaign, branding change, and YouTube post should reflect what *your* library provides to *your* users.

FURTHER READING

Marketing theory moves at a stately pace. Digital age marketing tools, however, evolve with almost unseemly speed. To reflect this dichotomy, the short resource list below includes both works that are "evergreen" and serials to keep you informed of the latest marketing trends and tools. In addition to the resources listed below, the American Association of Law Librarians and the American Library Association websites have regularly updated marketing pages.

American Association of Law Librarians. *Spectrum.* http://www.aallnet.org/main-menu/Publications/spectrum.

Crosby, Connie. 2010. *Effective blogging for libraries.* Chicago: Neal-Schuman.

Dempsey, Kathy. 2009. *The accidental library marketer.* Medford: Information Today.

Dowd, Nancy, Mary Evangeliste, and Jonathan Silberman. 2009. *Bite-sized marketing: Realistic solutions for the over-worked librarian.* Chicago: American Library Association.

Hastings, Robin M. 2010. *Microblogging and lifestreaming in libraries.* Chicago: Neal-Schuman.

Holloman, Christer, ed. 2012. *The social media MBA: Your competitive edge in social media strategy development & delivery.* Hoboken: Wiley.

Information Today, Inc. *Online: Exploring technology & resources for information professionals.* http://www.onlinemag.net.

———. *Marketing Library Services.* http://www.infotoday.com/mls/default.shtml.

International Legal Technology Association (ILTA). *Peer to peer: The quarterly magazine of ILTA.* http://www.iltanet.org/MainMenuCategory/Archives/PeertoPeerArchives.aspx.

Kanter, Beth, and Allison H. Fine. 2010. *The networked nonprofit: Connecting with social media to drive change.* San Francisco: Jossey-Bass/Wiley.

Kaushik, Anna, and Jagdish Arora. 2012. "Blogs on marketing library services." *Journal of Library & Information Technology* 32:186–192.

Kroski, Ellyssa. April 6, 2011. "Libraries to go: Mobile tech in libraries." http://www.slideshare.net/ellyssa/libraries-to-go-mobile-tech-in-libraries-presentation.

Lastufka, Alan, and Michael W. Dean. 2008. *YouTube: An insider's guide to climbing the charts.* Beijing; Sebastopol: O'Reilly Media.

"Library success: A best practices wiki." http://www.libsuccess.org.

Murphy, Joe. 2012. *Location-aware services and QR codes for libraries.* Chicago: Neal-Schuman.

Potter, Ned. 2012. *The library marketing toolkit.* London: Facet.

Schmidt, Aaron, and Amanda Etches. 2012. *User experience (UX) design for libraries.* Chicago: Neal-Schuman.

NOTES

1. Lynn Blumenstein, "West marketing message urges bypass of law librarians," *Library Journal* (August 25, 2009), http://www.libraryjournal.com/article/CA6685861.html?rssid=191.

2. Regis McKenna, "Marketing is everything," *Harvard Business Review* 69 (1991): 70.

3. McKenna, "Marketing is everything," 66.

4. McKenna, "Marketing is everything," 69.

5. See Richard P. Bagozzi, "What is a marketing relationship?" *Der Markt* 51 (1974): 64–69, for a seminal look at the theory. Also see Franklin S. Houston and Jule B. Gassenheimer, "Marketing and exchange," *Journal of Marketing* 51 (1987): 3–18.

6. Kristian Moller and Aino Halinen, "Relationship marketing theory: Its roots and direction," *Journal of Marketing Management* 16 (2000): 29–54.

7. Moller and Halinen, "Relationship marketing theory: Its roots and direction," 40.

8. Years ago, a nonprofit organization that I worked for asked me to create a day of programming for the general membership. I spent hundreds of hours and several thousand dollars to produce an outstanding lineup of speakers and panels. Membership turnout was low. I was crushed. One of the board members told me not to take it personally. He said, "You can put together the best program in the world, and if people don't want to come, you can't make them." I took that lesson to heart. Now, when I and a few others think something is a great idea, I ask if it addresses a real user want or need. If it is not something my users want or need, I am not wasting my resources on it.

9. Contra Costa County Library's website, QR pages, http://guides.ccclib.org/qr.

10. Julian Aiken, "Meet Monty," *Yale Law School Lillian Goldman Law Library* blog, September 19, 2012, http://library.law.yale.edu/blogs/news/2012/09/19/meet-monty.

11. According to January 20, 2013, phone conversation between Monty's owner, Julian Aiken, and the author, Mr. Aiken made a conscious decision to *not* make a Facebook page for Monty. The dog occasionally appears in the university and library system blog posts.

12. See Ball State University Libraries for an excellent example that includes a mobile card catalog, http://www.bsu.edu/libraries/mobile/.

13. Brent Nunnand Elizabeth Ruane, "Marketing gets personal: Promoting reference staff to reach users," *Journal of Library Administration* 51 (2011): 291–300.

14. Nunn and Ruane, "Marketing gets personal," 297.

15. See Gavan J. Fitzsimons et al., "Non-conscious influences on consumer choice," *Marketing Letters* 13 (2002): 269–279, for a discussion of how memory affects choices. The literature review's implication for libraries is that "out of sight, out of mind" is a very real danger.

16. My weekly e-mails are regularly opened by over one-third of the more than 1,400 recipients. Online resources mentioned in the e-mails show a spike in usage. In addition, the e-mails spark unrelated reference requests; there is a discernible "Hey, I can ask the library" effect. Tracking the response to my digital outreach is vital. It helps me determine if the outreach has crossed the line between informative and irritating. So if, for example, many people stopped opening my e-mails, I would send fewer of them.

17. Miroslaw Gorny and Jolanta Mazurek, "Key users of Polish digital libraries," *Electronic Library* 30 (2012): 544.

18. Gorny and Mazurek, "Key users of Polish digital libraries," 553.

19. Aaron Smith, "46% of American adults are smartphone users," *Pew Internet & American Life Project*, 2012, http://pewinternet.org/Reports/2012/Smartphones-Update-2012.aspx.

20. Aaron Smith, chart "Smartphone ownership within demographic groups, 2011–2012," *46% of American adults are smartphone users*, 4.

21. For a discussion on converting potential users, see "Convincing the swing vote: How to lure 'non-customer,'" *Knowledge@Wharton*, http://knowledge.wharton.upenn.edu/article.cfm?articleid=3084.

22. See Greg Lambert and Scott Preston, "Building a bridge between IT and library services," *Peer to Peer* 3 (2011): 62–64, for a discussion on library–IT collaboration.

23. Your marketing job is easier if they are current users, so try to bring them into the library fold.

24. For an excellent example of a complete marketing project, see the Contra Costa County Library's award-winning John Cotton Dana entry "Snap & Go! Library service access points anywhere you can stick a QR code," http://www.ebscohost.com/resources/john-cotton-dana/winners/2012/contra-costa-county-library.pdf.

25. See Camille D. Reynolds's June 2012 survey results, "Social media @ law firm libraries." The format is awkward to read, but the librarians' responses are instructive. http://www.slideshare.net/creynolds2009/survey-social-media-law-firm-libraries.

26. Beth Kanter, "Listening literacy for nonprofits," *@BrianSolis: Defining the Impact of Technology, Culture and Business* blog, August 1, 2009, http://www.briansolis.com/2009/08/listening-literacy-for-nonprofits%E2%80%A8/.

27. See Ian Cleary, "4 great tools for searching social media," on Razor Social for quick hints on using these tools. http://www.razorsocial.com/4-great-tools-for-searching-social-media/.

28. George Taoulstides and Meg Kribble, "Reaching out and meeting needs," *AALL Spectrum* 16 (2012): 12–13.

29. Joe J. Marquez, "Marketing in the library and beyond," *PNLA Quarterly* 76 (2011): 52–69.

30. Jordan Furlong, "Can a full-service law firm use Twitter?" http://www.stemlegal.com/strategyblog/2011/can-a-full-service-law-firm-use-twitter.

31. "Library success: A best practices wiki," http://www.libsuccess.org.

32. Cynthia Dudenhoffer, "Pin it! Pinterest as a library marketing and information literacy tool," *College & Research Library News* 73 (2012): 328–332.

33. Sarah Milstein, "Twitter for libraries (and librarians)," *Computers in Libraries* (May 2009).

34. The Mendik Mobile Application won the 2012 AALL Excellence in Marketing: Best Use of Technology Award. For more information, see the New York Law School's Mendik Mobile app page, http://www.nyls.edu/library/mendik_mobile_app/.

35. Temple Law Library, http://www.youtube.com/user/templelawlibrary.

36. Cornell Law Library, http://www.youtube.com/user/CornellLawLibrary.

37. Cat Casey and Alejandra Perez, "The rising tide of nonlinear review," *National Law Journal*, January 30, 2012, online edition, http://apps.us.hudson.com/legal/new/NLJ_RisingTide_HudsonLegal.pdf.

38. Google, "Google Maps API," https://developers.google.com/maps/.

39. Linda Absher, Adriene Lim, and Kerry Wu, "Library mashups for the virtual campus: Using Web 2.0 tools to create a new current awareness service," *Sailing into the future: Charting our destiny*, proceedings of the thirteenth National Conference of the Association of College and Research Libraries, March 29–April 1, 2007.

40. David Stuart, "Social media metrics," *Online: Exploring Technology and Resources for Information Professionals* 33 (2009): 22–24.

26

Competitive Intelligence

Jennifer Alexander
MCKENNA LONG

M. T. Hennessey
PAUL, WEISS, RIFKIND, WHARTON & GARRISON LLP

The practice of competitive intelligence (CI) is emerging as an integral part of doing business for many law firms, especially the larger firms. Many corporations have embraced CI for some time, making it a necessary part of their strategic and resource planning process. Recently, as competition among law firms for legal work has intensified, many of them have embraced CI for business development efforts, such as acquiring new clients and retaining and building new business with existing clients. In addition, law firms are using CI to gain deeper insight into their competitors as well as their own operations.

Librarians can take a significant role in the CI efforts at their law firms. Librarians already have many of the skills and training necessary for the CI position, as well as a devotion to customer service. Librarians interested in working in the CI field will need to cultivate other necessary competencies, such as data analysis, presentation skills, and creative thinking. It is a new and exciting field for law librarians, especially now that law firms are recognizing the importance of CI for their firms' futures.

Recognizing the primary role librarians play in research, some law firms have entrusted the responsibility of the CI function to their libraries. Hence, many law librarians have transitioned to the role of "CI librarian," offering not only legal research services but also managing and delivering nonlegal research, in-depth analysis, and so forth. In addition, the new CI librarian has an invaluable role assisting in the development of new business and supporting the law firm's strategic decisions.

WHAT CI IS AND IS NOT

First, what is CI? One of the premier CI organizations, the Strategic and Competitive Intelligence Professionals, also known as SCIP, provides a very general definition: "CI is a necessary, ethical business discipline for decision making based on understanding the competitive environment."[1]

This is a good start, but we are going to expand the definition to include researching, identifying, collecting, analyzing, and delivering insights into the strengths, weaknesses, and strategic initiatives of competitors as well as potential competitors, clients, and prospects. In addition, CI entails monitoring and reporting on significant developments within the business and economic environment that will likely affect the firm and/or its clients. It is important to note that CI doesn't involve espionage or breaking the law (more about this later); rather, it involves utilizing reliable resources—secondary, primary, and/or internal—to conduct research.

The goal of CI is to provide "actionable" information that will support the decision-making process in the law firm. For instance, CI can be used for a number of important decisions, including:

- Helping an attorney decide to pitch to a prospective client
- Deciding whether an industry sector should be targeted by the law firm for client acquisitions
- Choosing which vendor the firm should hire to upgrade its IT infrastructure
- Deciding whether the law firm should open an office or enter into a partnership in a particular city, country, or region

Importantly, CI is an ethical way to gather intelligence. Those working in CI use an array of information sources, such as publicly available secondary information and internal data (i.e., information stored in house, such as finance, accounting, knowledge management, practice and business development, client matters, and human resources) to analyze and provide intelligence on a company, industry, market, or similar group. There is no valid justification to resort to any unethical or illegal information gathering.

CI is not spying or being dishonest or underhanded. It does not consist of fraud, bribery, trespassing, deceit, eavesdropping, or stealing. To help adhere to ethical guidelines, everyone working in CI needs to be aware of rules and regulations relevant to the practice, such as the Economic Espionage Act of 1996 and other guidelines; the SCIP website has a Code of Ethics on its website.[2]

VALUE OF CI IN LAW FIRMS

The global economy and the economic downturn of 2008 have made gaining and obtaining clients very competitive for law firms. Companies are now closely scrutinizing the services provided by third parties, including law firms, much more deeply. Thus, CI can be an invaluable tool in a law firm's efforts to gain and/or retain business. For instance, CI librarians, working closely with a law firm's practice development units, senior management, and other personnel, coordinate and conduct specialized research and analysis to support those business development efforts.

In today's competitive market, law firms are often required to answer Request for Proposals (RFPs). RFPs are detailed applications where potential clients request that firms explain how well they know a potential client's business and identify how their experience and expertise stand out from other firms.

The RFP process is where the intelligence provided by a CI librarian is crucial; the CI librarian can provide extensive background information on a company, including corporate as well as legal issues, and provide insight on trends that are affecting the company and its industry sector. This information enables an attorney to be well prepared when he or she meets a potential client in person. In addition, many companies expect their outside counsel to know all about them and their industries, so the intelligence provided by the CI librarian can be crucial in a pitch.

In addition to aiding the RFP process, intelligence on a company can be used to assist in retaining and/or cross-selling between practice groups with existing clients. For instance, attorneys can reach out to their existing clients with new insights on their industry and markets, or on legal issues that might well impact their business. Attorneys can set themselves apart from the competition by showcasing their knowledge of the client's business and how they can help the company with legal issues or solve a specific problem.

The CI librarian is often charged with monitoring and reporting on legal and industry trends that are affecting clients to aid business development and marketing professionals and attorneys. CI also supports strategic and resource planning decisions by senior management within law firms. For instance, CI librarians, with their expertise in online resources and software, are able to provide insight into prospective software vendors and their products.

Law firms are constantly seeking to develop new business or grow their practices in new and existing markets. The CI function is integral to those efforts: for instance, it can help build an in-depth understanding of a new market, identify key competition, legal needs, recent market profitability, and economic climate to determine future growth opportunities.

Since the downturn in the economy, there has been an uptick in law firm partners making lateral moves, especially when the firms they are currently with are having financial challenges. Through monitoring these moves, CI librarians can track what firms are beefing up their partner ranks in certain practice areas or what firms are losing key legal personnel. CI research also can be used to assist recruiting of lateral hires or better understand the chances of a successful merger combination.

Table 26.1 highlights some of the critical capabilities and value that CI is able to contribute to a law firm.

It is imperative that the CI function in the library is properly staffed and has clear-cut responsibilities and focus. There is a lot of transparency in the legal marketplace, with a lot of information that needs to be collected and analyzed. Most of the larger law firms have many competitors—such as peer firms or specialized boutiques—as well as practice groups, usually in various locations, which have their own competitors. These practice groups also have some specific industry focus. In addition, clients—and potential clients—are numerous as well.

Table 26.1. Value of CI in a Law Firm

CAPABILITIES	VALUE	DELIVERABLES
Conducts research and analysis on companies, industries, and markets worldwide	Provides insight on potential client(s) or new developments affecting existing clients Identifies potential industries for firm to target Identifies market trends that affect the firm and its clients Assesses opportunities in new or existing markets	Company profiles (e.g., in-depth company profiles, two-page snapshots) Biographical sketches Industry/market screens for highly refined lists of ideas Targeted research In-depth industry reports
Monitors and reports on new developments that affect law firm's markets, clients, and competitive position	Assists with marketing and sales strategies Delivers relevant information in a timely fashion Disseminates information in understandable formats	News scans Periodic updates on companies, industries, and key issues/topics
Researches competitors' current positioning and future prospects	Offers insight into competitors' value propositions Assesses competitors' strengths and weaknesses	Competitor profiles Competitor tracking Report on lateral moves
Performs custom research and analysis for various departments (Finance, IT)	Assists decision making across the firm Supports strategic and resource planning	Executive briefings Ad hoc research
Provides expertise on an array of print and online resources	Availability of professionals to perform complex research	Skilled research support

CI RESEARCH PROCESS

The CI research process in law firms and other organizations includes a number of key steps, as noted in figure 26.1.

In general, when beginning a project, the CI librarian needs to define the issue (this is also called a needs assessment). This will help focus the research and deliverable, resulting in less wasted time and effort in the research process. This step is fairly straightforward for librarians, since they are well trained in the reference interview technique.

The next step is to collect the information from the relevant resources, which includes secondary (e.g., online financial databases, news), primary (e.g., in-house survey results gathered from clients), and internal (e.g., business development databases, billing data). This step also will determine what, if any, gaps exist in the resources available for the project or request.

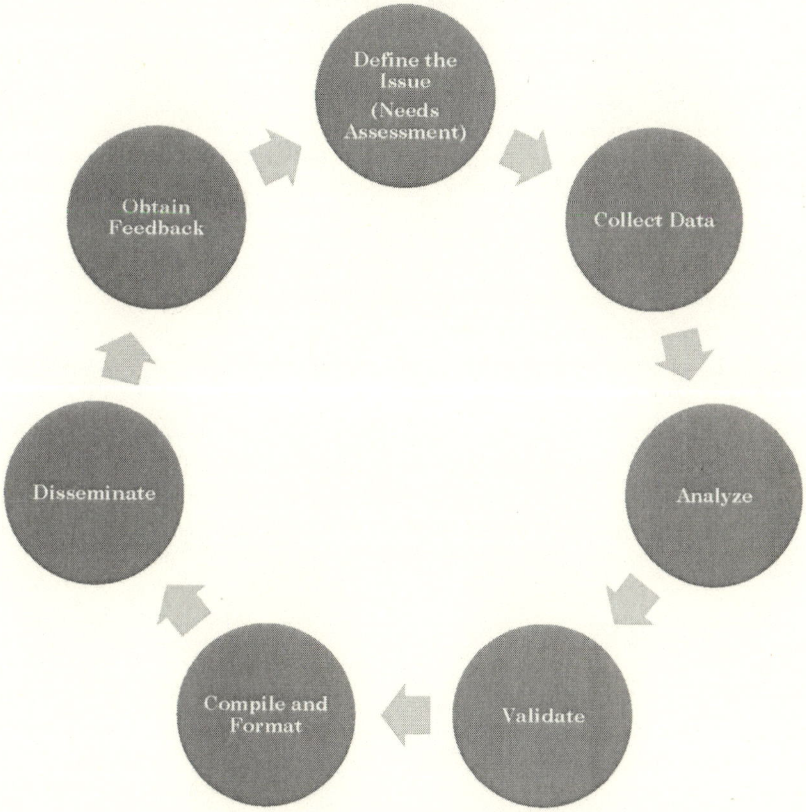

Figure 26.1 The CI Research Process.

The third step, analyzing the information, is one in which librarians have not been traditionally trained to perform. Although librarians do perform selected analysis for some requests (e.g., culling relevant articles to send to a requesting attorney), this is a step that requires additional training. The analysis needs to be validated—for example, is the assessment of the information justifiable?

After the information has been analyzed, the CI librarian writes the results of the analysis concisely and plainly, detailing the findings' relevance to the firm and the internal client. The analysis is then disseminated in a format that it easily understandable and customized for the client.

Finally, the CI librarian should obtain feedback on the analysis to help with future CI projects. In addition, the feedback can be used later to help show how it was a critical for the project (e.g., key to winning new business, identifying new lateral partner).

KEY CI DELIVERABLES

CI librarians in law firms provide an array of deliverables to their internal clients, as noted in table 26.1. These deliverables depend on the type of request; for instance, for an attorney meeting a prospective client that he knows absolutely nothing about, an in-depth company profile (time permitting) would be an appropriate deliverable.

For the most part, CI deliverables are customized and targeted to the specific needs and preferences of a requestor, depending on the type of request. As mentioned previously, the wrong format often will undermine any sort of insight and intelligence provided.

The type of request often determines the deliverable: Numeric data is often provided in an Excel spreadsheet, while a company profile is usually delivered in Microsoft Word or Adobe Acrobat. In addition, CI librarians often utilize a presentation product, such as Microsoft PowerPoint, when presenting data and information, since it can include graphs (pie charts, line graphs, bar charts) and images as well as text. Ideally, the CI librarian works closely with the person who requested the information to customize it to his or her specifications.

Creating Company Profiles

One of the most frequent requests of a CI librarian in the law firm is for intelligence on a company, usually for a pitch or new prospect. They also can be used as in-depth profiles of an existing client that is updated regularly with insights on news and developments.

Since time is money at law firms, attorneys do not want copious amounts of research with little or no focus or to spend unnecessary time on nonbillable work. Ideally, the company profile will be comprehensive as well as concise, not just a "data dump" (i.e., hundreds of pages of unstructured information).

For the most part, company profiles contain information that provides a clear picture of an organization—what it is (e.g., what does it do or produce? who is the current management team? where are its offices?), where it is going (e.g., opportunities for growth, strategic initiatives), and legal challenges and potential problems that the law firm can help to solve. The following is an outline of the key sections that a comprehensive company profile should include:

- *Executive Summary*: The first page of a profile will have the executive summary, which places key takeaways from the profile in a broader contact. The executive summary will highlight the requestor's focus (i.e., it will reflect the specific context of the request) and provide a conclusion or assessment. It should also identify opportunities for the law firm and highlight practice area strengths.
- *Company Overview*: This section provides a detailed, high-level description of the company, including its status (public or private, family owned), locations, the services or products it offers, and strategic initiatives.
- *Financials*: Brief summary of top- and bottom-line results (revenue and net income, respectively) from the latest quarterly and annual reports as well as financial data that are industry specific. Other pertinent data should be included as well, including segment data by product, geography, client types, etc.
- *Industry Trends*: Attorneys who are not familiar with a company's industry will often request a section on industry trends and outlook. This section summarizes the trends that will likely affect the company going forward as well as an analysis of its key competitors.
- *Legal Issues*: This section, depending on the request or client need, can cover litigation, corporate (M&A, IPO), real estate transactions, intellectual property (IP), or government affairs, to name a few. This section details what the company has been dealing with over a course of a year or two in various practice areas. For instance, has it been involved in litigation (especially in areas that the law firm is focused on)? How many M&A deals has the company been involved in and who was it legal advisor for those deals? Is it being investigated by any government agency? The section should also discuss possible future actions in this area.
- *Transactional Analysis*: For some company profiles, information on a company's recent transactions is required. This section provides an analysis of its deal making and aptitude for more (i.e., will they be looking to make more deals in areas where the firm has worked previously?).
- *News and Developments*: This section highlights what the company has been involved in over the recent past and the implications. For instance,

has the company been opening new offices outside its traditional markets (maybe in areas the law firm has no presence)? Has it been cutting costs or retrenching (this has implications for third-party providers)? Has there been a change in senior management?

- *Biographical Profiles*: The biographical profiles of C-level executives (CEO, CFO, COO) and the general counsel focus on work experience and educational background, as well as personal information not included in official biographies. Although the executive biographical profiles might not seem important, the information they provide gives an attorney confidence when meeting a potential client.

Significantly, researching, analyzing, and creating company profiles requires a lot of time and resources. Therefore, the end result should provide some sort of insight into the company as well as how the law firm can help it. But what if time is limited? Many online resources such as Standard & Poor's Capital IQ or Bloomberg Law (among others) have "ready-made" reports that can be customized and sent to internal clients. These are often the default deliverables to quick turnaround requests where the CI librarian has only a small window of opportunity to do the research and add insight to those types of reports.

CI TOOLS AND TECHNIQUES

There are numerous CI tools and techniques that can be utilized in a law firm. It is imperative that CI librarians become familiar with those techniques and develop expertise in the tools that they find most useful for their firm. Below are very brief summaries of some more popular and well known CI tools.

Key Intelligence Topics

Key intelligence topics (KITs) enable CI professionals to define and guide a research issue that is important for the law firm. KITs are essentially the "what keeps you up at night" issues or problems that need to be analyzed and solved.

KITs are a good starting point for a newly created CI function because the process involves conversations between CI librarians and senior management and attorneys. In this way, the CI analysis will directly address the specific intelligence needs of management. Some examples of KITs include senior management wanting to understand what the new technologies affecting the legal industry are and how they can respond with strategic and resource planning, or an attorney needs to find out what legal issues are likely to affect a certain client.

In addition, KITs provide insight into what resources that the CI function in the library will need for it to provide the appropriate level of support. The

KITs process also is a good outreach tool, introducing senior management and attorneys to the capabilities of the CI department.

SWOT

The SWOT analysis, which stands for strengths, weaknesses, opportunities, and threats, is a useful technique to evaluate a business or competitor and the external environment. Both strengths and weaknesses are the internal factors inherent in the business, while the opportunities and threats are the external factors.

SWOT analysis is an effective tool to portray a competitor or business because it can be easily explained and understood by everyone. It also is very focused on a business's or a competitor's status in an industry or their possible moves.

In addition, a SWOT analysis can be applied to many situations and intelligence issues. It can also show how a CI librarian's own firm stands in relation to its competitors that are being analyzed via the SWOT analysis.

Although it seems to be a relatively easy analytical tool to create, CI librarians should be aware of some of its drawbacks. For instance, it is easy to oversimplify some of the issues covered in the analysis; to overcome this, the SWOT analysis should have input from staff throughout the law firm to gain multiple perspectives. It also is sometimes difficult to identify strengths and weaknesses. For example, a law firm that has not expanded into a market where other firms have established offices might be one of its key strengths (focused in its own market) or a glaring weakness (unable to effectively compete outside its core market).

PEST and PESTLE

PEST analysis, which stands for political, economic, social, and technological, is a valuable tool for an environmental scan. PESTLE is a variation on PEST, adding legal and environmental to the analysis. Both tools are solely focused on the key external environmental influences on a business and how those issues could affect the law firm's strategy, marketing, or resource planning.

In addition to the PEST and PESTLE analysis, CI librarians often help to identify how a law firm should respond to those influences. For instance, CI can assist the law firm in investigating and assessing the regions where the firm might want to locate a new office or identify the new technological innovations that are affecting the legal industry.

Executive Profiling

Another important role for CI librarians in law firms is to compile personal profiles, whether to help the firm research and assess potential lateral hires from other firms or to educate an attorney about a senior executive or general counsel they will be meeting to pitch new business.

CI librarians often profile a company's general counsel for a firm attorney, most often when they are meeting for the first time. The executive profile will review the general counsel's past work experience—has she worked in a law firm? what are her areas of expertise? and so forth. Executive profiles also include other personal information, such as when and where they went to college, outside interests, and philanthropic efforts. (This information can be culled from LinkedIn profiles, nonprofit and college alumni websites, and the like.)

Executive profiles help to facilitate the relationship between an attorney and a potential client. This is especially important since many companies hire attorneys and law firms based on relationships.

Porter's Five Forces

No discussion of analytical tools in CI can ignore Michael Porter's Five Forces model. It is considered by many CI librarians as one of the best tools for understanding an industry.

According to Porter, the competitive environment in any industry is embodied in the following five forces:

- Threat of new potential entrants
- Threat of substitute products or services
- Bargaining power of suppliers
- Bargaining power of buyers
- Rivalry among existing competitors

According to Porter, "Understanding the competitive forces, and their underlying causes, reveals the roots of an industry's current profitability while providing a framework for anticipating and influencing competition (and profitability) over time."[3]

These five forces should be used as a conceptual background for identifying a law firm's competitive strengths and weaknesses and threats to and opportunities for the firm from its competitive environment. It also is a great brainstorming tool to discuss what the recent trends are in the legal marketplace and how they affect the firm.

There are other notable CI tools and techniques, such as customer segmentation analysis for business development, industry analysis, scenario planning/analysis, etc. For a full range of analytical models, the book *Strategic and Competitive Analysis* by Dr. Craig Fuller and Babette Bensoussan provides a clear and detailed explanation of many of the CI tools that have been discussed as well as many others. The authors also describe their system, called FAROUT (forward oriented, accurate, resource efficient, objective, useful, and timely), for determining the overall effectiveness of an analytical model.

SETTING UP A CI PROGRAM

Law firms that have decided to set up a CI unit usually turn to the library to house the function. However, some firms, with the research assistance provided by the library, house the CI unit in the business development area. We have set out some fundamental general steps below for setting up a CI unit in a law firm, with law librarians as the prime CI librarians.

In order to set up a CI function in the library, the head librarian must be sure that the existing library staff has the right skill set (this is discussed later in the chapter). In addition, the CI function must be properly staffed and have the right resources (e.g., critical sources for business research such as financial databases like Standard & Poor's Capital IQ, which provides multi-asset class and real-time data, research and analytics, and ThomsonONE, which contains a broad array of financial content, among others) and budget. Also, the firm needs to hire a CI librarian (if not already in place); importantly, the firm must set realistic expectations of what the CI librarian position will entail.

In some law firms, the CI librarian conducts legal reference in addition to doing CI work. This can be very challenging if not impossible; it can be very overwhelming to be expected to do the volume of work that comes with both roles. The CI librarian must be dedicated to the CI function.

The library must also institute a CI process, whereby the projects are monitored from the beginning to the end. When conducting CI, librarians research, gather, and compile data and information on the question at hand. These data and bits of information are analyzed and synthesized into executive summaries, spreadsheets, or PowerPoint presentations—all of which are time consuming. This tracking of requests and time, along with the outcome of CI projects, should be part of the CI process to help the library and the CI function to identify the value that it brings the law firm.

It also is important when establishing a CI unit to talk to the practice group chairs and senior management about the kind of CI research that is needed (e.g., key intelligence topics, or KITs). This will help set the agenda for the CI program as well as define the function. These conversations are also a good time to find out what sort of format is preferred for the finished work—intelligence is often discounted if not presented in a preferred format.

Once the statistics are gathered and key intelligence topics are determined, it's time to find a CI "champion" outside of the library (hopefully, this champion has already been identified since the law firm has decided to create the unit). In order to be successful and gain credibility within a law firm, the newly established CI function needs a "champion" to promote and support it. The champion should be a senior attorney or administrator who will help drive the process throughout the firm and assist with getting buy-in from other attorneys and senior management.

It is imperative that the CI librarian meet regularly with attorneys as well as senior management to discuss the accomplishments of the CI group. This also will help the CI librarian to identify resources in house, such as finance and information technology, which can be used for CI research. In addition, members of senior management who are not familiar with CI can become new internal clients.

As discussed previously, the CI librarian needs to create a process for the CI requests. The process should include best practices for request submissions and templates for deliverables (e.g., company profiles, alerts, and raw data). The process should also establish a way to prioritize projects and specify the ownership of the research.

Significantly, CI is not a "billable" activity in a law firm. Therefore, a monitoring and measuring process is needed to track the outcomes of the CI projects. This effort will show the value of the CI process to a firm's bottom line, whether it is helping to win new business with an existing client or identify new opportunities. As an example, the process should include following up on pitches that the CI function has been involved in and keeping track of any wins and feedback from attorneys. Obtaining feedback will help establish a clear and ongoing communication between end users and the CI librarians.

Lastly, it is important to sell, sell, sell the CI function consistently to the firm. Present CI capabilities at practice group meetings/retreats as well as new hire orientations. Presenting findings from the CI process, such as the CI support to win new clients, is imperative. Educating the firm about the function will help not only with marketing CI but also with setting expectations.

STAFFING THE CI FUNCTION

There are a number of important skills and attributes for a successful CI librarian in a law firm. For the most part, law librarians already have many of the skills needed for the position. They have training and experience in eliciting information (i.e., the reference interview), as well as the necessary research skills, including the ability to devise the best search strategy and proficiency at collecting information.

In addition, law librarians have an in-depth knowledge of resources—both online and print—and familiarity with the legal industry. These resources not only include legal databases like Lexis and Westlaw, but also many nonlegal services, such as financial databases mentioned earlier (Standard & Poor's Capital IQ, ThomsonONE) as well as specific industry services (e.g., Preqin for data and research on the private equity sector, and SNL Kagan for in-depth analysis and data on the media and communications business). Law librarians also know how to organize, manage, and share data, an important skill in a workplace that is inundated with data.

In addition to those practical skills, there are personal attributes that are necessary for CI librarians to be successful. They must be inquisitive and relentless in their search for reliable data and information. They also need to work well under pressure and be able to deliver bad news with a touch of optimism, since not all CI projects result in affirmative actionable intelligence (e.g., a move to a new region might not be in the law firm's best interest).

Law librarians do need to acquire and develop CI skills. Importantly, they must be able to assess and interpret information, draw conclusions, and present recommendations. They also must develop expertise in presenting ideas, whether in writing, graphics, statistical packages, PowerPoint, or other tools. Furthermore, law librarians need to develop an understanding of CI tools and techniques, such as SWOT analysis, KITs, PEST, and others (some key ones are discussed earlier in this chapter).

Law librarians also need to become familiar with their own law firms' focus and expertise. Knowledge of a firm's industry focus, key clients, and competitors is essential. Importantly, this knowledge will help ensure that the proper resources are in place, such as staff expertise in a particular industry and access to relevant databases.

Although it's not easy, it is critical that competitive intelligence remain free of bias, providing neutral results. Competitive intelligence is not intended to support an existing management decision. Good competitive intelligence should speak the truth and let management decide how it wants to proceed.

Ann Lee Gibson, in a presentation at the Ark Group/Managing Partner's fifth annual conference, themed "Competitive Intelligence in the Modern Law Firm," noted that very opinionated people often make poor analysts.[1] She also added if a current library staff does not pick up CI quickly, they most likely will never get it. As Gibson explained in her book, *Competitive Intelligence: Improving Law Firm Strategy and Decision Making*, effective CI librarians must be able to withhold judgment about what data mean while researching as well as being more of a creative thinker when analyzing and presenting the information.

CHALLENGES FOR ESTABLISHING CI IN LAW FIRMS

Though one could argue that setting up a CI program in a law firm is a necessity, many firms do not have an official CI role. Many are still becoming acquainted with CI and do not see the need for a full-time person or for dedicated resources to conduct CI.

Cost is a big issue as law firms struggle to control expenses while at the same time build business. In order to have a successful CI program, it needs to be properly staffed and have the appropriate resources. Some of those resources, such as databases like Standard & Poor's Capital IQ and ThomsonONE, can be very costly. Nonetheless, gathering information from many sources provides a solid basis for reporting and analysis.

In addition, some law firms struggle with the question of where to house the CI function—should it be in the library staffed by librarians, in business development or marketing, or somewhere else in the firm? In the 2012 American Lawyer Law Librarian survey, 56 percent of those polled said that the CI function at their firm was in the library. However, there are many firms where the function has been placed in the marketing department.

There is often a debate as to how well a law librarian can be a CI analyst. The advantages of having the position staffed in the library is that the person would most likely have a MLS, have research expertise, would be trained in online databases, and would be familiar with reference resources. Since analyzing data is not a requirement for a legal librarian, this skill set might not be readily found among the library staff. Placing the CI role in marketing with a nonlibrarian has its challenges. Though the marketing professional will be able to spot trends and create presentations utilizing the data found, he or she would not have the in-depth knowledge of online resources or the research and reference skills of a librarian. This CI staff would always have to depend on the library for assistance on projects.

Another major challenge for the CI unit is managing expectations. Attorneys can be very demanding and expect a quick turnaround for their requests. CI librarians need to be realistic on what they can deliver. Thus, educating the attorney and senior management on the CI function will help with managing expectations.

Lack of resources is a big issue in many law firm CI units. Electronic databases are very costly but necessary when conducting CI. If a law firm does not allot enough money for the CI librarian to obtain these online resources, the alternatives (such as searching the Internet for "free" information), though inexpensive, can be unreliable and often do not contain the data that are needed.

Arguably the biggest challenge is finding the right person to put into the CI role in a law firm. Law librarians traditionally have excellent research skills and knowledge of the legal industry but do not analyze the information that they are collecting or have the training to present their findings. A CI librarian is expected to conduct the research and then put the data in a concise format that contains insight. What has become a challenge for library directors has been finding a librarian who can conduct business and legal research as well as analyze data and place that data in a format that is concise and provides comprehensive overviews on clients, industries, and legal issues. Many law firms have turned to CI librarians outside the legal industry, people who have polished CI skills and corporate work experience but need to learn about the legal industry.

FUTURE TRENDS OF CI IN LAW FIRMS

With more organizations, including law firms, gathering information and intelligence on their own business and competitive landscape, CI librarians will most likely begin to work with "big data"—large and complex data sets that require expertise to organize and analyze the information for better decision making. Data can come from primary (internal), secondary, and proprietary sources. For the legal industry, large datasets can include internal financial and transactional data on clients combined with external data for pricing and profitability analyses or customer segmentation. CI librarians can support a law firm's efforts in this area, such as helping to determine what data are important for the firm, what tools are needed to access and make sense of the data, and how to transform this huge amount of data into meaningful insights.

More data and information vendors are targeting the legal marketplace with new products and services. CI librarians will need to keep up with new tools and online resources and decide what is essential to the CI function. For instance, traditional online legal vendors such as Westlaw and Lexis are enhancing their offerings, while new online legal services providers like Justia.com and Bloomberg Law are gaining traction. In addition, CI librarians need to become familiar with the many external sources that have expertise and primary information in a specific area.

With increasing globalization, law firms are setting up offices and/or partnerships outside their borders in an effort to support cross-border corporate finance and cross-border mergers and acquisitions practices. CI librarians can provide guidance to those expansion efforts. They also will need to provide support to those far-flung offices as well.

"Doing more with less" is a likely scenario going forward. More law firms are cutting expenses as demand for legal services remain flat or only moderately improve. According to the Citi Private Bank and Hildebrandt Consulting LLC report, *2013 Client Advisory*, the legal market will most likely remain challenging and volatile. However, those firms that will be successful in the market, notes the report, will have a clear, thought-out strategy for growth. They have targeted key practices, industries, clients, or regions in which they have prominence, and have developed a business plan around those targets. Importantly, those firms will need to align strategy with implementation.[5] This is where CI becomes an invaluable tool for a law firm's decision-making process.

CONCLUSION

In today's hyper-competitive environment, law firms are going to continue to contest every new opportunity and struggle to maintain existing clients. With this in mind, the CI function will continue to grow in importance in law firms. It has become an integral part of doing business at many larger law firms, and medium-sized firms are finding that CI offers certain advantages. It is essential that law firms understand their market, the environment that they operate in, and what they can offer to their clients and potential clients. It is now so easy for seemingly established clients to take their business elsewhere.

However, despite the benefit of having a strong CI function, law firms still struggle with utilizing their intelligence efforts. Recent research findings from ALM Legal Intelligence found that a small minority of law firms has "actively and strategically" aligned the CI functions with the firm's business objectives.[6] It seems that many of the firms that responded to the ALM Legal Intelligence survey that have CI programs use it on an ad hoc basis or it is not aligned with strategic initiatives.

We live in uncertain times. It is imperative that the CI function clarify any uncertainty in the market and business environment for the law firm. It also is imperative that the CI librarian promote the importance of CI in the law firm, especially with the growing consumerization of information and access to information. Notably, ALM Legal Intelligence's research found that the key issue to keeping the CI function from reaching the desired level of impact at most firms is the lack of awareness among attorneys.

As noted previously, the CI function requires specialized resources along with capable, well-trained staff that can gather information utilizing specialized resources and then analyze and present their assessments. CI librarians will need to create a value-added model that is recognized and is aligned with the firm's strategy, and to deliver CI services that have staying power.

FURTHER READING

Almendarez, Monica. October 27, 2008. "Weathering the storm with CI: Ensuring that law firms maximize their earning potential." *New Jersey Law Journal*: 25.

Applebaum, Zena. 2012. *Business intelligence for law firms*. Ark Group in association with Managing Partner.

Competitive Intelligence Magazine, Strategic and Competitive Intelligence Professionals (SCIP).

Fleisher, Craig S., and Babette E. Bensoussan. 2003. *Strategic and competitive analysis: Methods and techniques for analyzing business competition*. Upper Saddle River, N.J.: Prentice Hall.

Gibson, A. L. 2008. "How to create and use competitive intelligence: 45 tips for law firms." *Law Practice* 34, 2:47–52. Many law firms are still struggling to understand what CI is, and how it can benefit them. This primer explains how to reap the strategic benefits of CI.

Gibson, Ann Lee. 2012. *Competitive intelligence: Improving law firm strategy and decision-making*. Ark Group.

Lapachet, Jaye A. H. July 14, 2008. "Librarians can play key business-development role." *National Law Journal*: S3.

Porter, Michael E. 1980. *Competitive strategy: Techniques for analyzing industries and competitors*. New York: Free Press.

———. 2008. "The five competitive forces that shape strategy." *Harvard Business Review* 86, 1: 78–93.

Schweyer, Kitty. 2008. "Competitive intelligence resources in law firms—Kitty Schweyer, library and information resources manager at a major law firm, offers an overview of CI in law firms and evaluates the fee-based services for CI in the legal marketplace." *Searcher: The Magazine for Database Professionals* 16, 4:30.

Organizations

Strategic and Competitive Intelligence Professionals (SCIP)
Special Libraries Associate (SLA)—Competitive Intelligence Division
Association for Strategic Planning (ASP)
American Association of Law Libraries (AALL)—PLL/SIS Competitive Intelligence Group

Websites/Blogs

SCIP.org

SCIP's website provides information on CI and the CI industry for both nonmembers and members. It also has a publications section with featured I Books.

CI.SLA.org

SLA's Competitive Intelligence Division's site provides educational content (webinars, video) geared to CI librarians.

LLRX.com (law and technology resources for legal professionals)

Provides the CI page, "A Selective Resource Guide," for materials relating to CI, such as online tools for monitoring and tracking, and online people tracking sources, among others.

3 Geeks and a Law blog
A blog on topics and issues related to the legal industry, hosted by specialists in knowledge management, Internet marketing, and library science.

NOTES

1. SCIP website, http://www.scip.org/resources/content.cfm?itemnumber=601&navItemNumber=533.

2. SCIP Code of Ethics, http://www.scip.org/About/content.cfm?ItemNumber=578&navItemNumber=504.

3. M. E. Porter, "The five competitive forces that shape strategy," *Harvard Business Review* 86, 1 (2008): 78–93.

4. Ann Lee Gibson, "2011–2014: A bull market for law firm strategy and competitive intelligence," presentation at the Ark Group/Managing Partner's fifth annual conference, Competitive Intelligence in the Modern Law Firm, June 9, 2011.

5. Citi Private Bank and Hildebrandt Consulting LLC, "2013 Client Advisory."

6. ALM Legal Intelligence, "Competitive Intelligence: Getting the Edge—And Winning Clients," February 2013, 6.

Part VIII

PROFESSIONAL DEVELOPMENT AND THE FUTURE

27

Getting the Most from Major Associations, Publications, and Conferences

Holly M. Riccio
O'MELVENY & MYERS LLP

Professional development is an essential component of any law librarian's career: It not only enhances your existing knowledge and skills but also aids in building a foundation for future growth. The digital age has brought with it myriad ways to access, absorb, participate in, and reap the benefits of professional development opportunities. By combining the use of new technologies with more traditional approaches, law librarians can gain relevant skills and experience, expand their universe of networks and information sources, and develop name recognition within the law library community.

This chapter explores the wide spectrum of professional development opportunities for law librarians, from venerable professional associations to up-to-the-second social media, as well as best practices and tips for getting the most out of these tools. Finding the right mix of opportunities and information—and taking advantage of it—can fuel a lifetime of professional success and career advancement.

PROFESSIONAL ASSOCIATIONS

National Associations

Law librarians are served by two major national professional associations, the American Association of Law Libraries (AALL) and the Special Libraries Association (SLA). Each association provides professional development opportunities for members in a variety of formats, including annual meetings and conferences, webinars and online learning, print publications, and LISTSERVs and electronic communities.

American Association of Law Libraries

Established in 1906, AALL is the nation's oldest association for law librarians. It is dedicated to promoting the value of law librarians in all settings and equipping members with the tools to position themselves as essential to the mission of their organizations. AALL's work is carried out by its executive board, committees, special interest sections, caucuses, and chapters. Each of these entities offers a unique way to learn and grow through membership and participation.

The thirty-plus committees within AALL play integral roles in making law librarians leaders in legal information, research, and technology. AALL puts out a call for committee volunteers annually and makes committee assignments with an eye toward diversity and balance. Although the form for soliciting committee volunteers asks for members to list their top committee choices, librarians new to AALL may wish to leave those blank and ask to be placed wherever they can be most useful and effective.

AALL membership includes the opportunity to join special interest sections (SISs), which allows members to participate in smaller groups that deal with specialized areas of law librarianship. Some examples of SISs include Academic Law Libraries, Legal Information Services to the Public, Private Law Libraries, and Research Instruction and Patron Services. Joining and becoming actively involved in multiple SISs is a great way to keep abreast of current developments and trends in specific areas. Since the SISs have smaller memberships than AALL as a whole, they also provide a unique opportunity to assume leadership roles early on in your career. The SISs also provide educational opportunities targeted to their specific audience through newsletters, LISTSERVs, AALL communities and e-groups, webinars, and annual meeting programming.

Caucuses are informal, specialized groups of at least fifteen AALL members who share a common interest. Depending on their size, many caucuses also have executive boards and committee structures, which afford members another valuable way to get involved. Caucuses also provide educational opportunities aimed at their target audience in many of the ways that SISs do.

Chapters are made up of law librarians in specific geographic areas; AALL has more than thirty chapters throughout the United States. Chapters serve as a rich resource for information, providing access to education and events that may be specific to a certain state, city, or region. Many chapters play active roles in their communities, hosting various educational opportunities throughout the year, sponsoring outside speakers, and arranging for other types of networking and social events. Chapters also usually take the lead on government relations activities and present opportunities for members to get actively involved in influencing and advocating for legislation that affects law libraries.

Special Libraries Association

Founded in 1909, the Special Libraries Association acts as the international body representing the interests of thousands of information professionals in more than eighty countries worldwide. SLA's mission is to promote and strengthen members' careers through learning, advocacy, and networking initiatives. Its organizational structure includes a board of directors; chapters, divisions, and caucuses; committees and councils; and student groups. All these entities work together to represent the best interests of special librarians in corporate, academic, and government settings.

With SLA, various specialized committees and councils invite members to work together to address important topics and issues facing law librarianship. Volunteering for committee work in SLA is a valuable way to gain experience, but the majority of the association's work is carried out by its divisions and chapters.

Members of SLA can join one chapter and one division as part of their annual dues. SLA's fifty-six regional chapters all have elected officers, publish newsletters and meeting announcements, and hold a wide variety of programming throughout the year. SLA is also organized into divisions, each of which represents a different subject area, field, or type of information-handling approach. Like SLA chapters, divisions have elected officers and publish newsletters. They also participate in educational programming, but theirs is produced in conjunction with SLA's annual meeting. For areas of common interest that are not covered by a committee, chapter, or division, members can form a caucus. All these entities provide avenues for law librarians to get involved early in their careers—a step that can, over time, lead to even more enriching commitments. Starting out volunteering at the chapter or division level can jump-start your professional trajectory.

What Associations Have to Offer

Memberships in both AALL and SLA offer so much to your career, opening doors that extend beyond such tangible benefits as attendance at educational events and access to print publications. Each association is a respected leader in its field, and law librarians who take advantage of the opportunity to get involved in these associations are seen as forward-thinkers and leaders by their employers and colleagues. The opportunities that AALL and SLA offer law librarians to interact, network, and connect with their colleagues are invaluable, both for present success and future aspirations. The leadership, emotional, and social intelligence skills that law librarians can gain from embracing volunteer roles can over time be applied in the workplace and beyond.

Online Resources

Major Organizations' Websites

The website for AALL, called AALLNET (www.aallnet.org), contains detailed information about the benefits and opportunities the association provides members. In addition, members can log in to the site and gain access to an electronic version of the member directory as well as the AALL "My Communities" site. AALL's My Communities is the social networking component of AALLNET for special interest sections, caucuses, committees, and informal groups of members. Member groups can communicate via their e-groups, share documents through their resource libraries, post events, and create blogs.

The website for SLA (www.sla.org) is a comprehensive resource for members seeking to get involved or further their professional education. Members can access the SLA Community to search the online membership directory, connect with information professionals using a variety of formal and informal networks, and access SLA Connections, their electronic newsletter. SLA operates a host of blogs that members can utilize to keep up with topical information as well as share comments.

LinkedIn Groups

Both AALL and SLA have LinkedIn groups that invite members to connect online. Many of the subentities of each organization also have their own LinkedIn group presence. Joining these groups can help newer law librarians begin to network electronically with their colleagues. Also, since the membership of these LinkedIn groups is smaller than that of the overall association or its other entities, they can provide a unique opportunity to ask questions of colleagues and share information in a more informal environment.

Social Media Resources

Both AALL and SLA have official Twitter feeds and Facebook pages. In addition, many AALL and SLA entities, such as AALL's special interest sections and SLA's divisions, have their own Twitter and Facebook presence. Following Twitter feeds for both associations and their respective entities helps members keep an eye on upcoming educational events and learning opportunities. The same applies to the various Facebook pages, but the biggest advantage afforded by Facebook is in staying on top of events and activities at the local level, such as with AALL or SLA chapters. Many Facebook pages for local groups post information about upcoming local events and give members a chance to share photos and other information after the event has concluded.

CONFERENCES

Major Conferences

Both AALL and SLA hold annual meetings and conferences—the American Association of Law Libraries' annual meeting and conference is held every July and the Special Libraries Association annual conference is held in June. Throughout the year, law librarians can attend or otherwise take advantage of a host of other national and local conference and educational opportunities.

AALL Annual Meeting

The AALL annual meeting and conference comprises many educational and networking opportunities. Before the conference officially starts, AALL puts on half-day and daylong workshops on a variety of topics. Due to their length and size (attendance is usually limited), these can offer attendees an opportunity to gain a rich understanding of a specific topic and share their thoughts in a more intimate group setting. The conference itself consists of three days of programming, featuring sessions of varied lengths and employing a variety of learning styles and room settings. The special interest sections and caucuses are also afforded the opportunity to put on their own educational programs during the course of the conference. In addition to the educational content, attendees can take advantage of countless social and networking opportunities. Some of these are AALL-sponsored events open to all attendees, some are AALL-sponsored events intended for specific audiences, and some are organized and sponsored by exhibitors.

Law librarians new to the profession should consider attending the Conference for Newer Law Librarians (CONELL), which takes place the Saturday before the conference's official start. This event is intended to give new AALL members a wide breadth of exposure to AALL and its members, and many attendees make connections with colleagues that they nurture and leverage for their entire professional careers.

Another preconference event of particular interest to law firm librarians is the PLL Summit. This venture by the private law library community was created to explore ways to embrace change, demonstrate value, and learn about cutting-edge trends. Originally begun as a two-year project, the Summit has been presented for four years and is continuing to be a leading force in empowering law firm librarians to be change agents within their organizations.

SLA Conference

The SLA annual conference and INFO-EXPO invites attendees to participate in networking, communication, learning, and other developmental opportunities geared toward information professionals. In addition to the official conference programming, SLA offers division-sponsored preconference workshops that allow attendees to gain a more in-depth understanding of a specific topic. The conference consists of four days of programming encompassing topics and areas as varied as the SLA divisions that sponsor them. SLA's approach to new information professionals and conference attendees differs from that of AALL in that they hold events, such as luncheons and meet-ups, throughout the conference geared not only to connecting new attendees with each other but also to long-time members and veteran conference attendees.

Other Conferences

In addition to the AALL and SLA annual conferences, many other conferences take place throughout the year that can be of great benefit to law librarians. Some of them focus on specific types of libraries, such as academic libraries or private law libraries, and others are organized around a topic or concept that is wider in focus and, therefore, draw attendees from all kinds of library settings. The most important thing to keep in mind when deciding whether to attend any conference is how relevant the conference description and program information are to your job or career aspirations. But don't lock yourself in. Attending conferences not specifically aimed at law librarians can often yield some of the most innovative ideas and results—you just have to be open to seeing them.

- *Internet Librarian* (http://www.infotoday.com): This conference takes place in Monterey, California, each October and consists of five simultaneous tracks, numerous workshops, and ample networking opportunities. The intended audience is library and information professionals interested in technology, and the programming is geared to provide attendees insights, strategies, and practices that give them new tools for transforming the Internet, managing libraries and digital information, and enhancing the information-sharing and learning experiences of their library users.
- *AALS annual meeting* (http://www.aals.org/events_annualmeeting.php): The Association of American Law Schools is the professional organization for law teachers and faculty, and their annual meeting is the largest gathering of law faculty in the world.
- *Best Practices and Management Strategies for Law Firm Library, Research, and Information Services* (http://www.ark-group.com): This conference, established in 2007, is sponsored by the Ark Group and takes place in New York each February. A one-day educational event, it focuses on solving the key challenges facing law firms and the transformation taking place within the law library community.

- *Chapter Institutes:* Many AALL chapters hold their own annual educational and networking events, usually called Institutes. These can be wonderful opportunities to get the benefits of attending a conference in person without having to incur as many of the travel-related expenses, as chapter activities are closer to home and rarely exceed a day or two.

How to Get the Most Out of Conference Attendance

Although getting the most out of any conference will vary from person to person, everyone can benefit from some baseline tips and strategies.

Preconference Preparation

Start by doing your homework. Read through the conference agenda and decide what programs and events you want to attend. Look at the agenda with an open mind and do not always stick to programs from your library type or functional area—sometimes the best programs turn out to be the ones you might not have thought to attend. Consider using the online scheduler tool (if the conference has one) and share your schedule with others to start connecting over common interests and programs. Remember that elevator speech? It's not just for work anymore. Think about what you want new acquaintances to know about you so you can be prepared and make good first impressions. And even in the digital age, don't forget to pack plenty of business cards.

Be an Active Participant

Attendance at programs and events is just the first step in conference participation. In educational sessions, try to mix it up; rather than sitting with the same circle of acquaintances, take a leap and strike up a conversation with someone you do not know. If time has been left for questions at the end of the presentation, think of something to ask, and by all means, stand up and ask it. Talk to fellow attendees about what sessions they enjoyed the most and share your favorites with them. At social and networking events, make it a point to meet new people as well. Before the conference, when the call comes out for conference volunteers, consider volunteering for a task that will give you additional opportunities to interact with attendees, such as registration. To reinforce what you are learning about, consider live blogging or tweeting from the sessions you attend. In addition to documenting what you are learning, as you are learning it, social media allows you to connect with others, including those who may not be attending the conference in person.

Keep It Going after the Conference

After the conference is over, look at the agenda again and, based on the program descriptions and feedback you received from fellow attendees, download materials and recordings of sessions you didn't attend. Write up notes about the sessions you attended and what you got out of each of them to share with your co-workers. Volunteer to write an article about a session, possibly for your local AALL or SLA chapter newsletter. Regardless of your social media of choice (LinkedIn, Facebook, Twitter, etc.), take the time soon after the conference to connect with everyone you met or exchanged business cards with.

Keeping Up without Attending

Since many law librarians get the opportunity to attend only one conference a year, finding ways to learn from afar is an important skill. Many conferences will offer access to the educational content, such as audio of sessions and handouts or presentation slides, for a fee, after the conference has concluded. Once upon a time, such postconference materials were the only way to get the benefits of a conference without attending, but in the digital age, the options are endless. Consider trying some of the following tactics to keep up from your desk:

- See if any portion of the conference is being streamed or otherwise shared live.
- Find out the Twitter hash tag for a conference and follow it on a daily basis, as this can provide insight into what sessions might be worth following up on.
- See if there is a LinkedIn group that has been created for the conference and, if so, join it.
- Try contacting the speaker or speakers for a presentation of interest to see if they are willing to share their slides or materials with you.
- Scan law library-centric blogs for reviews of sessions by attendees.
- Check Slideshare (www.slideshare.com) for PowerPoint presentations shared by speakers.

Online Learning Opportunities

Although nothing can replace the benefits of attending conferences alongside colleagues, webinar technology has afforded associations the opportunity to provide more educational programming throughout the year and to many more members. Some webinars are free or, if not, available for a modest fee. AALL hosts many webinars throughout the year and then archives them in its online learning center, AALL2Go. Other AALL entities, such as SISs and chapters, may also hold webinars that cover topics tailored to their function area or library type or on geographical-specific issues. SLA also hosts numerous webinars during the year through its online learning community, Click University. The name, in case you were curious, is an acronym: Continuous Learning to Improve Career Knowledge.

The opportunities are plentiful for e-learning outside of those provided by professional associations. Many vendors sponsor training webinars to show how to get the most out of their products, and some will also host webinars discussing industry trends or other topics of interest to legal professionals. If you have an interest in a specific subject area (e.g., records management) or want to grow your leadership and communication skills, there are opportunities to do these through e-learning as well. Once you find and attend a webinar on a subject of interest and enjoy it, sign up to get information on future opportunities from that provider. Consider attending a webinar that sounds interesting just from the title and description—you may be pleasantly surprised to find that what you learn is both applicable to your job and valuable to your career.

PUBLICATIONS

What to Read

Association Publications

Both AALL and SLA serve their members with a wide variety of publications, both at the national and local levels, covering timely topics that are relevant to law librarianship. Scanning the tables of contents of association publications regularly, such as those listed here, and reading articles that are of particular interest can help you stay informed and keep current.

- *AALL Spectrum* (http://www.aallnet.org/main-menu/Publications/spectrum): AALL's monthly magazine publishes substantive articles on topics of interest to law librarians, as well as news about AALL, including its chapters, committees, and special interest sections. Members receive print copies; back issues are available on AALLNET.
- *Law Library Journal* (http://www.aallnet.org/main-menu/Publications/llj): This scholarly journal, published quarterly, has been the official publication of the AALL since 1908. Members receive print copies; back issues are available on AALLNET.
- *AALL E-Newsletter* (http://www.aallnet.org/main-menu/Publications/enews): This monthly electronic publication includes news items, important reminders about upcoming events, and deadlines that concern law librarians. All AALL members receive this automatically via e-mail.

- *Information Outlook* (http://www.sla.org/io/): The official publication of the Special Libraries Association provides information professionals with analyses of, and insights into, the trends, issues, and people affecting special libraries. Print publication ceased in 2013; a digital version, along with back issues, is now available to members online.
- *SLA News Connections* (www.sla.org/content/resources/lin/): These electronic newsletters keep members up to date on SLA happenings, including association news, professional development opportunities, and public policy issues.
- *SLA Chapter and Division Publications*
- *AALL Special Interest Section, Caucus, and Chapter Publications:* Many of these entities produce newsletters, most of them electronically, on a regular basis, and most have sections of their websites that offer both the current and archived editions.

Nonassociation Publications

Blogs of Interest. So many blogs are being written by and for law librarians and other legal professionals that keeping tabs on them all can be a challenge. How do you find out who is blogging and then determine what blogs are of interest to you? One way is to look at some of the lists that various groups have compiled, such as the AALL Computing Services SIS Law Library Blogs list (http://aallcssis.pbworks.com/w/page/1189465/Law%20Library%20Blogs) or the ABA Journal Blawg Directory (http://www.abajournal.com/blawgs/). You can also use Google Blogs (http://www.google.com/blogsearch) to search for relevant blogs and blog posts.

Sometimes the best way to find new blogs is through your colleagues, many of whom may have the same interests as you do. You can always use Twitter, Facebook, and LinkedIn to keep tabs on what your colleagues are reading and sharing, but consider using some of these other tools as well:

- Flipboard (http://flipboard.com/): This iOS app gathers links from Twitter, Facebook, and other online sources and auto-arranges them in a digital magazine format, providing an easy way to see what blogs your friends may be following or sharing.
- Paper.li (http://paper.li/): This tool creates a Web-based newspaper from the most popular links shared across your network of Twitter and Facebook friends.

The Law Library Twitter-verse. Making the most of Twitter for professional development takes concerted effort, but taking the plunge can pay dividends in the end. If you follow lots of Twitter feeds for nonwork interests, you can easily become overwhelmed with the amount of information flowing through your account night and day. Twitter offers users the option to create lists, akin to your own personally curated groups of Twitter users. You can then view a list timeline that will show you a stream of tweets from only the users on that list. Third-party Twitter management services can also be used to manage tweets and get the most out of this social media technology.

- HootSuite (http://hootsuite.com/): This desktop and Web-based service covers many social media networks and offers custom analytics information. It also allows you to post to multiple platforms at the same time and even schedule status updates and tweets to be posted on specific dates and times.
- Tweetcaster (https://tweetcaster.com): This clean, simple, and easy-to-edit option offers much more than the official Twitter app, including advanced searches, retweets with or without comments, content filters, and easy list-making to name a few. Available for iOS and Android.
- Echofon (http://www.echofon.com): This iOS app is a clean, streamlined Twitter client that allows you to post updates quickly and give you the functionality you need without making it too complicated. While other apps focus on throwing a lot of options at you, you can rely on Echofon's simplicity.

Figuring out whom to follow on Twitter is an organic process that evolves over time. Once you have identified some useful Twitter feeds, take a look at who they follow and who is following them, as this may bring to light others you may want to follow. Twitter will also offer suggestions on feeds you might be interested in, based on whom you are following, so mine those results regularly for possibilities as well. The AALL Computing Services SIS has also compiled a list of Law Libraries and Librarians on Twitter (http://aallcssis.pbworks.com/w/page/1189463/Law%20Libraries%20and%20Librarians%20on%20Twitter). Some words to the wise: Because of the speed and volume of information that Twitter generates, choose quality over quantity when selecting feeds to follow. A little Twitter goes a long way.

LISTSERVs. Although LISTSERVs, also called discussion forums, no longer qualify as modern technology, they still function as useful professional development tools for law librarians. Much LISTSERV traffic revolves around interlibrary loan requests, but forums are also used to announce upcoming professional development and educational events as well as articles, websites, and blogs of interest. Engaging discussions of hot topics can also occur on LISTSERVs from time to time. Some LISTSERVs are open for anyone to join, and others are closed to the specific group they serve or require approval from the LISTSERV administrator. To find LISTSERVs of interest, take a look at the following resources:

- AALLNET eGroup Discussion Lists: http://community.aallnet.org
- AALL Forums:
- http://share.aallnet.org/read/all_forums/?&max=100&skip=0&orderby=ListName+ASC
- AALL Chapter Discussion Forums: http://www.aallnet.org/main-menu/Member-Communities/discuss/list-index-chapter.html
- SLA Division Lists: http://www.sla.org/content/community/lists/divisionlists5979.cfm
- SLA Chapter Lists: http://www.sla.org/content/community/lists/chapterlists.cfm

Keeping It All Organized

How do you keep it together, organizing all the blogs, tweets, and other sites in a way that you can scan, evaluate, and read regularly? Since so many blogs and other online publishers syndicate their content as an RSS feed, using an RSS reader is a smart way to organize and manage your e-reading. The universe of RSS readers is vast, and finding the right one is very much a matter of personal taste, but here are some Web-based and desktop reader options you may want to consider:

- *Netvibes* (http://www.netvibes.com): This reader provides two different views: as a start page with customizable widgets or in reader view, similar to Google Reader's old interface, and users can toggle between the two views.
- *Feedly* (http://feedly.com/): This app has an extremely simple interface and also provides additional news and reading material based on topics you already subscribe to.
- *Newsblur* (http://www.newsblur.com): This Web-based reader is a good alternative for those with less than 64 feeds, as there is a monthly charge of $1 for unlimited feeds. Newsblur also has an optional feature that allows it to learn what is most interesting to you and highlight articles meeting your criteria.
- *Reeder* (http://reederapp.com/): This iOS RSS feed reader has a clean, easy-to-use interface and also integrates with services like Read It Later and Instapaper.

ACTIVITIES

Taking Ownership of Your Professional Development

Today more than ever, our professional and career success depends on our ability to take advantage of and manage our professional development opportunities.

The recent economic crisis has, to be sure, magnified this to some degree, but the concept isn't new at all. Pursuing the right balance of formal and informal training and development has always been one of the fundamental pillars of success and longevity in law librarianship. The key is to take advantage of every opportunity to build your skills and expand your network and relationships.

Activities to Consider

Conference Participation

Don't just attend conferences—help shape them! When the call for programs comes out, assess the conference theme and propose a program on a topic that interests you or that you can speak on. Remember that if you have a topic or idea but not the breadth or depth of knowledge to speak on it, you can ask colleagues or find other speakers to be part of your program. If your proposal is accepted, the experience of shepherding a program from idea to completion provides the opportunity to practice and enhance many skills, such as public speaking, project management, and networking.

Speaking at conferences, whether on your own or part of a larger panel, can help raise your profile within the profession and develop a name for yourself as an expert on a particular subject. When starting out, you may want to take advantage of as many speaking opportunities as possible, but as time goes on, think about your true passions and expertise and look for speaking engagements that allow you to refine them. Also, as scary as it may seem, take advantage of any opportunity to speak alongside amazing public speakers—the best way to perfect your presentation skills is to work with experts and learn from them firsthand.

Attend Whatever and Whenever

When beginning your career, make a priority of attending as many events as possible. Having face-to-face interaction with fellow law librarians and other legal professionals is one of the best ways to network for future career opportunities. Talking to others in the legal field can often enhance or alter your perspective on a current project or initiative. The competitive intelligence and collective wisdom of our profession can prove invaluable in your job: tapping the resources of the law librarian community shows your colleagues and, more importantly, your supervisors that you are enthusiastic and far-sighted about the profession's future.

Seek Out Writing/Publishing Opportunities

All AALL and SLA publications, both at the national and local levels, are interested in potential authors and new ideas for articles. If you have something you would like to write about, contact the publication's editor and volunteer to submit a draft. For those hesitant about writing, consider enlisting a coauthor. Writing on a regular basis in these publications will help establish your name and credibility over time. Also, since most publications are available electronically, your articles will begin to come up in search results from users outside the law library world. If the topic you want to write about lends itself to a wider or different audience, consider approaching editors of non-law-library-focused publications. The AALL Publishing Initiatives Caucus has compiled an annotated list of publications and publishers' contact information (http://www.aallnet.org/caucus/pic/publications.htm) that can help you identify potential publishing opportunities.

Blogging, Tweeting, and So Forth

If writing a full-fledged article seems intimidating, why not utilize social media outlets, such as blogs or Twitter? You probably already have a Twitter account, and setting up a blog is cheap and easy these days. (If you don't want to commit to blogging on your own, many law library blogs are happy to have guest bloggers—all you need to do is ask.) The main advantage of using these digital media outlets to communicate is that they allow for timely, quick commentary and responses to current legal issues that writing for a serial publication does not. However, there are other, more subtle advantages that can be gained from blogging or tweeting. For instance, social media can establish an online presence in the same space your professional colleagues, co-workers, and attorneys or professors inhabit and can provide a more informal way to connect with these groups as well.

The law librarians at Bryan Cave LLP recently started a Twitter account and began actively tweeting. In a guest blog post on 3 Geeks and a Law, one of the firm's librarians summed up the advantages they had seen:

> Twitter is a subtle way to develop relationships with attorneys by giving us an opportunity to casually interact with attorneys from all of our offices. Twitter is proving to be worth the small amount of time we devote to it. We encourage other law firm libraries to start tweeting. You may discover that it can help build relationships with your attorneys and co-workers!

Even more recently, the director of library services at Bose McKinney & Evans LLP, Cheryl Niemeier, became the first law librarian to start officially blogging on her firm's blog page. She raved about the benefits of her new role:

> Clearly, there is value in non-attorney posts on a law firm's page; it expands the reach of the firm's web site to a wider audience. Visitors to our site share our content via their own blogs, Twitter, or LinkedIn pages, further expanding our reach and enhancing the Firm's marketing efforts.

Whether you blog or tweet on your own behalf or in an official capacity for your organization, there is no denying that using these social media tools thoughtfully can enhance your professional name recognition and advance your career goals and objectives.

Maximizing Your Use of Social Media

In the digital age, an important component of your professional reputation is your online and social media presence. Using all the social media outlets you participate in—LinkedIn, Twitter, Facebook, and so on—as effectively as possible is an essential part of your professional persona now and will be increasingly so going forward. Learn how to use advanced features and functionality to your advantage to establish yourself as a knowledge leader. For example, with LinkedIn, you can join relevant groups and become an active participant in them by posting articles of interest or using the collective wisdom of the group's members to pose questions.

CONCLUSION

Although the technology that is available to provide and enhance educational opportunities in the digital age is forever changing, the concepts at the core of professional development are timeless. Law librarians need to continue to take an active role in their professional development and career growth and utilize all the tools and resources available to them. Whether that means dashing off 140-character missives or volunteering to help at an annual conference, taking advantage of the varied, evolving opportunities for law librarians is the first step in your pursuit of lifelong learning and professional growth.

FURTHER READING

AALL Spectrum. June 2008. "Member to member: What advice do you have for first-time conference attendees?" *12:30*. http://www.aallnet.org/main-menu/Publications/spectrum/Archives/Vol-12/pub_sp0806/pub-sp0806-more member.pdf.

Crosby, Connie. May 9, 2011. "6 reasons why your law library staff need to attend the annual professional conference." *Slaw: Canada's Online Legal Magazine*. http://www.slaw.ca/2011/05/09/6-reasons-why-your-law-library-staff-need-to-attend-the-annual-professional-conference/.

Crosby, Connie. January 15, 2006. "The tao of law librarianship: Do-it-yourself professional development." *LLRX*. http://www.llrx.com/columns/tao2.htm.

Crosby, Connie. March 20, 2008. "Twitter—What it is and how I am using it." Blog. http://conniecrosby.blogspot.com/2008/03/twitter-what-it-is-and-how-i-am-using.html.

Kroski, Ellyssa. January 28, 2013. "A library conference survival guide: 20 tips." *Open Education DataBase*. http://oedb.org/blogs/ilibrarian/2013/a-library-conference-survival-guide-tips/.

Lambert, Greg. June 2009. "Where do listservs fit in a social media world?" *AALL Spectrum* 13 (8–9): 13. http://www.aallnet.org/main-menu/Publications/spectrum/Archives/Vol-13/pub_sp0906/pub-sp0906-listservs.pdf.

"Tips for first-time attendees: Tips for making the most of an SLA conference." *Special Libraries Association*. http://sla2012.sla.org/general-information/tips-for-first-time-attendees/.

28

The Future of Law Librarianship

Scott D. Bailey
SQUIRE SANDERS (US) LLP

Julie Graves Krishnaswami
LILLIAN GOLDMAN LAW LIBRARY, YALE LAW SCHOOL

The law library has long ceased to be primarily a physical place; at its core, arguably, it has never been primarily a physical space. As the space was de-emphasized, service was elevated and linked to the mission of the parent institution, whether law school or firm. In providing service to their users, law libraries have evolved to be much more than an "as needed" destination. Their sphere of influence has expanded despite the reduction in their physical footprint and volume count. Law librarians have turned to the goal of accomplishing more—creating more productive scholars and lawyers—with less space, less money, fewer staff members, and without many other resources. In the future, law libraries will align to even more closely reflect the priorities of their users and institutions rather than being concerned with where they are "housed" and how to inevitably expand their physical presence.

The core valuable resource, the information management professional or the librarian, remains at the center of this service model. Users, whether students, faculty, attorneys, or partners, will continue to ask questions as they grapple with massive amounts of information and confront newer, more complicated, and specialized tasks. As a result, strategic plans will be played out in the libraries while law librarians—ever nimble—adjust their work to shifting priorities: analyzing, anticipating, spotting, and monitoring valuable trends as well as preserving institutional knowledge and securing them for future exploration. As law practice becomes increasingly specialized, globalized, and competitive, academic instruction will follow. Research courses will need to be sufficiently basic for new students, who are only beginning to learn legal vernacular as well as information literacy. Yet they must also be specialized, even with a global focus. Proving the value of these services, by counting what is difficult to quantify, and monetizing them where appropriate with increasing transparency will be a priority for librarians.

The benefits, opportunities, and challenges facing academic law libraries mirror law firm libraries and all other types of law libraries. First, law libraries of all kinds support changes in the practice and business of law, including the development of instructional programs to incorporate and support specialized, practice-specific research, greater analytic research skills, information and knowledge management, in a platform agnostic way. Next, being mindful of the future and the institution's strategic goals, law libraries will be responsive to the needs of their institutions, whether academic or firm, by controlling costs and promoting efficiencies, exploring new markets for legal services, supporting research, and providing instruction. Indeed, providing a high level of service has become bedrock of what we do. Finally, law libraries will combine forward-looking thinking and creativity to prepare for alternative roles, while continuing to engage in the nuts-and-bolts aspects of public service and teaching. This means branching out and developing subject specialties to support new practice areas or faculty interests, as well as developing high-level technological skills to build effective and interactive information portals, knowledge bases, and repositories. Librarians will take these new specialties and make them highly visible, developing lasting and strategic thought leadership, promotion, and institutional alignment. As with practice research support in the firm, instruction must be available on demand and just-in-time in a mobile environment and in a variety of modes to support the adult learner at his or her point of need.

SERVICE CONCEPTS: MISSION, GOALS, AND STRATEGIC PRIORITIES

Contemporary law librarians are flexible and innovative by nature, training, and necessity. The past twenty years have brought sweeping, transformative change to all areas of law librarianship. Nothing has been spared. Technology has changed everything from collection development practices to the use of space and reference and instruction. Research platforms and systems, such as Lexis, Westlaw, and West Digest System, which operated as the standards for decades, must be used alongside systems that operate with a completely different logic, that is still somewhat "fuzzy," not completely transparent in how it works, and more like Google than the classic research systems. And other databases and indexes are following suit. For the librarian, this means that as soon as she learns one system well, she has to adapt and learn another system. As a result, law librarians are constantly innovating and looking for opportunities to evolve and develop their skills, learning these new platforms because our users prefer them, and use them, but require help navigating them efficiently. Law librarianship joins several other industries in adaptability. Generally, increasing competition in the global marketplace has encouraged a customer service emphasis, in performance and title. Visibility and availability of support is a critical component of any sustained customer experience. New titles and job categories for librarians combined with the evolution of a solid history of service can promote and renew the value of the profession.

The pervasiveness of online chat in a support function to varying levels of effectiveness is an illustration of the direction of what consumers—the students, practitioners, associates, clients, or partners—expect. Yet, in the process of any given transaction, how often does a window pop up and ask whether help is needed? How often is "help" actually achieved with this method? Customers of all types quickly learn that all customer service is not equal just because the technology and infrastructure is similar. Without a knowledgeable, efficient, and well-educated provider end of help chat, the concept falls flat. The law librarian of the future needs to embody this helpful presence in an effective way, knowledgeable of the basics as well as the more advanced specialized fields. Likewise, to be credible practice support, she needs to be visible, available, responsive, efficient, and accurate. She is required to be the best, at all times and at all points of contact. Academic law librarians must couple teaching with service as they help students—so that they can later find the answer for themselves—even over chat. Whether bringing an inquiry to a traditional reference desk, using a mobile app, e-mail, or chat window, the service concepts must be carried through to a logical and helpful conclusion, and the librarian must meet the user where she finds her.

As for the concept of service in the large law firm library, increased competition from legal process outsourcing (LPO) and consultancies have entered the picture. With the rise of firms such as Integreon,[1] upper management has found the need to evaluate these options. Large firms in particular have been called upon to justify their costs and service levels in light of these alternatives. Yet an advantage an incumbent law librarian has is that she knows and understands her patrons and can customize the service levels to respond to and take into account the internal strategies and overarching mission of the firm. Where an outside firm or consultant must ramp up and get to know the firm at the outset of each project, the in-house librarian has a more insightful understanding of organization, strategy, personnel, and information-seeking behaviors (or idiosyncrasies). Even if a librarian is relatively new to a firm, if effective knowledge management has been exercised and she is an active contributor with access to the right information, there is no substitute for tailored contributions to the business.

Likewise, in the academic law library, the concept of service has moved beyond the reference department. All departments in the contemporary, academic law library must be focused on public service, with the mission of the institution driving the work. For instance, if the law school's mission is enhancing clinical and experiential programs, then librarians in each department have a role to play: Acquisition and collection development librarians must purchase the necessary monographs and practice materials; librarians with an expertise in electronic resources must make sure the appropriate databases are selected; access and circulation librarians must make sure that anyone who needs to access the library's materials can do so; and reference and instructional librarians must make sure they can teach research in the substantive areas, know how to find and use the best sources, author research guides and finding aids that meet the needs of clinicians and students. More importantly, all librarians must be seen as legitimate partners worthy of intellectual and substantive academic partnership.

To be sure, academic law libraries are frequently understaffed and overwhelmed with work as librarians have become some of the busiest and most in-demand professionals in a law school. Considering that most academic law librarians have responsibilities for teaching research courses, providing reference services (via chat, e-mail, telephone, or in person), collection development, cataloging or maintaining electronic records, writing research guides, and liaising with faculty to assist with their research, librarians must often determine what to stop doing.[2] Reflecting back on the institution's mission, strategic goals, and priorities can service as guide posts for making decisions about services.

CHANGE MANAGEMENT, CONTROLLING COSTS, AND PROMOTING EFFICIENCY IN A COMPETITIVE CLIMATE

Law firm practice has evolved due to economic pressures. Big law is consolidating and revenue is flattening due to downturns, globalization, and increased competition. The AmLaw100 is changing. Ten years ago, an average AmLaw100 firm had fewer offices and fewer attorneys but more revenue per attorney when factoring in inflation. Law school enrollment is down for the second year in a row by significant percentages. Firms and law schools are asking librarians to do more with less. As a result, the pressure to cut costs and provide more efficient services to clients is greater than ever before. An opportunity exists here for shrewd contract management in the law firm library as suppliers have also had to consolidate in the face of increasing competition and market pressure. More and more small vendors are being acquired by large publishers. In this developing environment, good relationships with a few core publishers are critical for today's librarian. Small firm, corporate, and solo practices are also affected by the shift in big law. Firms are increasing in size and diversifying their services alongside their suppliers.

The larger firms are in a race to the bottom in the regional markets for new business while the regionals markets are competing with solo practitioners for work that was formerly theirs for the taking. As a result, firm law librarians need to be mindful of the scale and flexibility of their contracts, with an eye to the business. Is a merger likely on the horizon? Are we hiring laterals primarily rather than promoting through a traditional summer associate track? Increasingly, summer associate classes are shrinking while lateral hires have increased. Research professionals must be aware of trends in the industry as this awareness informs the strategy of information managers. Likewise, as firms increase their geographic footprint, librarians must be mindful of opportunities to globalize contracts to achieve efficiencies of scale without losing the competitive edge by relying on too few suppliers with long-term contracts.

Typically, law firm library directors and managers are committing to short-term contracts and licenses that allow for more flexible licenses to better support law firms as they develop and change. Because legal research products have become increasingly user-friendly and retraining is straightforward, the costs associated with the implementation of a new platform have diminished. Firms are free to change platforms more frequently. The idea of a firm being exclusively a Westlaw or Lexis shop is no longer supported by training-challenge inertia or the lack of self-starting users. New entries into the marketplace such as the Bloomberg Law platform also give leverage to the savvy negotiator. Yet, due to increased competition and the pervasiveness of nondisclosure agreements, it is difficult to tell if there is an industry standardization for costs taking place. One general assumption is increased competition will drive prices down, but licensing value for proprietary content is still high. While Google Scholar is a starting place for certain primary research, the perception that "everything is free online" is dangerous for newer researchers, particularly in a profession where authority and editorial insight are at the heart of our work product. Insightful editorial content will always be valued, but the market is different. Targeting more specialized markets, the pressure is on for publishers to provide more useful material configurable in multiple ways for customer preference.

Similarly, law schools are critically evaluating their contracts, licensing agreements, and collective development policies and practices. Several factors have made it critical for the large law firm library to control costs: pressure from law firm management to spend less, competition from other departments within the firm, and outsourcing. Likewise, law school budgets are shrinking with the decreased enrollment and the redesign of legal education, in general. Whether in firms or academia, librarians must be prepared to justify spending practices and monitor their budgets closely.

Justifying the cost of legal services or proving the "value proposition," a practice heavily endorsed by many law firms and the Association of Corporate Counsel, has given rise to an increasing number of competitive pricing structures, notably Alternative Fee Arrangements (AFAs). Library and research service offerings should also promote those values, in the name of flexibility. For instance, the billable hour model for charging for research services is a powerful metric but only if it results in actual revenue value for the firm. If a library does not provide evidence of the value of its work to the firm and the client, the costs resulting from the librarians' work will be challenged by the client, and perhaps not recovered by the firm. This can cause the library to be viewed as a revenue drain.[3]

While some publications have heralded the demise of cost recovery, a substantial number of law firms continue to recover a significant portion of their online research expenditures by efficiently utilizing them and billing them to the client. Creative and efficient use of a variety of digital research subscriptions can result in a huge benefit for the client and can be recovered in some instances, if it is justified. Justifying these charges and promoting the value of our research resources, including staff, are a critical part of the future of law libraries.

Academic librarians, too, must be vigilant about scrutinizing budgets and spending practices. Databases frequently provide duplicative access to electronic resources. And many online resources may only be used to support the research of one or two faculty members. Librarians must compare costs of resources and books with the costs of labor involved in obtaining those materials. For instance, academic librarians frequently determine whether it makes financial sense to buy a new book or to spend staff time processing an interlibrary loan request for that book. Many academic law libraries have canceled their print reporters, law reviews, digests, and loose-leaf services, relying on the availability of these materials online. This, in turn, has freed up funds for purchasing more specialized collections, databases, and staff.

While some believe that all law firm research costs will eventually be subsumed into the hourly rates, questions remain: How does this impact those who believe that firms will increasingly employ AFAs that may not be dependent on hourly rates? How does this work in a flat fee or contingent or other discounted model? Flexibility will again be the order of the day in order to preserve firm revenue. Just as finance departments in law firms have been asked to find more options for a happy client retention agreement, so must the library be flexible in its method of responding to increasingly complex client billing demands.

Academic libraries must maintain this same flexibility. As librarians meet the needs of just-in-time users, they must also meet the needs of patrons who need specialty resources. Meeting these needs frequently requires a difficult calculus. For instance, when a faculty member researches legislation and legislative processes, does the law library support this research by licensing a resource like *Congressional Quarterly*, specifically designed for tracking legislation? Or does the library make do with the adequate but less sophisticated tracking tools built into the resources already licensed, like Lexis, Westlaw, or Bloomberg? A similar calculation is involved with student and faculty use of PACER, the electronic public access system for federal dockets and court documents. Students and faculty often need and want access to PACER but the fee structure—which charges per page—creates unpredictability in costs for the libraries that often pay the bill. Academic librarians must approach these questions and decisions with flexibility, the ability to be nimble, and change course when necessary, abandoning projects and products that no longer serve the institution's cohort, mission, or strategic priorities.

However a firm library responds to the downward curve of online cost recovery, it must stay focused on the primary goal of providing a valuable service to the firm and its clients. While much has been written about the decrease in the amount of cost recovery for legal research that firms can expect, little has been written about a responsive solution. Focusing on the service that a firm library provides is the way to counter the argument that the demise of recoveries points to the demise of the firm library. Monies unrecovered from billings can be made up for with efficiency of service and new billing methods to recapture the value of research resources and staff. While the commonly mentioned "solution" of billing research into the attorney's billable rates is pervasive, we must also consider this as an opportunity to advance the researcher's billable rate as a solution for lost cost recovery. Encouraging professional researchers with their vast array of specialties and levels of practice "embeddedness" to bill their time, even as a metric if it is not the firm's cultural practice, is a valuable way to offset this trend. Financial metrics that speak the firm's language can go a long way toward effective communication with management.

Special project billing from the library should mirror the firm's practices and provide complex ongoing research products for designated flat fees or retainers. As the complexity of research projects increases, direct billing for special research may be a possibility in many firms. While many firms already provide supervised client services with direct client contact, cost pressures indicate that more of this activity can take place in the future for a more profitable firm. Updating a fifty-state chart, legislative history, or global regulatory process paper for a client and providing it to them with a direct mechanism such as an extranet or other means of secure, remote access, can be another source of funds and value for the digital age library.

Use of subscription management applications that have a cost recovery or billing component can be an effective way to provide metrics. With the evolving nature of these products and the inclusion of workflow and collaborative solutions, they will demonstrate value better than tic marks at a reference desk ever could. The more comprehensive these tools are and the more of the research process they encompass, the more we will be able to seamlessly demonstrate our activities within the firm. The challenge with the existing systems is that they require some significant implementation time and a cultural shift to incorporate effectively. Ideally a system should work in real time and not interrupt the flow of the question for administrative benefit. Building a knowledge base of answers and perhaps even having them automatically updated would be a valuable resource for a firm.

These tools and statistics combined with billing applications can create a record of the research process that aids in the instruction of new employees, taking us beyond the traditional concepts of "cost-effective research." True cost-effective research that is tailored to the firm and the firm's contracts rather than based on general principles (like when to use an hourly versus transactional billing) can be informed by this knowledge base and will inform the research manager where problem areas may lie. Adapting to change and being a nimble and informed manager and instructor is an increasing part of the future of law libraries.

RESEARCH INSTRUCTION

Most firm librarians promote research training at the new associate orientation as well as when major upgrades occur with research platforms. Often regular training is offered by WebEx or in person by a vendor or librarian. Some larger firms have managers of continuing legal education (CLE) or training that is not in the library, but this can also be an opportunity for the library manager to offer added value to the firm. In the normal course of contract management, CLE and training products are routinely offered as incentives. Promoting this as a function of the library increases the profile of the librarian and contributes to her perception as cutting edge. Offering training and instruction on new products demonstrates that librarians are leading the way and are there to help. Obtaining buy-in from the partnership and key practice leaders for regular training opportunities will enhance the library's reputation, by providing a valuable "service within a service." In the global firm, providing on-demand training across time zones and on mobile devices will be essential.

Law firms and law schools share the challenge of instructing students and attorneys to use the right resource for the specific research at hand. Googling has trained everyone to always go to one tool, enter a list of words, and click through the first page or two results (and sometimes this technique will work well using Lexis Advance, WestlawNext, or Bloomberg Law). Research instructors must continue to teach students and attorneys to spend initial moments engaged in the research task or assignment to think about and conceptualize the research problem before venturing into the database. What is known about this issue? What do I already know? Unknown? What types of authority are required?

Next, students and practitioners must be reminded to think about the best type of source for the authority. There are many specialty resources or sources within large databases that can make searching for a particular type of material much more efficient than dumping words into a Google-like, mega-database. Students and attorneys should conceptualize their question and then ask, "Is there an app for that"—with "app" being a specific source.

Law librarians are the experts in legal research. As a result, research instruction will continue to be the bailiwick of reference librarians, despite some short-lived concern that legal writing instructors or vendors would take it over.[4] And so, like the law firm instruction practice, it will evolve accordingly. Students continue to require instruction in basic legal research skills—as they traditionally have—but they will also require researching instruction into new subject areas. But basic instruction must be supplanted by instruction in research management practices and basic information literacy skills, because it has been well documented that many college and law students today know how to Google and graze for information, but they lack the ability to search more deeply, analyze, and evaluate the content they find.[5] Moreover, WestlawNext, Lexis Advance, and Bloomberg Law, for example, have capitalized on this preference and created systems that encourage the use of Google-like search techniques, where students find relevant materials by weeding through thousands of results, of all types of materials, and then filtering, should they know what they are looking for.

Additionally, because supervising attorneys and partners expect students to leave law school "practice ready," contemporary research should move beyond informative, abstract, and basic. This is the new normal. Research instruction is an opportunity for students to become "practice ready" and begin to hone the skills they will be expected to have as attorneys. Law librarians must devise research problems that promote critical thinking. Law librarians teaching legal research must make the research class problem-based and built around the same constraints new associates will face in practice. For example, cost-effective research is a particular challenge facing new attorneys, and this is especially challenging to teaching in legal research courses where use of the research systems is free and unlimited. Yet students can be asked to research in certain databases or using only a certain number of sources. Likewise, students should be required to track and account for their time spent researching in the same contemporaneous way they will be required to do in practice, to begin to develop a comfort working under these types of constraints and conditions. Teaching students to become cost-effective researchers also means teaching them how to navigate the Internet and use Google effectively. The trend of clients being resistant to paying for research or firms absorbing the cost of research will likely continue and students must be able to use the Internet as efficiently as possible. Students should also understand how to manage mountains of information and stay abreast of changes in their field or practice areas. Research instruction needs to help students be comfortable with the uncertainties and constraints inherent in law practice. For instance, students need to learn how to balance the requirement to provide a client or partner with a coherent answer to a research question with the problems of time limits and the fact that as an attorney—unlike as a student—there will never be enough billable hours to read every authority.

With online research, digital books, and hyperlinked texts, students need instruction in "the context of legal information."[6] One of the problems with online research for novice researches such as law students is that they fail to understand the parts of texts as related to the whole: how they fit together, reference other parts of the text and sources, and when they should be used together. For example, when a student uses a keyword to search for, retrieve, and read a treatise online, she might find the keyword in isolation and often only read the one section of the treatise where the keyword appears—unless she takes an affirmative step to move forward and backward among sections or use the table of contents. In other words, students have to be taught that they are using a text, with the same features of a book or report, when they rely on the materials online. The other problem inherent in the lack of context is that the hierarchy of materials is obscured.[7] Today, sources lack organization in any meaningful way. As Professor Barbara Bintliff has explained, print legal resources brought structure and mechanization, which underscored and helped students understand the context and the material they were using. This is no longer the case with contemporary, Google-like research systems.[8] Helping students "understand the sources and forms of legal information in a holistic way, helping them discern the differences in pattern and structure of primary and secondary legal material" is now a learning objective of legal research instruction.[9]

Most importantly, law students must graduate with the ability to analyze a legal problem, synthesize the issue, and then draft a letter, memo, complaint, brief, contract, or other legal document. They must be able to interview a client and negotiate with an adversary. Other skills are required for more specialized practice, such as the ability to draft a comment to a proposed regulation, request or litigate for information under the Freedom of Information Act, or draft model legislation. And there are countless other skills new attorneys require. Yet, regardless of the specific task, problem solving "is the single intellectual skill on which all law practice is based."[10] The ability to locate and synthesize authority and then write well is critical. To be sure, locating the authority is no longer the critical mission in legal research—the new Google-like, "smart searching" systems have made finding the law relatively easily. But research instruction can serve as a training ground for teaching students these skills as students can be asked to submit not merely lists of sources located or a discussion of how the sources were located but also practice-like work product where they analyze the authority located.[11]

The types of authorities that librarians should focus their instruction on should also change. The dominance of cases, the lifeblood of legal education since Christopher Columbus Langdell, are being superseded by codes, statutes, and regulations. As Judge Judith Kaye of the New York Court of Appeals observed, statutory lawmaking "has . . . surpassed [the common law] as the preeminent source of law it once was."[12] Like no other time, statutory law and research is taking on a greater and more prominent role in the lawyer's work, and the dominance of statutory law has not gone unnoticed in the legal academy. Since the early 1980s, Judge Posner has called on law schools to improve instruction about legislation.[13] Law schools—Harvard, New York University, and Vanderbilt, among others—have been revising their curricula to require courses on regulation and legislation to reflect "the fact that regulations and statutes often play a more important role in the creation and elaboration of law than court decisions do."[14] Teaching students to work with statutes and regulations in a competent manner offers the additional benefits of more broadly preparing students for legal practice that will not necessarily include litigation. Whereas courts typically value clarity and predictability (as embodied in the concept of *stare decisis*), legislators and regulators strive for consensus, participation from various stakeholders, and information from experts and public interest groups. And this work requires somewhat different research competencies than searching for case law. It requires using the deep Web and using agency websites that are frequently cluttered and disorganized. This work also requires evaluating the information located. Staying current and abreast of changes in the field are paramount. Understanding how the various forms of authority work together and can be leveraged in research is also critical. Additionally, contemporary law students and practicing attorneys must be able to research and use foreign and international materials and nonlegal materials as these sources are routinely cited and judicially noted in court decisions.

SPACE: NO LONGER A FRONTIER?

For large law firms, where online subscriptions have long supplanted large print collections, this theme seems fully explored. Online resources have largely replaced the bulk of our print collections, and only print specialized materials and deskbooks—which reflect concessions for format preference—remain. Whether the "space" is a library of print materials that is now dispersed into practice areas, or on the desktop, or mobile device, or on a portal, or elsewhere in the future, space is still a major consideration and one that should be expertly addressed with effective management.

Due to the rising cost of print supplementation, office space or real estate, and the flattening of revenues, law firm libraries have kept their active print collections to a minimum. Although a current online subscription often costs the same as print subscription (despite the perceptions of many), pricing pressures from increased competition and the ubiquity of improving legal information sources that are lower in cost *should* drive prices of digital services down and their quality up. E-books and their circulation have received quite a bit of attention in the industry, but they have not been fully adopted by law firms. Regardless of the form of future digital services, many of the fundamental challenges of licensing and circulation persist. While an enterprise license for the firm or library will still likely be available, it is important to think about how to manage rights in this environment and communicate critical licensing terms to the users. As law firms produce materials and content for direct consumption by clients and provide a variety of licensed services to a variety of customers, careful management of copyright and subscription terms is required. This is a tremendous opportunity for the library manager to control costs and promote efficiency by managing and organizing the variety of digital subscriptions.

Academic law libraries are both physical and virtual. And like law firm libraries, academic library space is at a premium—but for different reasons. Students continue to flock to the libraries, even if they ignore the books on the shelves in favor of accessing material electronically. Instead, students want multiuse space suitable for study carrels or offices, stand-up desks, group study rooms, and storage space for books or law review materials. Within the walls of the library, students engage in a variety of tasks and projects from preparing for class, researching a law review note or seminar paper, to chatting and socializing with friends. So while all students may not be engaging in work *requiring* them to be in the law library, their continued presence in the physical space—a neutral, third space offering comfort, solitude, an opportunity for gathering with others—underscores the importance and the connection of the law library to law school.[15] Indeed, law schools acknowledge the importance of this space as they continue to renovate and build law libraries.[16]

The virtual space is no less important. As the card catalog or physical doors of the past, the library's website is the gateway to the materials maintained and the services offered by the library. Law librarians catalog material and create content such as research guides, blog posts, and lists of resources. Services, such as chat reference, the ability to reserve a room, or complete a tutorial, are available through the law library's website. As a law library's website is relied upon by students and faculty to access the collection, librarians and other staff frequently use the Web resources in the course of their reference and instructional work. This means that while librarians create and maintain content, they must simultaneously determine whether they are serving their primary users and to what extent are *they* using this content. Is it easy for those primary users to use? Talking with students about how they do research, conducting usability testing, or simply watching students use the library website can be illuminating. Law librarians may be shocked to discover that many law students access library databases through Google!

Although ceasing to collect reporters, loose-leafs, law reviews, and other material that is duplicative of licensed, electronic content and more cumbersome to use in print, law libraries continue to collect print materials. Monographs and codes are among the most popular. Users still read print—and prefer to, in many instances. And scholars continue to write books; faculty and students continue to rely on books in their research. Law libraries are showcasing monographs by sponsoring books talks or discussions, blogging about new books, and creating displays to show new faculty titles.[17] However, the law school's print collection must reflect the priorities and desires of the users. As Penny Hazelton thoughtfully observes,

> if law libraries ever forget their student users—that is, they fail to take into account students' needs in the academic environment—law students will not even darken the law library door. Law library space will be completely converted to other law school purposes, or our law libraries will be the cold, empty spaces the naysayers foresee in a future without books.[18]

Librarians should listen closely to their users and the needs of their users. If they are asking for materials, librarians should consider purchasing them; if they are asking for space for a particular use, that use should be carefully considered—with the presumption in favor of usage. We should closely monitor our physical and virtual spaces to confirm (1) whether they are being used, and (2) whether they are being used as intended. Too often librarians continue activities or practices that fail to resonate with users. There may be practices, such as blogging or creating research guides, they want to continue—or practices librarians think their patrons appreciate—but in reality, barely notice them.

EXPLORING NEW MARKETS FOR LEGAL SERVICES AND RESEARCH SERVICES

Academic and firm librarians are well prepared to leverage technology. They use technology to fill the holes within our institutions, as institutions grapple with using digital materials. Librarians regularly adjust their practices, whether work flow or pedagogy, to use collaborative platforms, redesign knowledge management to work paperlessly, save materials to the cloud, and interact with electronic platforms and materials in a myriad of individualized ways.

While other law firm administrative departments (including the library) have lost funding and reduced staffing levels to respond to flattening revenues, business development has typically not experienced the same treatment. Anecdotally, they are often seen as the "Hail Mary" of the law firm playbook and funds and staff are added, even in tough times. When there is less business, these professional colleagues are encouraged to strategically reach out and bring it in. The attitude toward this and the response by the law firm library is important to the future of the law firm as a whole.

Some library departments directly report to business development already; others are increasingly finding that more of their workload is based on business development or competitive intelligence needs of the firm. Seeing this change in the light of opportunity, the digital law firm library of the future can produce more integrated and responsive business development solutions. Other law library departments can contribute to project management or the evaluation of online resources for a practice group such as intellectual property (IP), where they may have significant experience.

In addition to these alternative roles for research services, other metrics and monetized services must be mined by the law firm for value. The law firm librarian or information professional should provide the vision. For example, where there is a need for international document authentication or visa applications, a law librarian can set up an intelligent process and service and monetize it appropriately. Librarians are natural managers of projects as well as organizers of information in logical and efficient ways, which is critical to the business of law. Thus, librarians will thrive in an environment where low-cost, high-efficiency projects are rewarded in this cost-pressure environment. Newsletter or current awareness projects can be marketed as a discrete standalone service or offered gratis as a loss leader for firm industrial business. Once contact is established, credibility builds with these services, and clients may be in direct contact with the law firm librarian of the future. Clients appreciate responsive work and keeping ahead of the curve while sorting out the sea of irrelevant material. Librarians can produce the customized, targeted information to the fore for those clients in an extranet, and this may become a trusted secret weapon. The partners and other stakeholders of a law firm appreciate this as well because it translates to satisfied clients and the may become willing to trust them with more business. It also simply reinforces that the firm is a credible knowledge leader in its desired area of work.

Law school libraries are not shying away from these new opportunities either. Indeed, the flexibility, innovation, and creativity of today's law librarians means they are leveraging high-level technology, teaching, and information management skills to take on new and unconventional projects to fill gaps and satisfy demand and need. With Duke Law School leading the way, law libraries are managing digital repositories to make faculty and student scholarship accessible and open to anyone with an Internet connection.[19] Law libraries are also approaching faculty projects by creating innovative resources. Following in the footsteps of university librarians, law librarians now create portals for the storage and access of primary documents or images from faculty published works.[20] Likewise, as law schools undergo a shift in the curriculum and focus away from doctrinal subjects to experiential or clinical training of specialized lawyers, the law library must find opportunities to align its work with emerging strategic goals of the contemporary law schools. And law library collections must bend and shift to follow the unique focus of law school, faculty, or mission. Law librarians must do this while continuing to engage in the bedrock, nuts-and-bolts work of teaching, reference, and faculty research.

CONCLUSION

To be sure, what is good for the institution is good for the law library—and this will be evolving as firms change and law schools follow. In the academic setting, law librarians will continue developing subject specialties, in whatever subjects are important for the institution and practice goals of the students and law firm environment. They will similarly use their high-level technological skills to spearhead projects that support faculty research in novel ways. The focus of basic and advanced legal research will continue to move from finding legal authority to understanding it within context and using it as means to teach analytic thinking and creative problem solving. Research instruction will also have to focus on information management and organization, and skills and principles imparted must be platform agnostic.

Law firm libraries will similarly evolve. As law school librarians become more closely integrated with faculty, law firm librarians will become more closely integrated with practice groups. Also like law school librarians, they will offer strategic support, with an emphasis on technological innovations. Projects will include knowledge management solutions, mobility, portal development, seamless service, monetized products, centralized resources, and search tools.

FURTHER READING

Balleste, Roy, Sonia Luna-Lamas, and Lisa Smith-Butler. 2007. *Law librarianship in the twenty-first century*. Lanham, Md.: Scarecrow Press.

Bintliff, Barbara, and Duncan E. Alford. 2011. *Teaching legal research*. London: Routledge.

Danner, Richard A., and Jules Winterton. 2011. *The IALL international handbook of legal information management*. Farnham, Surrey: Ashgate Pub. http://site.ebrary.com/id/10477049.

Goodenough, Oliver R., and Marc Lauritsen. 2012. *Educating the digital lawyer*. New Providence, N.J.: LexisNexis.

Lee, Leslie A., and Michelle M. Wu. 2007. *Beyond the books: People, politics, and librarianship*. Buffalo, N.Y.: W. S. Hein.

Mart, Susan Nevelow (ed.). 2013. *Boulder statements on legal research education: The intersection of intellectual and practical skills*. Buffalo, N.Y.: William S. Hein.

Mersky, Roy M., and Richard A. Leiter. 1991. *The spirit of law librarianship: A reader*. Littleton, Colo.: F. B. Rothman.

Palfrey, John G., and Urs Gasser. 2008. *Born digital: Understanding the first generation of digital natives*. New York: Basic Books.

Rubin, Edward L. 2012. *Legal education in the digital age*. Cambridge: Cambridge University Press.

Whisner, Mary. 2006. *Practicing reference: Thoughts for librarians and legal researchers*. Buffalo, N.Y.: William S. Hein.

NOTES

1. According to the website http://www.integreon.com, Integreon is a "provider of integrated legal, research and business support solutions to discerning professionals, including law firms and corporate law departments, leading corporations, financial services organisations and professional services firms."

2. Carl Yirka, "The Yirka question and Yirka answer: What should law libraries stop doing in order to address higher priority initiatives?" *AALL Spectrum* (July 1, 2008): 28–30, 32.

3. For instance, Bloomberg Law recently reported that 43 percent of law firms believe they are absorbing more research costs in 2012 than they did in 2010. "Law firms eating more costs," *Bloomberg Law*, http://about.bloomberglaw.com/2012/11/12/law-firms-eating-more-costs/.

4. For a summary of the contemporary models of legal research education in the United States, see S. Blair Kauffman, "Information literacy in law: Starting points for improving research competencies," *International Journal of Legal Information* (2010): 38.

5. John G. Palfrey and Urs Gasser, *Born digital: Understanding the first generation of digital natives* (New York: Basic Books, 2008).

6. Penny Hazelton, "Law students and the new law library," in *Legal education in the digital age*, edited by Edward L. Rubin (Cambridge: Cambridge University Press, 2012).

7. See, generally, Barbara Bintliff, "Context and legal research," *Law Library Journal* (2007): 99.

8. Bintliff, "Context and legal research," 259.

9. Hazelton, "Law students and the new law library," 174.

10. Myron Moskowvitz, "Behind the case method: It's time to teach with problems," *Journal of Legal Education* 42 (1992): 241, 245.

11. For more on teaching law students analytical skills required for legal practice, see Sarah Valentine, "Legal research as a fundamental skill: A lifeboat for students and law schools," *University of Baltimore Law Review* 39, 2 (2010): 175–255.

12. Eric Lane, "How to read a statute in New York: A response to Judge Kaye and Some more," *Hofstra Law Review* 28 (1999): 85, 86 quoting Judith S. Kaye. "State courts at the dawn of a new century: Common law courts reading statutes and constitutions," *N.Y.U. Law Review* 70, 1:18.

13. Richard A. Posner, "Statutory interpretation: In the classroom and in the courtroom," *University of Chicago Law Review* 50, 2 (1983): 800–822.

14. "Rethinking Langdell: Historic changes in 1L curriculum set state for new upper-level programs of study," http://www.law.harvard.edu/news/today/dec_hlt_langdell.php. Boston University, University of Colorado Law, Emory University School of Law, Ohio Northern University College of Law, and Vanderbilt University Law School also recently added a legislation course. See http://lawteaching.org/publications/ILTLchartoflegaleducationreform200905.pdf (Boston University, Ohio Northern University College of Law, Vanderbilt); http://www.coloradodaily.com/ci_15500723?source=most_viewed#axzz1GtYhDIA6 (University of Colorado Law). "Workshop on the future of the legal course book," 33 *Seattle University Law Rev.* 292, 301 (2010). Richard Posner, in *Statutory Interpretation: In the Classroom and in the Courtroom*, noted that the University of Chicago also offered a seminar in legislation.

15. For more on law libraries as a third space, see R. A. Danner, S. B. Kauffman, and J. G. Palfrey, "The twenty-first century law library," *Law Library Journal* 101, 2 (2009): 143–156.

16. Danner et al., "The twenty-first century law library," 143–156 (noting that Harvard and Yale Law Libraries underwent major renovation in the late 1990s). Other libraries that have been recently renovated include Stanford, the University of Chicago, Indiana University Law School, and Duke, among many others.

17. In this way, perhaps, libraries are taking a cue from public libraries, which began the trend of opening their doors and using their space for book talks or using their space to meet community—not necessarily library user—needs. See, for example, "Breakout no. 7: Libraries as community change agents," *Knight Foundation*, http://www.knightfoundation.org/press-room/other/breakout-7-libraries-community-change-agents/ (February 12, 2013); Deniz Koray, "Libraries help homeowners fight foreclosure," *American Libraries Magazine* (February 12, 2013), http//americanlibrariesmagazine.org/features/02122013/libraries-help-homeowners-fight-foreclosure.

18. Hazelton, "Law students and the new law library," 162.

19. For example, as of February 17, 2013, Yale Law School's Legal Scholarship Repository (http://digitalcommons.law.yale.edu/) contained 4,208 papers, 1,728,678 full-text downloads, and 1,057,512 downloads in the past year.

20. A recent example includes the Web portals created by Yale Law librarians to accompany *Lincoln's Code* by John Fabian Witt (http://documents.law.yale.edu/lincolns-code). The website contains images from the book, a bibliography, and related audio and video. Another example, also created by Yale Law librarians, was a website containing images and other material from the book *Representing Justice: Invention, Controversy, and Rights in City-States and Democratic Courtrooms* by Judith Resnick and Dennis Curtis (http://documents.law.yale.edu/representing-justice).

Index

List of Contributing Authors

Jennifer Alexander
Senior Business Analysis Manager
McKenna Long

Scott D. Bailey
Global Director of Research Services
Squire Sanders (US) LLP

Professor Rhea Ballard-Thrower
Director
Howard University Law Library

Theodora Belniak
Head of Collection Management
SUNY at Buffalo Law Library

Camille Broussard
Library Director
Professor of Law
Mendik Library/New York Law School

Molly (Mary) E. Brownfield
Manager of Library Services
Finnegan, Henderson, Farabow, Garrett & Dunner, LLP

Kim Clarke
Associate Vice-Provost (Libraries and Cultural Resources), Research Support, and Director of the Bennett Jones Law Library
University of Calgary

Kyle K. Courtney
Manager, Faculty Research and Scholarship
Harvard Law School

Valeri Craigle
Access Technologies Librarian
S. J. Quinney College of Law Library, University of Utah

Marcia L. Dority Baker
Access Services Librarian and Assistant Professor of Law Library
University of Nebraska College of Law, Schmid Law Library

Cheryl Kelly Fischer
Reference Librarian and Lecturer in Law
UCLA Law Library

Emily R. Florio
Manager of Libraries & Library Information Systems
Fish & Richardson

Edward T. Hart
Head of Technical Services
Lawton Chiles Legal Information Center, University of Florida

M. T. Hennessey
Business Intelligence Coordinator
Paul, Weiss, Rifkind, Wharton & Garrison LLP

Nadine R. Hoffman
Natural Resources, Energy & Environmental Law Librarian
Bennett Jones Law Library, University of Calgary

Emily Janoski-Haehlen
Associate Dean for Law Library Services
Valparaiso University Law School

Julie Graves Krishnaswami
Head of Instruction and Lecturer in Legal Research
Lillian Goldman Law Library, Yale Law School

Ellyssa Kroski
Director of Information Technology
The New York Law Institute

Steven A. Lastres
Director of Library & Knowledge Management
Debevoise & Plimpton LLP

Don MacLeod
KM Manager
Debevoise & Plimpton LLP

Sarah K. C. Mauldin
Director of Library Services
Smith, Gambrell & Russell, LLP

William R. Mills
Associate Librarian for Information & Technology Resources,
Professor of Legal Research
New York Law School Mendik Library

Ralph Monaco
Executive Director
The New York Law Institute

Catherine M. Monte
Chief Knowledge Officer
Fox Rothschild LLP

Carol Ottolenghi
Directory of Library Services
Office of the Ohio Attorney General

Andrew Plumb-Larrick
Acting Director
The Judge Ben C. Green Law Library
Case Western Reserve University

Holly M. Riccio
Director of Library Innovation, San Francisco Library Manager
O'Melveny & Myers LLP

Michael J. Robak
Associate Library Director, Leon E. Bloch Law Library,
and Director of Technologies
University of Missouri-Kansas City School of Law

Gitelle Seer
Senior Librarian Consultant
Visionary Information Solutions

Roger Vicarius Skalbeck
Associate Law Librarian for Electronic Resources and Services
Georgetown Law Center

Vicki Steiner
Reference Librarian
UCLA Law Library

Thomas J. Striepe
Faculty Services Librarian
University of Georgia Law Library

Mary Talley
Principal
TalleyPartners

Carol A. Watson
Director
University of Georgia Law Library

Jennifer Wertkin
Head of Public Services
The Arthur W. Diamond Law Library, Columbia University

Michelle M. Wu
Professor of Law and Law Library Director
Georgetown University Law Center

Editorial Board List